D0639946

MATHEMATICS FOR THE PHYSICAL SCIENCES

MATHEMATICS FOR THE PHYSICAL SCIENCES

by

Herbert S. Wilf, Ph.D.
Professor of Mathematics
University of Pennsylvania

DOVER PUBLICATIONS, INC.
NEW YORK

Published in Canada by General Publishing Company, Ltd., 30 Lesmill Road, Don Mills, Toronto, Ontario.
Published in the United Kingdom by Constable and Company, Ltd., 10 Orange Street, London WC2H 7EG.

This Dover edition, first published in 1978, is an unabridged republication of the work originally published by John Wiley & Sons, Inc., New York, in 1962. Corrections have been made on pages 27, 219 and 228 of the present edition.

International Standard Book Number:
0-486-63635-6
Library of Congress Catalog Card Number:
77-94162

Manufactured in the United States of America
Dover Publications, Inc.
180 Varick Street
New York, N.Y. 10014

To my mother and father

Preface

This book is based on a two-semester course in "The Mathematical Methods of Physics" which I have given in the mathematics department of the University of Illinois in recent years. The audience has consisted primarily of physicists, engineers, and other natural scientists in their first or second year of graduate study. Knowledge of the theory of functions of real and complex variables is assumed.

The subject matter has been shaped by the needs of the students and by my own experience. In many cases students who do not major in mathematics have room in their schedules for only one or two mathematics courses. The purpose of this book, therefore, is to provide the student with some heavy artillery in several fields of mathematics, in confidence that targets for these weapons will be amply provided by the student's own special field of interest. Naturally, in such an attempt, something must be sacrificed, and I have regarded as most expendable discussions of physical applications of the material being presented.

Again, in the short space allotted to each subject there is little chance to develop the theory beyond fundamentals. Thus in each chapter I have gone straight to (what I regard as) the heart of the matter, developing a subject just far enough so that applications can easily be made by the student himself. The exercises at the end of each chapter, along with their solutions at the back of the book, afford some further opportunities for using the theoretical apparatus.

The material herein is, for the most part, classical. The bibliographical references, particularly to journal articles, are given not so much to provide a jumping-off point for further research as to give the reader a feeling for the chronological development of these subjects and for the names of the men who created them.

Finally, I have, where possible, tried to say something about numerical methods for computing the solutions of various kinds of problems. These discussions, while brief, are oriented toward electronic computers and are intended to help bridge the gap between the "there exists" of a pure mathematician and the "find it to three decimal places" of an engineer.

I am indebted to Professor L. A. Rubel for permission to publish Theorem 7 of Chapter 3 here for the first time and to Professor R. P. Jerrard for some of the exercises in Chapter 7. To the well-known volume of Courant and Hilbert I owe the intriguing notion that, even in an age of specialization, it may be possible for physicists and mathematicians to understand each other.

HERBERT S. WILF

Philadelphia, Pennsylvania
March, 1962

Contents

MATHEMATICS FOR THE PHYSICAL SCIENCES

chapter 1

Vector spaces and matrices

1.1 VECTOR SPACES

A vector space V is a collection of objects $\mathbf{x}, \mathbf{y}, \ldots$ called vectors, satisfying the following postulates:

(I) If $\mathbf{x}$ and $\mathbf{y}$ are vectors, there is a unique vector $\mathbf{x} + \mathbf{y}$ in V called the sum of $\mathbf{x}$ and $\mathbf{y}$.

(II) If $\mathbf{x}$ is a vector and α a complex number, there is a uniquely defined vector $\alpha\mathbf{x}$ in V satisfying

(1) $\alpha(\mathbf{x} + \mathbf{y}) = \alpha\mathbf{x} + \alpha\mathbf{y}$ (2) $(\alpha\beta)\mathbf{x} = \alpha(\beta\mathbf{x})$

(3) $(\alpha + \beta)\mathbf{x} = \alpha\mathbf{x} + \beta\mathbf{x}$ (4) $1 \cdot \mathbf{x} = \mathbf{x}$

(5) $\mathbf{x} + \mathbf{y} = \mathbf{y} + \mathbf{x}$ (6) $\mathbf{x} + (\mathbf{y} + \mathbf{z}) = (\mathbf{x} + \mathbf{y}) + \mathbf{z}$

(III) There is a vector $\mathbf{0}$ in V satisfying

(7) $$\mathbf{x} + \mathbf{0} = \mathbf{0} + \mathbf{x} = \mathbf{x}$$

for every $\mathbf{x}$ in V, and, further, for every $\mathbf{x}$ in V there is a vector $-\mathbf{x}$ such that

(8) $$\mathbf{x} + (-\mathbf{x}) = \mathbf{0}.$$

We will use the notation $\mathbf{x} - \mathbf{y}$ to mean $\mathbf{x} + (-\mathbf{y})$, as might be expected.

(IV) If $\mathbf{x}$ and $\mathbf{y}$ are vectors in V, there is a uniquely defined complex number $(\mathbf{x}, \mathbf{y})$ called the "inner product" of $\mathbf{x}$ and $\mathbf{y}$ which satisfies

(9) $(\mathbf{x}, \mathbf{y}) = \overline{(\mathbf{y}, \mathbf{x})}$ (10) $(\alpha\mathbf{x}, \mathbf{y}) = \bar{\alpha}(\mathbf{x}, \mathbf{y})$

(11) $(\mathbf{x}, \mathbf{x}) \geqq 0$ (12) $(\mathbf{x} + \mathbf{y}, \mathbf{z}) = (\mathbf{x}, \mathbf{z}) + (\mathbf{y}, \mathbf{z})$

(13) $(\mathbf{x}, \mathbf{y} + \mathbf{z}) = (\mathbf{o}, \mathbf{y}) + (\mathbf{x}, \mathbf{z})$ (14) $(\mathbf{x}, \mathbf{x}) = 0$ if and only if $\mathbf{x} = \mathbf{0}$

We state at once that it is not our intention to develop here a purely axiomatic theory of vector spaces. However, in the remainder of this book we shall meet several vector spaces of different types, some of which will not "look like" vector spaces at all. It is most important to note that the only qualifications a system needs in order to be a vector space† are those just set forth, for only in this way can the true unity of such apparently diverse topics as finite dimensional matrices, Fourier series, orthogonal polynomials, integral equations, differential eigenvalue problems, and so on, be perceived. An enlightening exercise for the reader, for example, will be found in analyzing various results as they are proved for special systems, and asking whether or not the properties of the special system were used, or whether, as will more often happen, we have proved a general property of vector spaces.

Example 1. The set of ordered n-tuples of complex numbers $(\alpha_1, \alpha_2, \ldots, \alpha_n)$ is a vector space V_n (Euclidean n-space) if we define for any vectors

$$\mathbf{x} = (\alpha_1, \ldots, \alpha_n), \; \mathbf{y} = (\beta_1, \ldots, \beta_n),$$

$$\mathbf{x} + \mathbf{y} = (\alpha_1 + \beta_1, \ldots, \alpha_n + \beta_n) \tag{15}$$

$$\gamma \mathbf{x} = (\gamma\alpha_1, \gamma\alpha_2, \ldots, \gamma\alpha_n) \tag{16}$$

$$(\mathbf{x}, \mathbf{y}) = \sum_{i=1}^{n} \bar{\alpha}_i \beta_i. \tag{17}$$

The complex numbers $\alpha_1, \alpha_2, \ldots, \alpha_n$ are called the *components* of the vector $\mathbf{x}$, and postulates (I)–(IV) are easily verified here by direct calculation. For example, to prove (11),

$$(\mathbf{x}, \mathbf{x}) = \sum_{i=1}^{n} \bar{\alpha}_i \alpha_i = \sum_{i=1}^{n} |\alpha_i|^2 \geqq 0. \tag{18}$$

Example 2. The class of functions $f(x)$ of a real variable x, on the interval $[a, b]$ of the real axis for which

$$\int_a^b |f(x)|^2 \, dx < \infty \tag{19}$$

forms a vector space, each vector now being a function satisfying (19). Here the sum of two vectors $f(x)$, $g(x)$ is the vector $f(x) + g(x)$, and the inner product is defined by

$$(f, g) = \int_a^b \overline{f(x)} g(x) \, dx \tag{20}$$

for which the postulates can again easily be shown to be satisfied. This is the space $\mathscr{L}^2(a, b)$, which is of particular interest, for example, in quantum mechanics.

† Our terminology is not conventional. Actually axioms I–III define a vector space, whereas, with axiom IV the structure is sometimes called a "unitary space." We use the term "vector space" for simplicity.

Example 3. Let $w(x) \not\equiv 0$ be a fixed, real-valued, integrable function, defined and non-negative on the interval $[a, b]$ of the real axis. Consider the set of all polynomials

$$f(x) = a_0 + a_1x + \cdots + a_nx^n \tag{21}$$

with real coefficients, and of degree $\leqq n$, for some fixed n. This class forms a vector space if addition of two vectors (polynomials) is defined in the obvious way, and if the inner product is given by

$$(f, g) = \int_a^b f(x)g(x)w(x)\,dx. \tag{22}$$

It is in this vector space that we will develop the theory of orthogonal polynomials in the next chapter.

1.2 SCHWARZ'S INEQUALITY AND ORTHOGONAL SETS

Theorem 1. (*Schwarz's inequality*). *Let* $\mathbf{x}$, $\mathbf{y}$ *be vectors in a vector space* V. *Then*

$$|(\mathbf{x}, \mathbf{y})|^2 \leqq (\mathbf{x}, \mathbf{x})(\mathbf{y}, \mathbf{y}) \tag{23}$$

the sign of equality holding if and only if there is a complex number α *such that* $\mathbf{x} = \alpha\mathbf{y}$ (i.e., *if* $\mathbf{x}$ *and* $\mathbf{y}$ *are proportional*).
Proof. Let λ be any real number. By (11),

$$(\mathbf{x} + \lambda(\mathbf{y}, \mathbf{x})\mathbf{y}, \quad \mathbf{x} + \lambda(\mathbf{y}, \mathbf{x})\mathbf{y}) \geqq 0. \tag{24}$$

Hence by (9), (12), and (13),

$$0 \leqq (\mathbf{x}, \mathbf{x}) + 2\lambda\,|(\mathbf{x}, \mathbf{y})|^2 + \lambda^2\,|(\mathbf{x}, \mathbf{y})|^2\,(\mathbf{y}, \mathbf{y}) \tag{25}$$

for all real λ.

Thus the discriminant of this quadratic polynomial is not positive, that is,

$$|(\mathbf{x}, \mathbf{y})|^4 - (\mathbf{x}, \mathbf{x})\,|(\mathbf{x}, \mathbf{y})|^2\,(\mathbf{y}, \mathbf{y}) \leqq 0.$$

If $(\mathbf{x}, \mathbf{y}) \neq 0$, we get

$$|(\mathbf{x}, \mathbf{y})|^2 \leqq (\mathbf{x}, \mathbf{x})(\mathbf{y}, \mathbf{y}) \tag{26}$$

whereas if $(\mathbf{x}, \mathbf{y}) = 0$, (26) is obvious. Finally, suppose the sign of equality holds in (26). Then in (25) we have a quadratic polynomial with zero discriminant, which therefore is zero for some real value of λ, say λ_0. Referring to (14) and (24) we see that

$$\mathbf{x} + \lambda_0(\mathbf{y}, \mathbf{x})\mathbf{y} = \mathbf{0} \tag{27}$$

which is to say that x is proportional to y. Conversely, if $\mathbf{x} = \beta\mathbf{y}$, substitution in (23) shows at once that the sign of equality holds.

Two vectors $\mathbf{x}$ and $\mathbf{y}$ are said to be *orthogonal* if

$$(\mathbf{x}, \mathbf{y}) = 0. \tag{28}$$

The *length* of a vector $\mathbf{x}$ is defined by

$$\|\mathbf{x}\| = (\mathbf{x}, \mathbf{x})^{1/2} \tag{29}$$

and is always a non-negative real number. In terms of the length, Schwarz's inequality (23) reads

$$|(\mathbf{x}, \mathbf{y})| \leqq \|\mathbf{x}\| \, \|\mathbf{y}\|. \tag{30}$$

A finite or infinite sequence of vectors $\mathbf{x}_1, \mathbf{x}_2, \mathbf{x}_3, \ldots$ is called an *orthogonal set* if

$$(\mathbf{x}_i, \mathbf{x}_j) = 0 \qquad (i \neq j; \; i, j = 1, 2, 3, \ldots) \tag{31}$$

and an *orthonormal* set if, in addition to (31), we have also

$$\|\mathbf{x}_i\| = 1 \qquad (i = 1, 2, \ldots). \tag{32}$$

The two conditions (31) and (32) are frequently combined in the form

$$(\mathbf{x}_i, \mathbf{x}_j) = \delta_{ij} \qquad (i, j = 1, 2, \ldots) \tag{33}$$

where δ_{ij}, the "Kronecker delta," is defined by

$$\delta_{ij} = \begin{cases} 1 & \text{if } i = j \\ 0 & \text{if } i \neq j. \end{cases} \tag{34}$$

A vector $\mathbf{x}$ of length unity is said to be *normalized.*

Now let $\mathbf{f}$ be an arbitrary vector in a vector space V, and let† $\mathbf{x}_1, \mathbf{x}_2, \mathbf{x}_3, \ldots$ be an orthonormal set in V. The numbers

$$\gamma_\nu = (\mathbf{x}_\nu, \mathbf{f}) \qquad (\nu = 1, 2, \ldots) \tag{35}$$

are called the *Fourier coefficients* of $\mathbf{f}$ with respect to the set $\mathbf{x}_1, \mathbf{x}_2, \ldots$. These coefficients are of considerable importance in applications. As an example, consider the following approximation problem: let n be a fixed integer, $\mathbf{f}$ a given vector of a vector space V, and $\mathbf{x}_1, \ldots, \mathbf{x}_n$ an orthonormal set lying in V. It is required to find numbers $\alpha_1, \alpha_2, \ldots, \alpha_n$ for which the vector

$$\mathbf{h} = \alpha_1 \mathbf{x}_1 + \cdots + \alpha_n \mathbf{x}_n \tag{36}$$

is the best possible approximation to $\mathbf{f}$ in the sense that $\|\mathbf{f} - \mathbf{h}\|$ is as small as possible.

† For existence see § 1.3.

To solve this problem, we have

$$
\begin{aligned}
(37) \quad \|\mathbf{f} - \mathbf{h}\|^2 &= (\mathbf{f} - \mathbf{h}, \mathbf{f} - \mathbf{h}) \\
&= (\mathbf{f} - \alpha_1\mathbf{x}_1 - \cdots - \alpha_n\mathbf{x}_n, \mathbf{f} - \alpha_1\mathbf{x}_1 - \cdots - \alpha_n\mathbf{x}_n) \\
&= (\mathbf{f}, \mathbf{f}) - \alpha_1(\mathbf{f}, \mathbf{x}_1) - \cdots - \alpha_n(\mathbf{f}, \mathbf{x}_n) \\
&\quad - \bar{\alpha}_1(\mathbf{x}_1, \mathbf{f}) - \cdots - \bar{\alpha}_n(\mathbf{x}_n, \mathbf{f}) + |\alpha_1|^2 + \cdots + |\alpha_n|^2 \\
&= (\mathbf{f}, \mathbf{f}) - \alpha_1\bar{\gamma}_1 - \bar{\alpha}_1\gamma_1 - \cdots - \alpha_n\bar{\gamma}_n - \bar{\alpha}_n\gamma_n \\
&\quad + |\alpha_1|^2 + \cdots + |\alpha_n|^2 \\
&= (\mathbf{f}, \mathbf{f}) + (|\alpha_1|^2 - 2\mathrm{Re}\,\alpha_1\bar{\gamma}_1) + \cdots + (|\alpha_n|^2 - 2\mathrm{Re}\,\alpha_n\bar{\gamma}_n) \\
&= (\mathbf{f}, \mathbf{f}) + |\alpha_1 - \gamma_1|^2 + \cdots + |\alpha_n - \gamma_n|^2 - |\gamma_1|^2 - \cdots - |\gamma_n|^2.
\end{aligned}
$$

Now, remembering that $\mathbf{f}, \gamma_1, \ldots, \gamma_n$ are fixed, and only $\alpha_1, \ldots, \alpha_n$ are at our disposal, it is plain that the choice of $\alpha_1, \ldots, \alpha_n$ which minimizes the "least squares error" $\|\mathbf{f} - \mathbf{h}\|^2$ is

$$
(38) \qquad \alpha_\nu = \gamma_\nu = (\mathbf{x}_\nu, \mathbf{f}) \qquad (\nu = 1, 2, \ldots, n).
$$

Furthermore, if we make this optimal choice of the α_ν as the Fourier coefficients of $\mathbf{f}$, (37) shows clearly that

$$
0 \leq (\mathbf{f}, \mathbf{f}) - |\gamma_1|^2 - \cdots - |\gamma_n|^2
$$

or

$$
(39) \qquad \sum_{\nu=1}^{n} |\gamma_\nu|^2 \leqq (\mathbf{f}, \mathbf{f}).
$$

This inequality, known as Bessel's inequality, is seen to be a property of the vector $\mathbf{f}$ and the set $\mathbf{x}_1, \ldots, \mathbf{x}_n$ only, and therefore expresses a general property of Fourier coefficients.

It may happen that a given orthonormal set $\mathbf{x}_1, \mathbf{x}_2, \mathbf{x}_3, \ldots$ has the property that every vector $\mathbf{f}$ in the space V can be approximated arbitrarily closely by taking n, the number of vectors used from the set, large enough.

More precisely, let $\mathbf{x}_1, \mathbf{x}_2, \mathbf{x}_3, \ldots$ be an orthonormal set with the property that if $\varepsilon > 0$ and an arbitrary vector $\mathbf{f}$ of V are given, there is an n for which the vector (36) with (38) implies

$$
\|\mathbf{f} - \mathbf{h}\| < \varepsilon.
$$

We then say that $\mathbf{x}_1. \mathbf{x}_2, \ldots$ is a *complete* orthonormal set. The following theorems are now clear:

Theorem 2. *Let* $\mathbf{x}_1, \mathbf{x}_2, \ldots$ *be a complete orthonormal set in a vector space* V *and let* $\mathbf{f}$ *be a vector of* V. *Then*

$$
(40) \qquad \sum_{\nu=1}^{\infty} |(\mathbf{x}_\nu, \mathbf{f})|^2 = (\mathbf{f}, \mathbf{f}) \qquad (\textit{Parseval's identity}).
$$

Theorem 3. *(The Riemann-Lebesgue Lemma). If* $\mathbf{x}_1, \mathbf{x}_2, \ldots$ *is an infinite orthonormal set and* $\mathbf{f}$ *is any vector of* V*, then*

$$|(\mathbf{x}_\nu, \mathbf{f})| \to 0 \qquad (\nu \to \infty). \tag{41}$$

Since the series on the left side of (39) obviously converges, its terms must approach zero.

1.3 LINEAR DEPENDENCE AND INDEPENDENCE

The vectors $\mathbf{x}_1, \mathbf{x}_2, \ldots, \mathbf{x}_n$ are said to be *linearly dependent* if there are constants $\alpha_1, \ldots, \alpha_n$ not all zero, such that

$$\alpha_1\mathbf{x}_1 + \cdots + \alpha_n\mathbf{x}_n = \mathbf{0}. \tag{42}$$

Otherwise the vectors are *linearly independent.*

Let $\mathbf{x}_1, \mathbf{x}_2, \ldots, \mathbf{x}_n$ be linearly independent. We wish to transform the set $\mathbf{x}_1, \ldots, \mathbf{x}_n$ into a new set $\mathbf{y}_1, \ldots, \mathbf{y}_n$ having the properties: (i) $\mathbf{y}_1, \ldots, \mathbf{y}_n$ is an orthonormal set, (ii) each $\mathbf{y}_i$ is a linear combination of the $\mathbf{x}_j$ ($j = 1, \ldots, n$). This may be accomplished by the following procedure, called the *Gram-Schmidt process.*

First, take

$$\mathbf{y}_1 = \frac{\mathbf{x}_1}{\|\mathbf{x}_1\|}. \tag{43}$$

Then, clearly $\|\mathbf{y}_1\| = 1$. Next, assume

$$\mathbf{y}_2' = \mathbf{x}_2 - \lambda_1\mathbf{y}_1$$

and determine the constant λ_1, such that $(\mathbf{y}_2', \mathbf{y}_1) = 0$, i.e., take

$$\lambda_1 = (\mathbf{y}_1, \mathbf{x}_2).$$

Since $\mathbf{x}_1, \mathbf{x}_2$ are linearly independent, $\mathbf{y}_2' \neq \mathbf{0}$, and we set

$$\mathbf{y}_2 = \frac{\mathbf{y}_2'}{\|\mathbf{y}_2'\|}$$

In general, if $\mathbf{y}_1, \mathbf{y}_2, \ldots, \mathbf{y}_k$ have been constructed, write

$$\mathbf{y}'_{k+1} = \mathbf{x}_{k+1} - \sigma_1\mathbf{y}_1 - \cdots - \sigma_k\mathbf{y}_k \tag{44}$$

and determine the constants $\sigma_1, \ldots, \sigma_k$ so that

$$(\mathbf{y}'_{k+1}, \mathbf{y}_j) = 0 \qquad (j = 1, 2, \ldots, k) \tag{45}$$

that is, choose

$$\sigma_j = (\mathbf{y}_j, \mathbf{x}_{k+1}) \qquad (j = 1, 2, \ldots, k). \tag{46}$$

As before, $\mathbf{y}'_{k+1} \neq \mathbf{0}$, and taking $\mathbf{y}_{k+1} = \mathbf{y}'_{k+1}/\|\mathbf{y}'_{k+1}\|$, we have constructed the next vector in the set.

A vector space V is said to be of *dimension n* if it contains n linearly independent vectors, but every $n + 1$ vectors are linearly dependent. A space which for every integer n contains n linearly independent vectors is said to be *infinite dimensional*. By virtue of the Gram-Schmidt process we see that the dimension of a vector space is also the length of the longest orthonormal set contained in the space.

A set of vectors $\mathbf{x}_1, \mathbf{x}_2, \ldots, \mathbf{x}_n$ is said to *span* a vector space V if every vector of V is a linear combination of $\mathbf{x}_1, \mathbf{x}_2, \ldots, \mathbf{x}_n$, that is, if $\mathbf{f}$ is an arbitrary vector of V, there exist complex numbers $\alpha_1, \alpha_2, \ldots$ such that

$$\mathbf{f} = \alpha_1\mathbf{x}_1 + \alpha_2\mathbf{x}_2 + \cdots. \tag{47}$$

A set of vectors $\mathbf{x}_1, \mathbf{x}_2, \ldots$ is said to form a *basis* for a vector space V if (i) the set spans the space and (ii) the set is linearly independent.

1.4 LINEAR OPERATORS ON A VECTOR SPACE

A linear operator on a vector space V is a rule which assigns to each vector $\mathbf{f}$ of V a unique vector $T\mathbf{f}$ of V, in such a way that

$$T(\alpha\mathbf{f} + \mathbf{g}) = \alpha T\mathbf{f} + T\mathbf{g} \tag{48}$$

for every pair of vectors $\mathbf{f}$, $\mathbf{g}$ in V and every complex number α.

Example 1. For Euclidean n-space, the operator which associates with

$$\mathbf{x} = (\alpha_1, \alpha_2, \ldots, \alpha_n)$$

the vector

$$T\mathbf{x} = (\alpha_1, \alpha_1 + \alpha_2, \alpha_1 + \alpha_2 + \alpha_3, \cdots, \alpha_1 + \alpha_2 + \cdots + \alpha_n)$$

is a linear operator.

Example 2. In $\mathscr{L}^2(a, b)$ the rule which associates with the vector $f(x)$ the vector

$$Tf(x) = \int_a^x f(y)\, dy \qquad (a \leq x \leq b)$$

is a linear operator.

Henceforth the term "operator" will invariably refer to a linear operator on the space in question.

The *identity operator I* is the operator which assigns to any vector $\mathbf{f}$ the vector $\mathbf{f}$ itself, i.e.,

$$I\mathbf{f} = \mathbf{f} \qquad (\text{all } \mathbf{f} \text{ in } V). \tag{49}$$

This is clearly linear. Two operators T, U are said to be *equal* if their effect on every vector of V is the same, that is, $T = U$ means

$$Tf = Uf \qquad (\text{all } \mathbf{f} \text{ in } V). \tag{50}$$

The product TU of two operators T and U is defined by

$$(TU)\mathbf{f} = T(U\mathbf{f}). \tag{51}$$

In general, we do not have $TU = UT$. If $TU = UT$, however, we say that T *commutes* with U, and, in any case, the commutator $[T, U]$ of two operators is

$$[T, U] = TU - UT \tag{52}$$

so that two operators commute if and only if their commutator is the zero operator.

Let T be an operator on V. There may or may not be an operator U on V such that

$$UT = TU = I.$$

If there is such a U, we say that U is the *inverse* of T, and write $U = T^{-1}$. Hence

$$T^{-1}T = TT^{-1} = I. \tag{53}$$

The operator T^{-1}, when it exists, "undoes" the work of T in the sense that if $\mathbf{f}$ is any vector of V, we have

$$T^{-1}(T\mathbf{f}) = (T^{-1}T)\mathbf{f} = I\mathbf{f} = \mathbf{f}. \tag{54}$$

An operator which has an inverse will be called *nonsingular*, otherwise the operator is *singular*. A simple property of the inverse operator is

Theorem 4. *The inverse of a product is the product of the inverses in reverse order*, i.e.,

$$(ST)^{-1} = T^{-1}S^{-1} \tag{55}$$

if S and T are nonsingular.

Proof.

$$(ST)(T^{-1}S^{-1}) = S(TT^{-1})S^{-1} = SIS^{-1} = SS^{-1} = I$$
$$(T^{-1}S^{-1})(ST) = T^{-1}(S^{-1}S)T = T^{-1}IT = T^{-1}T = I$$

which was to be shown.

1.5 EIGENVALUES AND HERMITIAN OPERATORS

Let T be an operator on a vector space V. Among all the vectors of V, there may be some nonzero vectors which, when operated on by T, do not have their "direction" changed, but only their magnitude. More precisely, there may exist a nonzero vector $\mathbf{f}$ and a complex number λ such that

$$T\mathbf{f} = \lambda\mathbf{f}. \tag{56}$$

Any such vector $\mathbf{f}$ is called an *eigenvector* (*characteristic vector, proper vector*) of the operator T, and for any such $\mathbf{f}$, the number λ in (56) is called the *eigenvalue* (*characteristic value, proper value*) of T corresponding to the eigenvector $\mathbf{f}$.

Example 1. Let V be the vector space of all odd trigonometric polynomials

$$f(x) = a_1 \sin x + a_2 \sin 2x + \cdots + a_n \sin nx$$

and let

$$Tf(x) = -f''(x). \tag{57}$$

An eigenvector of this operator, according to (56), is a function $f(x)$ of V for which

$$-f''(x) = \lambda f(x).$$

Hence this operator has infinitely many independent eigenvectors

$$f_n(x) = A_n \sin nx \qquad (n = 1, 2, \ldots) \tag{58}$$

where the A_n are arbitrary constants, and the eigenvalues of T are the numbers $1, 4, 9, \ldots, n^2, \ldots$, the eigenvalue corresponding to the nth eigenvector being n^2.

Example 2. The space V being the same as in Example 1, consider the operator S given by

$$Sf(x) = 3f(x).$$

Clearly every vector in the space V is an eigenvector of S, yet S has only one eigenvalue, $\lambda = 3$.

Let T be a linear operator. The *adjoint operator* T^* is the operator having the property that

$$(\mathbf{x}, T\mathbf{y}) = (T^*\mathbf{x}, \mathbf{y}) \tag{59}$$

for every pair of vectors $\mathbf{x}$, $\mathbf{y}$ in V. We are not stating here that every operator has an adjoint (although this is true in a "complete vector space" or Hilbert space, which is a space satisfying all of our axioms in addition to having the property that every Cauchy sequence of vectors, $\|\mathbf{f}_n - \mathbf{f}_m\| \to 0$, has a limit vector $\mathbf{f}$ in the space) but merely that if an operator T^* satisfying (59) exists, it is called the adjoint of T. Clearly, from (59), $(T^*)^* = T$, for every operator T.

An operator T is *Hermitian*, or *self-adjoint*, if it is its own adjoint, i.e., if $T^* = T$, or equivalently, if

$$(\mathbf{x}, T\mathbf{y}) = (T\mathbf{x}, \mathbf{y}) \tag{60}$$

for every $\mathbf{x}$, $\mathbf{y}$ in V.

Theorem 5. *Let the operators T, U possess adjoints T^*, U^*, respectively. Then the adjoint of TU exists and is U^*T^*.*

Proof. Let x and y be arbitrary vectors of V. Then

$$(x, TUy) = (T^*x, Uy) = (U^*T^*x, y).$$

Theorem 6. *Let T be a self-adjoint operator and x an arbitrary vector of V. Then $(\mathbf{x}, T\mathbf{x})$ is a real number.*

Proof.

$$(\mathbf{x}, T\mathbf{x}) = (T^*\mathbf{x}, \mathbf{x}) = (T\mathbf{x}, \mathbf{x}) = \overline{(\mathbf{x}, T\mathbf{x})}.$$

Theorem 7. *The eigenvalues of a Hermitian operator are real.*

Proof. If

$$A\mathbf{x} = \lambda\mathbf{x}$$

then

$$(\mathbf{x}, A\mathbf{x}) = \lambda(\mathbf{x}, \mathbf{x}),$$

whence the result, since $(\mathbf{x}, A\mathbf{x})$ and $(\mathbf{x}, \mathbf{x})$ are both real.

Theorem 8. *Let $\mathbf{x}$ and $\mathbf{y}$ be eigenvectors of a Hermitian operator T, belonging to distinct eigenvalues λ_1, λ_2, respectively. Then $\mathbf{x}$ and $\mathbf{y}$ are orthogonal.*

Proof. Our hypotheses are:

$$\text{(i)}\ A\mathbf{x} = \lambda_1\mathbf{x}$$
$$\text{(ii)}\ A\mathbf{y} = \lambda_2\mathbf{y}$$
$$\text{(iii)}\ \lambda_1 \neq \lambda_2$$
$$\text{(iv)}\ A = A^*.$$

Taking the inner product of (i) with $\mathbf{y}$ and of (ii) with $\mathbf{x}$,

$$(\mathbf{y}, A\mathbf{x}) = \lambda_1(\mathbf{y}, \mathbf{x})$$
$$(\mathbf{x}, A\mathbf{y}) = \lambda_2(\mathbf{x}, \mathbf{y}).$$

Hence

$$\lambda_2(\mathbf{x}, \mathbf{y}) = (\mathbf{x}, A\mathbf{y}) = (A^*\mathbf{x}, \mathbf{y}) = (A\mathbf{x}, \mathbf{y}) = \overline{(\mathbf{y}, A\mathbf{x})} = \overline{\lambda_1(\mathbf{y}, \mathbf{x})} = \lambda_1(\mathbf{x}, \mathbf{y})$$

and by (iii), $(\mathbf{x}, \mathbf{y}) = 0$, which was to be shown.

1.6 UNITARY OPERATORS

An operator U is said to be *unitary* if it possesses an inverse U^{-1}, an adjoint U^*, and these are equal:

$$U^{-1} = U^* \qquad \text{or} \qquad UU^* = U^*U = I. \tag{61}$$

An operator U is *isometric* if it preserves all inner products, i.e.,

$$(\mathbf{x}, \mathbf{y}) = (U\mathbf{x}, U\mathbf{y}) \qquad \text{all } \mathbf{x}, \mathbf{y} \text{ in } V. \tag{62}$$

In particular, an isometric operator preserves the length of every vector, since

$$\|Ux\|^2 = (Ux, Ux) = (x, x) = \|x\|^2.$$

Thus an isometry may be thought of as a generalized rotation of the vector space V.

Theorem 9. If U* exists, then *U is isometric if and only if it is unitary.*

Proof. If U is unitary, then

$$(U\mathbf{x}, U\mathbf{y}) = (\mathbf{x}, U^*U\mathbf{y}) = (\mathbf{x}, I\mathbf{y}) = (\mathbf{x}, \mathbf{y})$$

and U is isometric.

Conversely, if U is isometric,

$$(U\mathbf{x}, U\mathbf{y}) = (\mathbf{x}, U^*U\mathbf{y}) = (\mathbf{x}, \mathbf{y})$$

for every $\mathbf{x}$, $\mathbf{y}$; hence if we set $S = U^*U - I$, we have $(\mathbf{x}, S\mathbf{y}) = 0$ for every $\mathbf{x}$, $\mathbf{y}$.

Taking, in particular, $\mathbf{x} = S\mathbf{y}$, we find

$$(S\mathbf{y}, S\mathbf{y}) = 0$$

and therefore $S\mathbf{y} = \mathbf{0}$. Since $\mathbf{y}$ was arbitrary, $S = 0$, which was to be shown.

1.7 PROJECTION OPERATORS

An operator P is a projection operator if (i) P is Hermitian and (ii) $P^2 = P$.

Example 1. In Euclidean n-dimensional space, the operator P which associates with the vector $\mathbf{x} = (\alpha_1, \alpha_2, \ldots, \alpha_n)$ the vector $P\mathbf{x} = (\alpha_1, 0, 0, \ldots, 0)$ is a projection. Condition (ii) is obviously satisfied, and equation (60) can be verified by a trivial calculation.

Example 2. In the space of trigonometric polynomials

$$f(x) = a_1 \sin x + a_2 \sin 2x + \cdots + a_n \sin nx$$

with inner product

$$(\mathbf{f}, \mathbf{g}) = \frac{1}{2\pi} \int_0^{2\pi} f(x) g(x)\, dx$$

the operator which carries $f(x)$ into $Pf(x) = a_3 \sin 3x$ is a projection.

Theorem 10. *Let P be a projection operator and* $\mathbf{x}$ *an arbitrary vector. Then we can write*

$$\mathbf{x} = \mathbf{y} + \mathbf{z}$$

where $P\mathbf{y} = \mathbf{y}$, $P\mathbf{z} = \mathbf{0}$.

Proof. Consider the identity

$$\mathbf{x} = P\mathbf{x} + (I - P)\mathbf{x} = \mathbf{y} + \mathbf{z}$$

then

$$P\mathbf{y} = P(P\mathbf{x}) = P^2\mathbf{x} = P\mathbf{x} = \mathbf{y}$$

and

$$P\mathbf{z} = P(I - P)\mathbf{x} = (P - P^2)\mathbf{x} = (P - P)\mathbf{x} = \mathbf{0}. \qquad \text{QED}$$

Thus, if P is a given projection operator, consider the two vector spaces: (a) the set $\mathscr{M}$ of all vectors $P\mathbf{x}$, for $\mathbf{x}$ in V, and (b) the set $\mathscr{M}_\perp$ of all vectors $(I - P)\mathbf{x}$ for $\mathbf{x}$ in V. These two spaces are orthogonal to each other in that if $\mathbf{x}_1$ is in $\mathscr{M}$, $\mathbf{x}_2$ in $\mathscr{M}_\perp$,

$$(\mathbf{x}_1, \mathbf{x}_2) = (P\mathbf{y}_1, (I - P)\mathbf{y}_2) = (\mathbf{y}_1, (P - P^2)\mathbf{y}_2) = (\mathbf{y}_1, \mathbf{0}) = 0$$

and, by the theorem above, these two spaces span V in the sense that any vector of V has the form $\mathbf{x} = \mathbf{y} + \mathbf{z}$, $\mathbf{y}$ in $\mathscr{M}$, $\mathbf{z}$ in $\mathscr{M}_\perp$.

The space $\mathscr{M}$ is the space onto which P projects; $\mathscr{M}_\perp$ is the *orthogonal complement* of $\mathscr{M}$. For any vector $\mathbf{x}$, the vector $P\mathbf{x}$ is the projection of $\mathbf{x}$ onto the space $\mathscr{M}$.

1.8 EUCLIDEAN n-SPACE AND MATRICES

Euclidean n-dimensional space is the space of vectors

$$\mathbf{x} = (\alpha_1, \alpha_2, \ldots, \alpha_n)$$

of ordered n-tuples of complex numbers α_ν ($\nu = 1, 2, \ldots, n$) (the components of $\mathbf{x}$), with addition defined by

$$\begin{aligned}\mathbf{x} + \mathbf{y} &= (\alpha_1, \alpha_2, \ldots, \alpha_n) + (\beta_1, \beta_2, \ldots, \beta_n) \\ &= (\alpha_1 + \beta_1, \alpha_2 + \beta_2, \ldots, \alpha_n + \beta_n)\end{aligned}$$

and the inner product

$$(\mathbf{x}, \mathbf{y}) = \sum_{\nu=1}^{n} \bar{\alpha}_\nu \beta_\nu. \tag{63}$$

The symbol $(\mathbf{x})_i$ will denote the ith component α_i, of the vector $\mathbf{x}$.

Now, let T be an operator† which carries E_m into E_n, that is, if $\mathbf{x}$ is a vector in E_m, then $T\mathbf{x}$ is a vector in E_n. Consider the m vectors $\mathbf{e}_1, \mathbf{e}_2, \ldots, \mathbf{e}_m$ in E_m, where the ith component of $\mathbf{e}_j$ is δ_{ij}. That is,

$$\mathbf{e}_1 = (1, 0, 0, \ldots, 0), \quad \mathbf{e}_2 = (0, 1, 0, \ldots, 0), \ldots, \quad \mathbf{e}_m = (0, 0, \ldots, 0, 1)$$

and let $\mathbf{f}_1, \mathbf{f}_2, \ldots, \mathbf{f}_n$ be the analogous vectors in E_n,

$$(\mathbf{f}_j)_i = \delta_{ij} \qquad (i, j = 1, 2, \ldots, n).$$

† Here we broaden the notion of linear operator to allow T to carry one space into another. The generalized definition of linearity is obvious.

The vectors $\mathbf{e}_1, \ldots, \mathbf{e}_m$ are clearly a basis for E_m, the vectors $\mathbf{f}_1, \ldots, \mathbf{f}_n$ a basis for E_n. For the given operator T, the vectors

$$\mathbf{h}_i = T\mathbf{e}_i \qquad (i = 1, 2, \ldots, m)$$

are in E_n, and therefore are a linear combination of $\mathbf{f}_1, \mathbf{f}_2, \ldots, \mathbf{f}_n$, say

$$\mathbf{h}_i = \sum_{j=1}^{n} \tau_{ji}\mathbf{f}_j \qquad (i = 1, 2, \ldots, m). \tag{64}$$

Now let $\mathbf{x} = (\alpha_1, \alpha_2, \ldots, \alpha_m)$ be an arbitrary vector of E_m. Since

$$\mathbf{x} = \sum_{\nu=1}^{m} \alpha_\nu \mathbf{e}_\nu$$

we have, by linearity of T,

$$T\mathbf{x} = \sum_{\nu=1}^{m} \alpha_\nu T\mathbf{e}_\nu \tag{65}$$

$$= \sum_{\nu=1}^{m} \alpha_\nu \mathbf{h}_\nu.$$

Substituting (64) in (65),

$$T\mathbf{x} = \sum_{\nu=1}^{m} \alpha_\nu \sum_{j=1}^{n} \tau_{j\nu}\mathbf{f}_j$$

$$= \sum_{j=1}^{n} \left\{ \sum_{\nu=1}^{m} \tau_{j\nu}\alpha_\nu \right\} \mathbf{f}_j.$$

Recalling the definition of the vectors $\mathbf{f}_j$, we see that the ith component of the vector $T\mathbf{x}$ is

$$(T\mathbf{x})_i = \sum_{\nu=1}^{m} \tau_{i\nu}\alpha_\nu \qquad (i = 1, 2, \ldots, n). \tag{66}$$

Thus the action of the operator T is completely known (known on any vector) if we only know the numbers $\tau_{i\nu}$ $(i = 1, 2, \ldots, n;\ \nu = 1, 2, \ldots, m)$. If we visualize the numbers τ_{ij} as arranged in a rectangular array having n rows and m columns

$$\begin{pmatrix} \tau_{11} & \tau_{12} & \tau_{13} & \cdots & \tau_{1m} \\ \tau_{21} & \tau_{22} & \tau_{23} & \cdots & \tau_{2m} \\ \cdot & \cdot & \cdot & & \cdot \\ \cdot & \cdot & \cdot & & \cdot \\ \cdot & \cdot & \cdot & & \cdot \\ \tau_{n1} & \tau_{n2} & \tau_{n3} & \cdots & \tau_{nm} \end{pmatrix} \tag{67}$$

then this array is called an $n \times m$ matrix. Since, by (66), this matrix fully represents, or describes, the operator T, we use the same letter T to denote the matrix. We have shown

Theorem 11. *Every linear operator T carrying E_m into E_n is representable, in the sense* (66), *by a matrix of n rows and m columns. Conversely, any $n \times m$ matrix is such an operator if used in the manner* (66).

The equation

$$T = (\tau_{i\nu}) \qquad (i = 1, 2, \ldots, n; \ \nu = 1, 2, \ldots, m)$$

means that T is the matrix which has the number $\tau_{i\nu}$ in row i, column ν.

1.9 MATRIX ALGEBRA

Let A, B be two operators, each carrying E_m into E_n, and represented by the matrices a_{ij}, b_{ij} $(i = 1, 2, \ldots, n; \ j = 1, 2, \ldots, m)$, respectively. If $\mathbf{x}$ is any vector, $\mathbf{x} = (\alpha_1, \ldots, \alpha_m)$, of E_m, we have

$$\begin{aligned}[(A + B)\mathbf{x}]_i &= (A\mathbf{x})_i + (B\mathbf{x})_i \\ &= \sum_{\nu=1}^{m} a_{i\nu}\alpha_\nu + \sum_{\nu=1}^{m} b_{i\nu}\alpha_\nu \\ &= \sum_{\nu=1}^{m} (a_{i\nu} + b_{i\nu})\alpha_\nu.\end{aligned}$$

Comparing with (66), we see that the sum of two operators A, B is represented by the matrix

$$\text{(68)} \qquad \mathbf{A} + \mathbf{B} = (a_{ij} + b_{ij}) \qquad (i = 1, 2, \ldots, n; \ j = 1, 2, \ldots, m).$$

Our next task is to discover how to multiply two matrices. Some care is necessary here, however, for if A and B are defined as above, the operator AB is meaningless, since if $\mathbf{x}$ is in E_m, $B\mathbf{x}$ is in E_n, and A is not defined on the vector $B\mathbf{x}$.

Therefore, let A carry E_m into E_p, and B carry E_p into E_n. The operator BA (not AB!) is meaningful, and carries E_m into E_n. It should be representable by a certain matrix BA of n rows and m columns. We wish to express the elements of this matrix in terms of the elements of the matrices

$$\begin{aligned} A &= (a_{ij}) \qquad (i = 1, 2, \ldots, p, j = 1, \ldots, m) \\ B &= (b_{ij}) \qquad (i = 1, 2, \ldots, n, j = 1, 2, \ldots, p)\end{aligned}$$

representing the operators A, B, respectively.

To do this, let $\mathbf{x} = (x_i)$ be a vector of E_m. Then

$$(A\mathbf{x})_i = \sum_{j=1}^{m} a_{ij}x_j \qquad (i = 1, 2, \ldots, p).$$

Applying the operator B to the vector $A\mathbf{x}$ (in E_p), we find

$$\begin{aligned}((BA)\mathbf{x})_i &= \sum_{k=1}^{p} b_{ik}(A\mathbf{x})_k \\ &= \sum_{k=1}^{p} b_{ik} \sum_{j=1}^{m} a_{kj}x_j \\ &= \sum_{j=1}^{m} \left\{ \sum_{k=1}^{p} b_{ik}a_{kj} \right\} x_j.\end{aligned}$$

Comparing this with the prototype, equation (66), we see that the operator BA is represented by the matrix

$$BA = \left(\sum_{k=1}^{p} b_{ik}a_{kj} \right) \qquad (i = 1, \ldots, n;\ j = 1, \ldots, m). \tag{69}$$

Hence the product of an $n \times p$ matrix by a $p \times m$ matrix is an $n \times m$ matrix.

Example: $(n = 3, p = 2, m = 2)$.

$$AB = \begin{pmatrix} 1 & 2 \\ -1 & 0 \\ 1 & 1 \end{pmatrix} \begin{pmatrix} 1 & 4 \\ 2 & -1 \end{pmatrix} = \begin{pmatrix} 1 \cdot 1 + 2 \cdot 2 & 1 \cdot 4 + 2 \cdot (-1) \\ (-1) \cdot 1 + 0 \cdot 2 & (-1) \cdot 4 + 0 \cdot (-1) \\ 1 \cdot 1 + 1 \cdot 2 & 1 \cdot 4 + 1 \cdot (-1) \end{pmatrix}$$

$$= \begin{pmatrix} 5 & 2 \\ -1 & -4 \\ 3 & 3 \end{pmatrix}$$

1.10 THE ADJOINT MATRIX

Let $A = (a_{ij})$ $(i = 1, 2, \ldots, m;\ j = 1, 2, \ldots, n)$ be an $m \times n$ matrix. Consider the $n \times m$ matrix $B = (b_{ij})$, where

$$b_{ij} = \overline{a_{ji}} \qquad (i = 1, \ldots, n;\ j = 1, \ldots, m). \tag{70}$$

Let $\mathbf{x} = (\alpha_1, \ldots, \alpha_m)$, $\mathbf{y} = (\beta_1, \ldots, \beta_n)$ be arbitrary vectors, respectively, in E_m and E_n. Then

$$\begin{aligned}(\mathbf{x}, A\mathbf{y}) &= \sum_{i=1}^{m} \overline{\alpha_i}(A\mathbf{y})_i \\ &= \sum_{i=1}^{m} \overline{\alpha_i} \sum_{k=1}^{n} a_{ik}\beta_k \\ &= \sum_{i=1}^{m} \sum_{k=1}^{n} \bar{\alpha}_i a_{ik}\beta_k\end{aligned}$$

while

$$(B\mathbf{x}, \mathbf{y}) = \sum_{i=1}^{n} \overline{(B\mathbf{x})_i}\beta_i = \sum_{i=1}^{n} \overline{\left(\sum_{k=1}^{m} \overline{a_{ki}}\alpha_k\right)}\beta_i$$
$$= \sum_{i=1}^{n} \sum_{k=1}^{m} a_{ki}\overline{\alpha_k}\beta_i \equiv (\mathbf{x}, A\mathbf{y}).$$

Hence the matrix B defined by (70) has the property of the operator adjoint to A, that is, $B = A^*$. In symbols,

$$(A^*)_{ij} = \overline{(A)}_{ji}. \tag{71}$$

The matrix A^* is variously referred to as the adjoint matrix, the conjugate transposed matrix, or the Hermitian conjugate matrix. Unfortunately, the name "adjoint matrix" is also applied sometimes to another matrix which we shall encounter presently under the name of "adjugate."

A square ($n \times n$) matrix, therefore, is Hermitian if

$$a_{ij} = \overline{a_{ji}} \qquad (i, j = 1, \ldots, n).$$

The transpose of a matrix A is the matrix A^T given by

$$(A^T)_{ij} = (A)_{ji} \qquad (i = 1, \ldots, m; j = 1, \ldots, n).$$

We see that

$$A^* = \overline{(A^T)}.$$

A square matrix A is *symmetric* if $A = A^T$. For a square matrix with *real* elements to be Hermitian it is necessary and sufficient that it be symmetric.

Theorem 12. *Let A, B be $n \times n$ Hermitian matrices. In order that AB be Hermitian it is necessary and sufficient that A and B commute*, $[A, B] = 0$.

Proof. If AB is Hermitian, then

$$(AB)^* = B^*A^* = BA = AB.$$

Conversely, if $AB = BA$, then

$$(AB)^* = (BA)^* = A^*B^* = AB$$

which was to be shown.

1.11 THE INVERSE MATRIX

Let $A = (a_{ij})$ $(i, j = 1, \ldots, n)$ be a square matrix. We wish to know under what circumstances A has an inverse A^{-1}. For given i, j, the *cofactor* of a_{ij} is defined as the determinant of the matrix A after striking out the ith row and jth column, multiplied by $(-1)^{i+j}$. We assume that the elementary properties of determinants are known. Denoting the cofactor of the element

a_{ij} by the symbol a^{ij}, the familiar Laplace expansion of the determinant by cofactors has the form

$$\sum_{j=1}^{n} a_{ij}a^{ij} = \det(A) \qquad (i = 1, 2, \ldots, n). \tag{72}$$

Now let $i \neq k$, and consider

$$\sum_{j=1}^{n} a_{ij}a^{kj} \tag{73}$$

i.e., the sum of products of elements of one row by cofactors of *another* row. A moment's reflection will show that (73) is exactly the expansion (72) of the determinant of a *modified matrix*, obtained from the given matrix A by deleting the kth row, and substituting for that row, the ith row of A. But this last matrix has two identical rows, hence its determinant must vanish; which is to say that (73) is zero if $i \neq k$. Combining this result with (72), we have found

$$\sum_{j=1}^{n} a_{ij}a^{kj} = \delta_{jk} \cdot \det(A) \qquad (i, k = 1, \ldots, n). \tag{74}$$

If $\det A \neq 0$, we may define the matrix

$$(A^{-1})_{ij} = \frac{a^{ji}}{\det(A)} \qquad (i, j = 1, \ldots, n) \tag{75}$$

and reference to (74) shows that

$$(AA^{-1})_{ij} = \sum_{k=1}^{n} a_{ik}(A^{-1})_{kj}$$

$$= \sum_{k=1}^{n} a_{ik} \frac{a^{jk}}{\det(A)} = \delta_{ij}$$

or

$$AA^{-1} = I \tag{76}$$

where I, the unit matrix, has ones on the diagonal and zeros elsewhere. A trivial modification of the above argument shows that also

$$A^{-1}A = I$$

and hence the matrix A^{-1} defined by (75) is indeed the inverse of A. The matrix whose elements are a^{ij} is the *adjugate* of A, written adj A, and (75) states that

$$A^{-1} = \frac{1}{\det(A)} (\text{adj } A)^T.$$

Hence the calculation of the inverse of a matrix by this method involves the following steps:

(1) Replace each element of A by its cofactor.
(2) Transpose the resulting matrix.
(3) Divide each element by the determinant of A if this is not zero.

We accept without proof (here) the relation

$$\det(AB) = (\det A)(\det B) \tag{77}$$

valid for square matrices A, B, and deduce as a consequence.

Theorem 13. *In order that the square matrix A be nonsingular, it is necessary and sufficient that $\det A \neq 0$. If the inverse exists, it is uniquely given by* (75).

We have already shown the sufficiency. If A^{-1} exists, however, and $\det A = 0$, an absurdity results from using (77) on both sides of (76), which proves the necessity. As to uniqueness, if A has two inverses, say B and C, then

$$AB = AC = I,$$

and multiplying from the left by, say, B, we find $B = C$.

Theorem 14. *Suppose A is nonsingular. Then so is A^* and*

$$(A^*)^{-1} = (A^{-1})^* \tag{78}$$

Proof.

$$A^*(A^{-1})^* = (A^{-1}A)^* = I^* = I$$
$$(A^{-1})^*A^* = (AA^{-1})^* = I^* = I$$

1.12 EIGENVALUES OF MATRICES

In this and the following sections all matrices, unless otherwise specified, will be square ($n \times n$).

Suppose $\mathbf{x}$ is an eigenvector of A corresponding to the eigenvalue λ. Then

$$A\mathbf{x} = \lambda\mathbf{x}.$$

In component form,

$$\sum_{j=1}^{n} a_{ij}x_j = \lambda x_i \qquad (i = 1, \ldots, n)$$

or equivalently,

$$\sum_{j=1}^{n} (a_{ij} - \lambda\delta_{ij})x_j = 0 \qquad (i = 1, \ldots, n). \tag{79}$$

This is a system of n linear, algebraic, *homogeneous* equations in n unknowns. Such a system invariably admits the solution

$$x_1 = x_2 = \cdots = x_n = 0.$$

In order to get a *nontrivial* solution, however, it is necessary for the determinant of the coefficients to vanish, i.e.,

$$\det(A - \lambda I) = \begin{vmatrix} a_{11} - \lambda & a_{12} & \cdots & a_{1n} \\ a_{21} & a_{22} - \lambda & \cdots & a_{2n} \\ \cdot & \cdot & & \cdot \\ \cdot & \cdot & & \cdot \\ \cdot & \cdot & & \cdot \\ a_{n1} & a_{n2} & \cdots & a_{nn} - \lambda \end{vmatrix} = 0 \tag{80}$$

It is clear that (80) is a polynomial equation of degree n in λ, the *characteristic equation* of the matrix A. The polynomial

$$\phi(\lambda) = \det(A - \lambda I) \tag{81}$$

is the *characteristic polynomial* of A. If λ is any one of the n roots $\lambda_1, \lambda_2, \ldots, \lambda_n$ of the characteristic equation of A, then λ is an eigenvalue of A, and conversely. If λ_i is an eigenvalue of A, we may take $\lambda = \lambda_i$ in (79), and the resulting equations will be redundant, by virtue of (80). Hence, by choosing one of the components of the vector $\mathbf{x}$ arbitrarily, the remaining components of $\mathbf{x}$ may be found (possibly not uniquely) from (79). In general, corresponding to any given eigenvalue λ_i, we may expect to find $p \leqq n$ independent vectors $\mathbf{x}$ satisfying (79) with $\lambda = \lambda_i$. In this case we say that the ith eigenstate of the matrix exhibits p-fold *degeneracy* or *multiplicity*. If $p = 1$, the state is nondegenerate.

Example 1. $A = I$, the identity matrix. Here

$$\phi(\lambda) = \det(I - \lambda I) = \det((1 - \lambda)I) = (1 - \lambda)^n.$$

The matrix has exactly one eigenvalue, $\lambda = 1$. Corresponding to this eigenvalue, we may choose *any* n independent vectors $e_1, \ldots, e_n$ as eigenvectors, since $Ie_k = e_k = 1 \cdot e_k$ for any e_k. The eigenvalue $\lambda = 1$, therefore, has degeneracy (or multiplicity) n.

Example 2.

$$A = \begin{pmatrix} 1 & 2 \\ 2 & 3 \end{pmatrix}$$

Here,

$$\phi(\lambda) = \det(A - \lambda I) = \det\begin{pmatrix} 1 - \lambda & 2 \\ 2 & 3 - \lambda \end{pmatrix} = \lambda^2 - 4\lambda - 1$$

The eigenvalues are $\lambda_1 = 2 + \sqrt{5}$, $\lambda_2 = 2 - \sqrt{5}$. For λ_1, we seek a vector $\mathbf{x} = (\alpha_1, \alpha_2)$ such that

$$Ax = \begin{pmatrix} 1 & 2 \\ 2 & 3 \end{pmatrix}\begin{pmatrix} \alpha_1 \\ \alpha_2 \end{pmatrix} = \lambda_1\begin{pmatrix} \alpha_1 \\ \alpha_2 \end{pmatrix} = (2 + \sqrt{5})\begin{pmatrix} \alpha_1 \\ \alpha_2 \end{pmatrix} = \begin{pmatrix} \alpha_1 + 2\alpha_2 \\ 2\alpha_1 + 3\alpha_2 \end{pmatrix}$$

Hence $\alpha_1 + 2\alpha_2 = (2 + \sqrt{5})\alpha_1$; $2\alpha_1 + 3\alpha_2 = \alpha_2(2 + \sqrt{5})$.

Choosing $\alpha_2 = 1$, either equation gives $\alpha_1 = \dfrac{\sqrt{5} - 1}{2}$; hence the eigenvector corresponding to λ_1 is

$$\mathbf{x}_1 = c_1\begin{pmatrix} \sqrt{5} - 1 \\ 2 \end{pmatrix}$$

where c_1 is an arbitrary constant; Similarly, corresponding to λ_2, we find

$$\mathbf{x}_2 = c_2\begin{pmatrix} -(\sqrt{5} + 1) \\ 2 \end{pmatrix}$$

Since this matrix A is Hermitian (symmetric and real), we should expect real eigenvalues and orthogonal eigenvectors, as is indeed the case.

The *trace* (spur) of a matrix $A = (a_{ij})_1{}^n$ is

$$\text{(82)} \qquad \operatorname{Tr}(A) = \sum_{i=1}^{n} a_{ii}$$

the sum of the diagonal elements of A.

Theorem 15. *Let the eigenvalues of A be $\lambda_1, \lambda_2, \ldots, \lambda_n$. Then*

$$\text{(83)} \qquad \lambda_1\lambda_2 \ldots \lambda_n = \det A$$

$$\text{(84)} \qquad \lambda_1 + \lambda_2 + \cdots + \lambda_n = \operatorname{Tr} A.$$

Proof. One finds by expanding the characteristic polynomial $\phi(\lambda)$, in (80),

$$\begin{aligned} \phi(\lambda) &= (-1)^n[\lambda^n - (a_{11} + a_{22} + \cdots + a_{nn})\lambda^{n-1} + \cdots + (-1)^n(\det A)] \\ &= (-1)^n[(\lambda - \lambda_1)(\lambda - \lambda_2)\cdots(\lambda - \lambda_n)] \end{aligned}$$

and the result follows by matching the coefficients of λ^{n-1} and λ^0.

Generally we may write

$$\text{(85)} \qquad \phi(\lambda) = (-1)^n[\lambda^n - \alpha_1\lambda^{n-1} + \alpha_2\lambda^{n-2} + \cdots + (-1)^n\alpha_n]$$

and we see that

$$\alpha_\nu = \sigma_\nu(\lambda_1, \lambda_2, \ldots, \lambda_n) \qquad (\nu = 1, 2, \ldots, n)$$

where σ_ν is the νth elementary symmetric function formed from its arguments. Thus

$$\text{(86)} \qquad \begin{aligned} \alpha_1 &= \lambda_1 + \lambda_2 + \cdots + \lambda_n \\ \alpha_2 &= \lambda_1\lambda_2 + \lambda_1\lambda_3 + \cdots + \lambda_1\lambda_n + \lambda_2\lambda_3 + \cdots + \lambda_2\lambda_n + \cdots + \lambda_{n-1}\lambda_n \\ &\vdots \\ \alpha_n &= \lambda_1\lambda_2 \ldots \lambda_n \end{aligned}$$

1.13 DIAGONALIZATION OF MATRICES

Let $\mathbf{x}_1, \mathbf{x}_2, \ldots, \mathbf{x}_n$ denote the eigenvectors of a given matrix A, corresponding to eigenvalues $\lambda_1, \ldots, \lambda_n$, respectively. Then

$$A\mathbf{x}_j = \lambda_j \mathbf{x}_j \qquad (j = 1, 2, \ldots, n)$$

We define a matrix P by placing in the jth column of P the components of the vector $\mathbf{x}_j$. That is

$$(P)_{ij} = (\mathbf{x}_j)_i \qquad i, j = 1, 2, \ldots, n). \tag{87}$$

Further, we define a diagonal matrix Λ (i.e., a matrix whose off-diagonal elements vanish) by placing λ_i in the ith position on the diagonal,

$$(\Lambda)_{ij} = \lambda_i \delta_{ij} \qquad (i, j = 1, 2, \ldots, n). \tag{88}$$

Then

$$\begin{aligned}(AP)_{ij} &= \sum_{k=1}^{n} a_{ik}(P)_{kj} = \sum_{k=1}^{n} a_{ik}(x_j)_k \\ &= \lambda_j (x_j)_i\end{aligned}$$

and

$$\begin{aligned}(P\Lambda)_{ij} &= \sum_{k=1}^{n} (P)_{ik}(\Lambda)_{kj} = \sum_{k=1}^{n} (x_k)_i \lambda_k \delta_{kj} \\ &= \lambda_j (x_j)_i = (AP)_{ij}.\end{aligned}$$

Thus we have shown that

$$AP = P\Lambda. \tag{89}$$

Suppose now that the eigenvectors $\mathbf{x}_1, \mathbf{x}_2, \ldots, \mathbf{x}_n$ are linearly independent. Then the columns of P are linearly independent, and the determinant of P is not zero. Hence P has an inverse P^{-1}, and multiplying (89) from the right by P^{-1}, we find

$$A = P\Lambda P^{-1}. \tag{90}$$

This representation of A as the product of a nonsingular matrix P, a diagonal matrix Λ, and P^{-1} is called the diagonal form of A, and finding the matrices P, Λ (i.e., the eigenvectors and eigenvalues of A) is referred to as *diagonalizing* the matrix A. A matrix which has n linearly independent eigenvectors is said to be *diagonalizable* (diagonable). The importance of diagonalizing a matrix will appear in the next few sections. The matrix P which occurs in (90), defined by (87), is called the *polar matrix* (modal matrix) of A.

Two matrices A, B are said to be *similar* if there is a nonsingular matrix P such that

$$A = P^{-1}BP. \tag{91}$$

Theorem 16. *Similar matrices have the same eigenvalues.*
Proof. If A and B are similar, let their characteristic polynomials be $\phi(\lambda)$, $\psi(\lambda)$, respectively. Then

$$\begin{aligned}\phi(\lambda) &= \det(A - \lambda I) = \det(P^{-1}BP - \lambda I) \\ &= \det(P^{-1}BP - \lambda P^{-1}P) \\ &= \det[P^{-1}(B - \lambda I)P] \\ &= (\det P^{-1})\det(B - \lambda I)(\det P) \\ &= \det(B - \lambda I) \\ &= \psi(\lambda)\end{aligned}$$

so that their characteristic equations are identical.

We have already proved

Theorem 17. *An arbitrary diagonalizable matrix A is similar to a diagonal matrix Λ.*

We see from these theorems that if we are so fortunate as to be given a matrix in diagonal form (90), then the eigenvalues of A can be read off by inspection, for they are the same as the eigenvalues of the diagonal matrix Λ, namely the numbers on the main diagonal of Λ.

Theorem 18. *Let A be Hermitian. Then A has a polar matrix U which is unitary, and hence the diagonal form of A is*

$$A = U^*\Lambda U. \tag{92}$$

In particular, every Hermitian matrix is diagonalizable.
Proof. First, suppose that A has *distinct* eigenvalues $\lambda_1, \lambda_2, \ldots, \lambda_n$. Then we know that the eigenvectors of A are orthogonal and, *a fortiori*, independent. Hence A is diagonalizable. If these eigenvectors are normalized, and the polar matrix P is constructed as in (87), then the relation

$$P^*P = I$$

which says that P is unitary, is an immediate consequence of the assertion that the columns of P are an orthonormal set, the definition (87) of P, and the laws of matrix multiplication.

If the eigenvalues of A are not distinct, we can perturb the elements of A by adding a Hermitian matrix δA such that $A + \delta A$ has distinct eigenvalues. The previous argument now applies and shows that the modal matrix $U + \delta U$ of $A + \delta A$ exists and may be taken as a unitary matrix, whence by making $\delta A \to 0$, the conclusion persists.

This last proof must be regarded as unsatisfactory in that it intrudes ideas of analysis (continuity of the zeros of a polynomial, limits, etc.) into the domain of algebra—unnecessarily, for purely algebraic proofs can be given. These are, however, complicated, and we have chosen to proceed as above.

1.14 FUNCTIONS OF MATRICES

Let A be a diagonalizable matrix. Then

$$A = P^{-1}\Lambda P$$

for some nonsingular matrix P and diagonal matrix Λ. Hence

$$\begin{aligned} A^2 &= (P^{-1}\Lambda P)(P^{-1}\Lambda P) = P^{-1}\Lambda(PP^{-1})\Lambda P \\ &= P^{-1}\Lambda^2 P \end{aligned}$$

and, in general, for any positive integer k,

$$A^k = P^{-1}\Lambda^k P.$$

We notice at once that calculating Λ^k is a trivial matter, for

$$(\Lambda^k)_{ij} = \lambda_i^{\,k}\delta_{ij} \qquad (i, j = 1, 2, \ldots, n). \tag{93}$$

Next, if

$$f(z) = c_0 + c_1 z + \cdots + c_m z^m \tag{94}$$

we have, by the above,

$$\begin{aligned} f(A) &= P^{-1}\{c_0 I + c_1\Lambda + \cdots + c_m\Lambda^m\}P \\ &= P^{-1}f(\Lambda)P \end{aligned} \tag{95}$$

valid for every polynomial $f(z)$. An obvious limiting argument shows that (95) persists for entire functions (functions regular throughout the plane) $f(z)$. We conclude that the calculation of polynomial or entire functions of a diagonalizable matrix is a simple matter once the eigenvalues and modal matrix (eigenvectors) are known. The matrix $f(\Lambda)$, by (93), is clearly

$$(f(\Lambda))_{ij} = f(\lambda_i)\delta_{ij} \qquad (i, j = 1, \ldots, n). \tag{96}$$

A remarkable consequence follows from (95) and (96) if we choose, for the polynomial $f(z)$, the characteristic polynomial of A, $\phi(\lambda)$. For then, reference to (96) shows that $\phi(\Lambda) = 0$ (the zero matrix), and (95) gives the result

$$\phi(A) = 0$$

which is to say that "a matrix satisfies its own characteristic equation." This result, known as the Cayley-Hamilton theorem, has been proved here only for diagonalizable matrices. It actually is true for all matrices, and we state

Theorem 19. (*Cayley-Hamilton*): *Let $\phi(\lambda)$ be the characteristic polynomial of A. Then $\phi(A)$ is the zero matrix.*

As an application, we note a second method of finding the inverse of a nonsingular matrix A. Indeed, if

$$\phi(\lambda) = \lambda^n - c_1\lambda^{n-1} + \cdots + (-1)^n c_n = 0$$

is the characteristic equation of A, we have

$$A^n - c_1 A^{n-1} + \cdots + (-1)^n c_n I = 0$$

and multiplying by A^{-1},

$$(-1)^{n+1} c_n A^{-1} = A^{n-1} - c_1 A^{n-2} + \cdots + (-1)^{n+1} c_{n-1} I \tag{97}$$

which, since $c_{n_1} = \det A \neq 0$, determines A^{-1} as a linear combination of I, $A, \ldots, A^{n-1}$ with coefficients obtained from the characteristic polynomial of A.

Example. Let

$$A = \begin{pmatrix} 3 & 1 \\ 1 & 3 \end{pmatrix} \tag{98}$$

The characteristic polynomial of A is

$$\phi(\lambda) = \lambda^2 - 6\lambda + 8.$$

Hence

$$A^2 - 6A + 8I = \begin{pmatrix} 10 & 6 \\ 6 & 10 \end{pmatrix} - 6\begin{pmatrix} 3 & 1 \\ 1 & 3 \end{pmatrix} + 8\begin{pmatrix} 1 & 0 \\ 0 & 1 \end{pmatrix} = \begin{pmatrix} 0 & 0 \\ 0 & 0 \end{pmatrix}$$

as required by Theorem 19. The eigenvalues of A are $\lambda_1 = 2$, $\lambda_2 = 4$, with corresponding eigenvectors

$$\mathbf{x}_1 = \begin{pmatrix} 1 \\ -1 \end{pmatrix} \qquad \mathbf{x}_2 = \begin{pmatrix} 1 \\ 1 \end{pmatrix}$$

Thus we may take, for the modal matrix of A,

$$P = \begin{pmatrix} 1 & 1 \\ -1 & 1 \end{pmatrix}$$

and find that the diagonal form of A is

$$A = \begin{pmatrix} 1 & 1 \\ -1 & 1 \end{pmatrix}\begin{pmatrix} 2 & 0 \\ 0 & 4 \end{pmatrix}\begin{pmatrix} \frac{1}{2} & -\frac{1}{2} \\ \frac{1}{2} & \frac{1}{2} \end{pmatrix}$$

If $f(z)$ is any polynomial or entire function of z,

$$f(A) = \begin{pmatrix} 1 & 1 \\ -1 & 1 \end{pmatrix}\begin{pmatrix} f(2) & 0 \\ 0 & f(4) \end{pmatrix}\begin{pmatrix} \frac{1}{2} & -\frac{1}{2} \\ \frac{1}{2} & \frac{1}{2} \end{pmatrix}$$

Note, for example, the labor saved by writing

$$A^{300} = \begin{pmatrix} 1 & 1 \\ -1 & 1 \end{pmatrix} \begin{pmatrix} 2^{300} & 0 \\ 0 & 4^{300} \end{pmatrix} \begin{pmatrix} \frac{1}{2} & -\frac{1}{2} \\ \frac{1}{2} & \frac{1}{2} \end{pmatrix}$$

1.15 THE COMPANION MATRIX

We have already seen that the characteristic equation of a given matrix is a polynomial equation of degree n, n being the order of the matrix. Conversely, suppose

$$\phi(z) = z^n + a_1 z^{n-1} + \cdots + a_{n-1} z + a_n \tag{99}$$

is a given polynomial of degree n; can we find an $n \times n$ matrix whose characteristic polynomial is $\phi(z)$? We observe that the answer is not unique, since if A is any such matrix, $P^{-1}AP$ is another, for any nonsingular matrix P. The answer to the question, however, is always in the affirmative, and the proof is of special interest because it gives an explicit (and easy) construction of a matrix with the desired property.

Theorem 20. *Every polynomial of degree n is the characteristic polynomial of an $n \times n$ matrix.*

Proof. Consider the matrix

$$A = \begin{pmatrix} -a_1 & -a_2 & -a_3 & \cdots & -a_{n-1} & -a_n \\ 1 & 0 & 0 & \cdots & 0 & 0 \\ 0 & 1 & 0 & \cdots & 0 & 0 \\ \cdot & \cdot & & & & \cdot \\ \cdot & \cdot & & & & \cdot \\ \cdot & \cdot & & & & \cdot \\ 0 & 0 & 0 & \cdots & 1 & 0 \end{pmatrix} \tag{100}$$

We claim that $\phi(z)$ of (99) is the characteristic polynomial of A. Indeed,

$$\det(\lambda I - A) = \left| \begin{pmatrix} a_1 + \lambda & a_2 & a_3 & \cdots & a_n \\ -1 & \lambda & 0 & \cdots & 0 \\ 0 & -1 & \lambda & \cdots & 0 \\ \cdot & \cdot & \cdot & & \cdot \\ \cdot & \cdot & \cdot & & \cdot \\ \cdot & \cdot & \cdot & & \cdot \\ 0 & 0 & 0 & \cdots & \lambda \end{pmatrix} \right|$$

To evaluate this determinant, multiply the first column by λ and add to the second, getting

$$\det(\lambda I - A) = \left| \begin{pmatrix} a_1 + \lambda & a_2 + a_1\lambda + \lambda^2 & a_3 & \cdots & a_n \\ -1 & 0 & 0 & \cdots & 0 \\ 0 & -1 & \lambda & \cdots & 0 \\ \cdot & \cdot & \cdot & & \cdot \\ \cdot & \cdot & \cdot & & \cdot \\ \cdot & \cdot & \cdot & & \cdot \\ 0 & 0 & 0 & \cdots & \lambda \end{pmatrix} \right|$$

Now, multiply the second column by λ and add to the third, etc. The final result of these elementary column operations, which do not change the value of the determinant, is

$$\det(\lambda I - A) = \begin{vmatrix} * & * & * & \cdots & \phi(\lambda) \\ -1 & 0 & 0 & \cdots & 0 \\ 0 & -1 & 0 & \cdots & 0 \\ \cdot & \cdot & \cdot & & \cdot \\ \cdot & \cdot & \cdot & & \cdot \\ \cdot & \cdot & \cdot & & \cdot \\ 0 & 0 & 0 & \cdots & 0 \end{vmatrix} = \phi(\lambda)$$

where $\phi(\lambda)$ is the given polynomial (99), and the proof is complete.

The matrix A defined in (100) is called the *companion matrix* of the polynomial $\phi(z)$, and will be of considerable importance in later applications.

1.16 BORDERING HERMITIAN MATRICES

Let A be an $n \times n$ matrix, $\mathbf{u}$, $\mathbf{v}$ (column) vectors, and α a scalar. The matrix

$$\tilde{A} = \begin{pmatrix} A & \mathbf{u} \\ \mathbf{v}^* & \alpha \end{pmatrix} \tag{101}$$

is of order $n + 1$. The process which builds the matrix $\tilde{A}$ from A is called *bordering*, for obvious reasons. If A is Hermitian, then $\tilde{A}$ will also be Hermitian if and only if $\mathbf{u} = \mathbf{v}$, in which case

$$\tilde{A} = \begin{pmatrix} A & \mathbf{u} \\ \mathbf{u}^* & \alpha \end{pmatrix} \tag{102}$$

We are interested here in discovering what happens to the eigenvalues and eigenvectors of a Hermitian matrix when it is bordered.

The notation $\mathbf{x} = (\mathbf{y}, \beta)$, where $\mathbf{y}$ is an n-vector and β a scalar means the $(n + 1)$-vector whose first n components are those of $\mathbf{y}$ and whose last component is β.

Now suppose $\mathbf{x} = (\mathbf{y}, \beta)$ is an eigenvector of $\tilde{A}$. Then

$$\begin{pmatrix} A & \mathbf{u} \\ \mathbf{u}^* & \alpha \end{pmatrix} \begin{pmatrix} \mathbf{y} \\ \beta \end{pmatrix} = \lambda \begin{pmatrix} \mathbf{y} \\ \beta \end{pmatrix}$$

Carrying out the multiplication, we get two equations (one vector and one scalar)

$$A\mathbf{y} + \beta\mathbf{u} = \lambda\mathbf{y} \tag{103}$$

$$(\mathbf{u}, \mathbf{y}) + \alpha\beta = \lambda\beta. \tag{104}$$

Now, suppose the diagonal form of A is known, say,

$$A = U \Lambda U^*, \tag{105}$$

and let $\mathbf{y} = U\mathbf{v}$, where $\mathbf{v}$ is to be determined. Then, from (103),

$$AU\mathbf{v} + \beta\mathbf{u} = \lambda U\mathbf{v}$$

or

$$U \Lambda \mathbf{v} + \beta\mathbf{u} = \lambda U\mathbf{v}.$$

Multiplying by U^*, and remembering that U is unitary,

$$\Lambda\mathbf{v} + \beta U^*\mathbf{u} = \lambda\mathbf{v}$$

which is to say that

$$\mathbf{v} = \beta(\lambda I - \Lambda)^{-1}U^*\mathbf{u} \tag{106}$$

giving the eigenvector, if the eigenvalue is known. Now, using (104),

$$(\mathbf{u}, U\mathbf{v}) = (\lambda - \alpha)\beta$$

and hence

$$(\mathbf{u}, U(\lambda I - \Lambda)^{-1}U^*\mathbf{u}) = (\lambda - \alpha). \tag{107}$$

This is an algebraic equation which determines the eigenvalues λ of $\tilde{A}$. More explicitly, if the ith column of U (ith eigenvector of A) is $\mathbf{x}_i$, equation (107) can be written

$$\sum_{i=1}^{n} \frac{|(\mathbf{u}, \mathbf{x}_i)|^2}{\lambda - \lambda_i} = \lambda - \alpha \tag{108}$$

which brings out quite clearly its algebraic nature. In particular, by plotting, as a function of λ, the left and right sides of (108), it is easy to see that an eigenvalue of $\tilde{A}$ lies between each pair of eigenvalues of A, one to the right of all of them, and one to the left of all of them. If A has a multiple eigenvalue λ, repeated, say, p times, then $\tilde{A}$ has the eigenvalue λ repeated $p - 1$ times.

Theorem 21. *Let A be a Hermitian matrix with diagonal form* (105). *If $\tilde{A}$ is given by* (101), *then the eigenvalues of $\tilde{A}$ are the roots of the algebraic equation* (108). *If λ is any of these, the eigenvector $\mathbf{x} = (\mathbf{y}, \beta)$ corresponding to λ is given by* (106), *where $\mathbf{y} = U\mathbf{v}$. The eigenvalues of A separate those of $\tilde{A}$.*

1.17 DEFINITE MATRICES

Let A be a Hermitian matrix. We know from Theorem 6 that if $\mathbf{x}$ is any vector, then $(\mathbf{x}, A\mathbf{x})$ is a real number. We say that the matrix A is *positive definite* if for every vector $\mathbf{x} \neq \mathbf{0}$, we have

$$(\mathbf{x}, A\mathbf{x}) > 0. \tag{109}$$

The matrix is *non-negative definite* (respectively, *negative definite, negative semidefinite*) if for every $\mathbf{x} \neq \mathbf{0}$, we have $(\mathbf{x}, A\mathbf{x}) \geqq 0$ (respectively, <0, $\leqq 0$). If none of these four alternatives holds (i.e., for some $\mathbf{x} \neq \mathbf{0}$, $(\mathbf{x}, A\mathbf{x}) > 0$ and for others $(\mathbf{x}, A\mathbf{x}) < 0$), the matrix is *indefinite*. The discussion below will be phrased in terms of positive definite matrices, for concreteness, but each assertion will have an obvious analogue in the other three cases.

Theorem 22. *Let A be Hermitian. For A to be positive definite it is necessary and sufficient that all the eigenvalues of A be positive.*
Proof. If A is positive definite, the equation

$$A\mathbf{x} = \lambda\mathbf{x}$$

gives

$$(\mathbf{x}, A\mathbf{x}) = \lambda(\mathbf{x}, \mathbf{x})$$

and λ is clearly positive.

Conversely, suppose all eigenvalues $\lambda_1, \lambda_2, \ldots, \lambda_n$ of A are positive. Let $\mathbf{x} \neq \mathbf{0}$ be an arbitrary vector. Since A is Hermitian, it has n orthogonal eigenvectors $\mathbf{e}_1, \mathbf{e}_2, \ldots, \mathbf{e}_n$, which we may suppose normalized. Then

$$\mathbf{x} = (\mathbf{x}, \mathbf{e}_1)\mathbf{e}_1 + (\mathbf{x}, \mathbf{e}_2)\mathbf{e}_2 + \cdots + (\mathbf{x}, \mathbf{e}_n)\mathbf{e}_n \tag{110}$$

$$A\mathbf{x} = (\mathbf{x}, \mathbf{e}_1)\lambda_1\mathbf{e}_1 + (\mathbf{x}, \mathbf{e}_2)\lambda_2\mathbf{e}_2 + \cdots + (\mathbf{x}, \mathbf{e}_n)\lambda_n\mathbf{e}_n \tag{111}$$

$$(\mathbf{x}, A\mathbf{x}) = |(\mathbf{x}, \mathbf{e}_1)|^2 \lambda_1 + \cdots + |(\mathbf{x}, \mathbf{e}_n)|^2 \lambda_n > 0, \tag{112}$$

which was to be shown.

Corollary 22. For any Hermitian matrix A and vector $\mathbf{x} \neq \mathbf{0}$, we have

$$\lambda_n \leqq \frac{(\mathbf{x}, A\mathbf{x})}{(\mathbf{x}, \mathbf{x})} \leqq \lambda_1 \tag{113}$$

where λ_n is the smallest and λ_1 is the largest eigenvalue of A.

Proof. From (111) we have

$$\lambda_1 - \frac{(\mathbf{x}, A\mathbf{x})}{(\mathbf{x}, \mathbf{x})} = \frac{\lambda_1(\mathbf{x}, \mathbf{x}) - (\mathbf{x}, A\mathbf{x})}{(\mathbf{x}, \mathbf{x})}$$

$$= \frac{1}{(\mathbf{x}, \mathbf{x})}\{(\lambda_1 - \lambda_2)\,|(\mathbf{x}, \mathbf{e}_1)|^2 + \cdots + (\lambda_1 - \lambda_n)\,|(\mathbf{x}, \mathbf{e}_n)|^2\}$$

$$\geqq 0$$

which proves the right hand inequality (113), the left now being obvious.

Theorem 23. *For a Hermitian matrix A to be positive definite it is necessary and sufficient that the coefficients of the characteristic equation of A alternate in sign.*

Proof. First, since A is Hermitian, its eigenvalues are real. Thus, if the coefficients alternate in sign, the characteristic equation clearly cannot have a negative root. Hence all the eigenvalues of A are positive, and by the previous theorem, A is positive definite. Conversely, if A is positive definite, all the eigenvalues of A are positive, and the alternation in sign is evident by inspection of (85) and (86).

If A is a given matrix, the *principal submatrices* $A^{(m)}$ of A are the $m \times m$ matrices formed from the first m rows and columns of A, for each $m = 1, 2, \ldots, n$. The *discriminants* Δ_m of a matrix A are the determinants

$$\Delta_1 = \det A^{(1)} = a_{11}; \quad \Delta_2 = \det A^{(2)}; \quad \ldots; \quad \Delta_n = \det A^{(n)} = \det A.$$

Theorem 24. *For the Hermitian matrix A to be positive definite it is necessary and sufficient that each of its principal submatrices be positive definite.*

Proof. Since $A = A^{(n)}$ is itself one of these submatrices, the condition is clearly sufficient. Conversely, if A is positive definite, let $\mathbf{x}^{(m)} = (\alpha_1, \ldots, \alpha_m)$ be any vector $\neq \mathbf{0}$ in E_m. Then $\mathbf{x} = (\alpha_1, \alpha_2, \ldots, a_m, 0, 0, \ldots, 0)$ is in E_n and it is obvious that

$$(\mathbf{x}^{(m)}, A^{(m)}\mathbf{x}^{(m)}) = (\mathbf{x}, A\mathbf{x}) > 0$$

which shows that each $A^{(m)}$ is also positive definite.

We come, finally, to a criterion for definiteness which does not require either a knowledge of the eigenvalues of A or the labor involved in finding the characteristic equation of A.

Theorem 25. *For a Hermitian matrix A to be positive definite it is necessary and sufficient that*

$$\Delta_1 > 0, \Delta_2 > 0, \ldots, \Delta_n > 0. \tag{114}$$

Proof. If A is positive definite, then so is $A^{(m)}$ for $1 \leqq m \leqq n$. Hence $\Delta_m = \det A^{(m)}$, being the product of the eigenvalues of $A^{(m)}$, is positive, and the condition is necessary.

Conversely, if (114) holds, the theorem is plainly true if $n = 1$ (i.e., if A is a 1×1 matrix). Inductively, suppose (114) is sufficient for matrices of order $1, 2, \ldots, k$, and let A be a $(k + 1) \times (k + 1)$ matrix for which (114) holds, with $n = k + 1$. By the inductive hypothesis, $A^{(k)}$ is positive definite. Suppose A is not. Then A has a negative eigenvalue. Since $\det A = \Delta_{k+1} = \lambda_1\lambda_2 \ldots \lambda_{k+1} > 0$, A must have two negative eigenvalues at least. By Theorem 21 of the preceding section, $A^{(k)}$ has an eigenvalue between these two, which must be negative—a contradiction—and the proof is complete by induction.

1.18 RANK AND NULLITY

If A is nonsingular, then clearly $A\mathbf{x} = \mathbf{0}$ implies $\mathbf{x} = \mathbf{0}$, which is to say that A carries no nonzero vector into zero. On the other hand, if A is singular, there will be nonzero vectors which are annihilated by A. Indeed, since A is singular, it has the eigenvalue zero, and any eigenvector corresponding to this eigenvalue has the required property. Hence, for the given matrix A, there is a space $\mathcal{N}$, called the *null space* of A, which consists of all the vectors $\mathbf{x}$ such that $A\mathbf{x} = \mathbf{0}$. This space contains, with any vectors $\mathbf{x}$, $\mathbf{y}$, the vectors $\mathbf{x} \pm \mathbf{y}$, $\alpha\mathbf{x} + \beta\mathbf{y}$, etc., and hence is itself a vector space, a *subspace* of E_n. The dimension ν of this space $\mathcal{N}$ is called the *nullity* of A.

A complementary notion is that of the *rank* r of a matrix A. This is defined as the dimension of the image of E_n under A. More precisely, let $\mathcal{R}$ denote the set of all vectors $A\mathbf{x}$ as $\mathbf{x}$ runs through all nonzero vectors of E_n. $\mathcal{R}$ is also a subspace of E_n, and its dimension is the rank of A.

Theorem 26. *Let A be an $n \times n$ matrix. Then*

$$\text{(rank of } A) + \text{(nullity of } A) \equiv r(A) + \nu(A) = n. \tag{115}$$

Proof. Let the vectors $\mathbf{e}_1, \ldots, \mathbf{e}_\nu$ form a basis for $\mathcal{N}$. Then we can enlarge this to a basis $\mathbf{e}_1, \ldots, \mathbf{e}_\nu, \mathbf{e}_{\nu+1}, \ldots, \mathbf{e}_n$ for all of E_n. We claim that the vectors $A\mathbf{e}_{\nu+1}, \ldots, A\mathbf{e}_n$ are a basis for $\mathcal{R}$. First, these vectors obviously belong to $\mathcal{R}$.

Next, if $\mathbf{x}$ is any vector of E_n,

$$\mathbf{x} = \sum_{s=1}^{n} \alpha_s \mathbf{e}_s$$

for some scalars $\alpha_1, \ldots, \alpha_n$. Hence

$$A\mathbf{x} = \sum_{s=1}^{n} \alpha_s A\mathbf{e}_s = \sum_{s=\nu+1}^{n} \alpha_s A\mathbf{e}_s$$

and therefore any vector of $\mathscr{R}$ can be expressed as a linear combination of $A\mathbf{e}_{\nu+1}, \ldots, A\mathbf{e}_n$ (they span $\mathscr{R}$). Finally, these vectors are linearly independent, for suppose

$$c_{\nu+1}A\mathbf{e}_{\nu+1} + c_{\nu+2}A\mathbf{e}_{\nu+2} + \cdots + c_nA\mathbf{e}_n = 0$$
$$= A(c_{\nu+1}\mathbf{e}_{\nu+1} + \cdots + c_n\mathbf{e}_n).$$

Then the vector $c_{\nu+1}\mathbf{e}_{\nu+1} + \cdots + c_n\mathbf{e}_n$ belongs to $\mathscr{N}$. Hence it can be expressed as a linear combination of $\mathbf{e}_1, \mathbf{e}_2, \ldots, \mathbf{e}_\nu$

$$c_{\nu+1}\mathbf{e}_{\nu+1} + \cdots + c_n\mathbf{e}_n = c_1\mathbf{e}_1 + \cdots + c_\nu\mathbf{e}_\nu.$$

But this is a linear dependence among the $\mathbf{e}_j$ $(j = 1, \ldots, n)$, contradicting their independence, unless $c_1 = \cdots = c_n = 0$. Thus the set $A\mathbf{e}_{\nu+1}, \ldots, A\mathbf{e}_n$ is a basis for $\mathscr{R}$. Hence the dimension of $\mathscr{R}$, or rank of A, is $n - \nu$, which was to be shown.

A particularly interesting class of matrices is the class of matrices of rank one. Here, the image of E_n under A is of dimension one, which is to say that A carries every vector into a scalar multiple of a certain fixed vector. These matrices are completely characterized by

Theorem 27. *For a matrix A to be of rank one, it is necessary and sufficient that there exist vectors* $\mathbf{u}$, $\mathbf{v}$ *such that*

$$A = \mathbf{u}\mathbf{v}^*. \tag{116}$$

Proof. Suppose such a $\mathbf{u}$ and $\mathbf{v}$ exist. Then, if $\mathbf{x}$ is an arbitrary vector

$$A\mathbf{x} = \mathbf{u}\mathbf{v}^*\mathbf{x} = (\mathbf{v}, \mathbf{x})\mathbf{u}$$

and A is of rank one. Conversely, if $r(A) = 1$, by (115), $\nu(A) = n - 1$, and we may take vectors $\mathbf{e}_1, \ldots, \mathbf{e}_{n-1}$ as an orthonormal basis for the null space $\mathscr{N}$ of A. Adjoining a vector $\mathbf{v}$, we get an orthonormal basis $\mathbf{e}_1, \mathbf{e}_2, \ldots, \mathbf{e}_{n-1}, \mathbf{v}$ for all of E_n. If $\mathbf{x}$ is an arbitrary vector, we have

$$\mathbf{x} = (\mathbf{e}_1, \mathbf{x})\mathbf{e}_1 + \cdots + (\mathbf{e}_{n-1}, \mathbf{x})\mathbf{e}_{n-1} + (\mathbf{v}, \mathbf{x})\mathbf{v}.$$

Hence

$$A\mathbf{x} = (\mathbf{v}, \mathbf{x})A\mathbf{v} \equiv (\mathbf{v}, \mathbf{x})\mathbf{u}$$

where $\mathbf{u} = A\mathbf{v}$. If $\mathbf{u}$ and $\mathbf{v}$ are so defined, then the matrix $\mathbf{u}\mathbf{v}^*$ must be identical with A, since its effect on every vector is the same, which was to be shown.

Equation (116) shows that

$$(A)_{ij} = u_i\bar{v}_j \qquad (i, j = 1, 2, \ldots, n) \tag{117}$$

and therefore a square $n \times n$ matrix of rank one has only $2n$ independent elements rather than n^2.

Suppose we have a matrix A whose inverse A^{-1} is known. Consider now the matrix $\tilde{A}$ obtained from A by changing exactly one element, say a_{pq}, to $\tilde{a}_{pq}$. The difference between A and $\tilde{A}$ is a matrix, all of whose elements vanish except one. Such a matrix is easily seen, from (116), to be of rank one. This state of affairs persists even if a whole row or column of A is changed. We are thus led to inquire into the relationship between A^{-1} and $\tilde{A}^{-1}$ when $A - \tilde{A}$ is of rank one.

Theorem 28.† *Let A be nonsingular, and let* **u** *and* **v** *be vectors. Then*

$$(A + \mathbf{uv}^*)^{-1} = A^{-1} - \frac{A^{-1}\mathbf{uv}^*A^{-1}}{1 + (\mathbf{v}, A^{-1}\mathbf{u})} \tag{118}$$

in the sense that if either side exists, then so does the other and they are equal.

Proof. First, suppose $A = I$, the identity matrix. Then

$$(I + \mathbf{uv}^*)\left[I - \frac{\mathbf{uv}^*}{1 + (\mathbf{v}, \mathbf{u})}\right] = I - \frac{\mathbf{uv}^*}{1 + (\mathbf{v}, \mathbf{u})} + \mathbf{uv}^* - \frac{(\mathbf{v}, \mathbf{u})\mathbf{uv}^*}{1 + (\mathbf{v}, \mathbf{u})} = I$$

and (118) is proved in this case. In the general case,

$$(A + \mathbf{uv}^*)^{-1} = [A(I + A^{-1}\mathbf{uv}^*)]^{-1} = [I + (A^{-1}\mathbf{u})\mathbf{v}^*]^{-1}A^{-1}$$

which is of the type already considered, and the result follows by applying the formula already proved.

Next, let A be a Hermitian matrix, with eigenvalues $\lambda_1, \lambda_2, \ldots, \lambda_n$ and eigenvectors $\mathbf{e}_1, \mathbf{e}_2, \ldots, \mathbf{e}_n$, which we assume to be orthonormal. Consider the matrix

$$B = A - \lambda_1\mathbf{e}_1\mathbf{e}_1^*. \tag{119}$$

If $j \geqq 2$, we have

$$\begin{aligned} B\mathbf{e}_j &= A\mathbf{e}_j - \lambda_1\mathbf{e}_1\mathbf{e}_1^*\mathbf{e}_j = \lambda_j\mathbf{e}_j - \lambda_1(\mathbf{e}_1, \mathbf{e}_j)\mathbf{e}_1 \\ &= \lambda_j\mathbf{e}_j \end{aligned}$$

while

$$B\mathbf{e}_1 = A\mathbf{e}_1 - \lambda_1\mathbf{e}_1(\mathbf{e}_1, \mathbf{e}_1) = \lambda_1\mathbf{e}_1 - \lambda_1\mathbf{e}_1 = \mathbf{0}.$$

Hence the matrix B has eigenvalues $\lambda_2, \lambda_3, \ldots, \lambda_n$, 0 with corresponding eigenvectors $\mathbf{e}_2, \mathbf{e}_3, \ldots, \mathbf{e}_n, \mathbf{e}_1$. The subtraction (119) is of particular value, in practice, if one has available some technique which is capable of finding the largest eigenvalue and corresponding eigenvector of a Hermitian matrix. Applying such a technique to A, we would find λ_1, $\mathbf{e}_1$; subtracting as in (119), we would then form B and apply the method to it, finding λ_2, $\mathbf{e}_2$, etc. The end result of this process would be the Hermitian matrix

$$A - \lambda_1\mathbf{e}_1\mathbf{e}_1^* - \cdots - \lambda_n\mathbf{e}_n\mathbf{e}_n^*$$

† Sherman and Morrison [1], Woodbury [1].

which has only the eigenvalue zero, of multiplicity n. Hence this must be the zero matrix, and we have discovered that

$$A = \lambda_1 \mathbf{e}_1 \mathbf{e}_1^* + \cdots + \lambda_n \mathbf{e}_n \mathbf{e}_n^*. \tag{120}$$

This, however, is the second time we have discovered this fact, for, as the reader should verify, (120) is merely a rewriting of (92).

1.19 SIMULTANEOUS DIAGONALIZATION AND COMMUTATIVITY

We know that a Hermitian matrix A can always be written

$$A = U\Lambda_a U^* \tag{121}$$

where Λ_a is diagonal and U is unitary. Suppose B is another Hermitian matrix having the same modal matrix U, say

$$B = U\Lambda_b U^* \tag{122}$$

where Λ_b is diagonal. Then we say that the matrices A and B are *simultaneously diagonalized* by the unitary matrix U. If this happens, then

$$\begin{aligned} AB - BA &= U\Lambda_a U^* U\Lambda_b U^* - U\Lambda_b U^* U\Lambda_a U^* \\ &= U\Lambda_a\Lambda_b U^* - U\Lambda_b\Lambda_a U^* \\ &= U(\Lambda_a\Lambda_b - \Lambda_b\Lambda_a)U^* \\ &= 0 \end{aligned}$$

since diagonal matrices always commute. Hence, if A and B are simultaneously diagonalizable, then $[A, B] = 0$. The converse is also true, and we state

Theorem 29. *Let A and B be Hermitian matrices. For A and B to be simultaneously diagonalizable it is necessary and sufficient that A and B commute.*

Proof. We have already shown the necessity. Suppose now that $AB = BA$. Suppose further that A has the distinct eigenvalues $\lambda_1, \ldots, \lambda_n$ and normalized eigenvectors $\mathbf{e}_1, \ldots, \mathbf{e}_n$, and that B has the distinct eigenvalues $\mu_1, \ldots, \mu_n$, with normalized eigenvectors $\mathbf{f}_1, \ldots, \mathbf{f}_n$. Then, for any j,

$$\begin{aligned} A(B\mathbf{e}_j) = (AB)\mathbf{e}_j = (BA)\mathbf{e}_j &= B(A\mathbf{e}_j) \\ = B(\lambda_j \mathbf{e}_j) &= \lambda_j(B\mathbf{e}_j). \end{aligned}$$

Thus $B\mathbf{e}_j$ is the eigenvector of A corresponding to the eigenvalue λ_j; therefore $B\mathbf{e}_j$ is proportional to $\mathbf{e}_j$, say

$$B\mathbf{e}_j = \sigma_j \mathbf{e}_j$$

But this means that $\mathbf{e}_j$ is one of the eigenvectors of B. Hence the eigenvectors $\mathbf{f}_1, \mathbf{f}_2, \ldots, \mathbf{f}_n$ of B are merely a rearrangement of $\mathbf{e}_1, \mathbf{e}_2, \ldots, \mathbf{e}_n$, and by renumbering them, if necessary, A and B have precisely the same eigenvectors, and the desired conclusion follows. If A or B (or both) fails to have distinct eigenvalues, we may reach the conclusion by, say, the same perturbation argument that was used in the proof of Theorem 18.

If, in Theorem 29, the word "matrices" is replaced by "operators" and "diagonalizable" by "measurable with arbitrary precision," we have a fundamental theorem of quantum mechanics known as the uncertainty principle. The reasons for permitting this substitution of words belong to the domain of physics and will not concern us here.

1.20 THE NUMERICAL CALCULATION OF EIGENVALUES

Let A be a Hermitian matrix with eigenvalues $|\lambda_1| > |\lambda_2| \geqq |\lambda_3| \geqq \cdots \geqq |\lambda_n|$ and orthonormal eigenvectors $\mathbf{e}_1, \mathbf{e}_2, \ldots, \mathbf{e}_n$. Let $\mathbf{x}$ be an arbitrary nonzero vector. Define a sequence of vectors $\mathbf{x}_0, \mathbf{x}_1, \mathbf{x}_2, \ldots$ by

$$\begin{cases} \mathbf{x}_0 = \mathbf{x} \\ \mathbf{x}_{\nu+1} = A\mathbf{x}_\nu \qquad (\nu = 0, 1, 2, \ldots) \end{cases} \tag{123}$$

and a sequence of numbers

$$\sigma_\nu = \frac{(\mathbf{x}_\nu, \mathbf{x}_{\nu+1})}{(\mathbf{x}_\nu, \mathbf{x}_\nu)} = \frac{(\mathbf{x}_\nu, A\mathbf{x}_\nu)}{(\mathbf{x}_\nu, \mathbf{x}_\nu)} \qquad (\nu = 0, 1, 2, \ldots). \tag{124}$$

We will show that

$$\lim_{\nu\to\infty} \sigma_\nu = \lambda_1 \tag{125}$$

if $(\mathbf{x}, \mathbf{e}_1) \neq 0$.

Indeed, we have

$$\mathbf{x} = (\mathbf{x}, \mathbf{e}_1)\mathbf{e}_1 + \cdots + (\mathbf{x}, \mathbf{e}_n)\mathbf{e}_n$$

and thus,

$$\begin{aligned} \mathbf{x}_\nu = A^\nu\mathbf{x} &= (\mathbf{x}, \mathbf{e}_1)A^\nu\mathbf{e}_1 + \cdots + (\mathbf{x}, \mathbf{e}_n)A^\nu\mathbf{e}_n \\ &= \lambda_1{}^\nu(\mathbf{x}, \mathbf{e}_1)\mathbf{e}_1 + \cdots + \lambda_n{}^\nu(\mathbf{x}, \mathbf{e}_n)\mathbf{e}_n. \end{aligned}$$

It follows that

$$\begin{aligned} \sigma_\nu = \frac{(\mathbf{x}_\nu, A\mathbf{x}_\nu)}{(\mathbf{x}_\nu, \mathbf{x}_\nu)} &= \frac{\lambda_1^{2\nu+1}\,|(\mathbf{x}, \mathbf{e}_1)|^2 + \cdots + \lambda_n^{2\nu+1}\,|(\mathbf{x}, \mathbf{e}_n)|^2}{\lambda_1^{2\nu}\,|(\mathbf{x}, \mathbf{e}_1)|^2 + \cdots + \lambda_n^{2\nu}\,|(\mathbf{x}, \mathbf{e}_n)|^2} \\ &= \lambda_1 \left\{ \frac{\lambda_1^{2\nu}\,|(\mathbf{x}, \mathbf{e}_1)|^2 + \cdots + \dfrac{\lambda_n^{2\nu+1}}{\lambda_1}\,|(\mathbf{x}, \mathbf{e}_n)|^2}{\lambda_1^{2\nu}\,|(\mathbf{x}, \mathbf{e}_1)|^2 + \cdots + \lambda_n^{2\nu}\,|(\mathbf{x}, \mathbf{e}_n)|^2} \right\} \end{aligned}$$

When ν is large, the first term in the denominator dominates it, and the first term in the numerator is dominant there; hence the expression in braces tends to unity as $\nu \to \infty$, and $\sigma_\nu \to \lambda_1$. If the vectors $\mathbf{x}_\nu$ are normalized after each iteration, then it is obvious that the $\mathbf{x}_\nu$ converge to the dominant eigenvector of A. Having found this dominant eigenvalue and eigenvector, equation (119) may be used to find the remaining eigenvalues and eigenvectors.

If the smallest eigenvalue is desired, rather considerable labor is involved in the above process, and it is preferable to develop other methods which converge to the smallest eigenvalue directly. One way, of course, is to consider A^{-1}, whose largest eigenvalue is the reciprocal of the smallest eigenvalue of A, and apply the process (123) to it. Even this labor can be avoided, however, by constructing a slight variant of (123) in the following manner. With an arbitrary initial vector $\mathbf{x}$, take

$$\begin{aligned} \mathbf{x}_0 &= \mathbf{x} \\ \mathbf{x}_{\nu+1} &= (A - \eta_\nu I)\mathbf{x}_\nu \qquad (\nu = 0, 1, 2, \ldots) \end{aligned} \tag{126}$$

where the number η_ν is to be determined. Estimating the eigenvalue by (124), again we find

$$\begin{aligned} \sigma_\nu &= \frac{(\mathbf{x}_{\nu+1}, A\mathbf{x}_{\nu+1})}{(\mathbf{x}_{\nu+1}, \mathbf{x}_{\nu+1})} \\ &= \frac{(A\mathbf{x}_\nu - \eta_\nu \mathbf{x}_\nu, A^2\mathbf{x}_\nu - \eta_\nu A\mathbf{x}_\nu)}{(A\mathbf{x}_\nu - \eta_\nu \mathbf{x}_\nu, A\mathbf{x}_\nu - \eta_\nu \mathbf{x}_\nu)} \\ &= \frac{\tau_3 - 2\eta_\nu\tau_2 + \eta_\nu^2\tau_1}{\tau_2 - 2\eta_\nu\tau_1 + \eta_\nu^2\tau_0} \end{aligned} \tag{127}$$

where we have written $\tau_i = (\mathbf{x}_\nu, A^i\mathbf{x}_\nu)$, $(i = 0, 1, 2, 3)$. To converge to the largest eigenvalue most rapidly we should choose η_ν, at each stage, so as to maximize (127); to converge to the smallest eigenvalue we need only to minimize (127).† The choices of η_ν which accomplish these objectives are the two roots of the quadratic equation

$$\begin{vmatrix} 1 & \eta & \eta^2 \\ \tau_0 & \tau_1 & \tau_2 \\ \tau_1 & \tau_2 & \tau_3 \end{vmatrix} = 0 \tag{128}$$

The complete iterative process consists, then, in choosing an arbitrary vector $\mathbf{x}_0$, and, in general, if $\mathbf{x}_\nu$ has been determined, calculating τ_0, τ_1, τ_2, τ_3, and taking for η_ν either the larger or smaller of the roots of (128) in (126), depending on whether the largest or smallest eigenvalue of A is sought.

† Hestenes and Karush [1].

1.21 APPLICATION TO DIFFERENTIAL EQUATIONS

Let A be a square matrix. Then the series

$$e^{At} = I + At + \frac{A^2t^2}{2!} + \cdots \tag{129}$$

converges uniformly on any finite interval of t, as does the derived series, and hence by differentiation

$$Ae^{At} = \frac{d}{dt}(e^{At}) = e^{At}A \quad (!!) \tag{130}$$

the equation meaning that

$$\frac{d}{dt}\{(e^{At})_{ij}\} = (Ae^{At})_{ij} = (e^{At}A)_{ij} \qquad (i, j = 1, \ldots, n).$$

Hence consider the system of n linear ordinary differential equations in n unknown functions $y_1(t), \ldots, y_n(t)$

$$\begin{cases} \dfrac{dy_i}{dt} = \displaystyle\sum_{j=1}^{n} a_{ij}y_j(t) & (i = 1, 2, \ldots, n) \\ y_i(0) \text{ given} \end{cases} \tag{131}$$

where the a_{ij} are constants.

If we define the vector $\mathbf{y}(t) = (y_1(t), \ldots, y_n(t))$, (131) is

$$\begin{cases} \mathbf{y}'(t) = A\mathbf{y}(t) \\ \mathbf{y}(0) \text{ given} \end{cases} \tag{132}$$

whose solution is plainly

$$\mathbf{y}(t) = e^{At}\mathbf{y}(0). \tag{133}$$

Hence the solution of the initial value problem (132) is equivalent to finding the matrix e^{At}. If $A = P\Lambda P^{-1}$ is diagonalizable, (133) becomes

$$\mathbf{y}(t) = Pe^{\Lambda t}P^{-1}\mathbf{y}(0). \tag{134}$$

Another way of looking at (132) if A is diagonalizable is to make the change of dependent variable $\mathbf{u}(t) = P^{-1}\mathbf{y}(t)$. Then (132) takes the form

$$\mathbf{u}'(t) = \Lambda\mathbf{u}(t) \tag{135}$$

in which the equations are *uncoupled*, and hence may be solved separately, the final result being, of course, (134).

As an example, suppose $\mathbf{y}(t)$ satisfies $\mathbf{y}'(t) = A\mathbf{y}(t)$ for $0 \leqq t \leqq t_0$ and $\mathbf{y}'(t) = B\mathbf{y}(t)$ for $t_0 \leqq t \leqq t_1$, where A, B are different, diagonalizable, constant matrices, and the solution vector is required to be everywhere continuous.

For $0 \leqq t \leqq t_0$,

$$\mathbf{y}(t) = e^{At}\mathbf{y}(0) \tag{136}$$

while for $t_0 \leqq t \leqq t_1$,

$$\mathbf{y}(t) = e^{Bt}\mathbf{c} \tag{137}$$

where $\mathbf{c}$ is a vector of constants. Joining these solutions at $t = t_0$,

$$e^{Bt_0}\mathbf{c} = e^{At_0}\mathbf{y}(0)$$

or

$$\mathbf{c} = e^{-Bt_0}e^{At_0}\mathbf{y}(0) \qquad (\neq e^{(A-B)t_0}\mathbf{y}(0)).$$

Hence in $t_0 \leqq t \leqq t_1$ the solution is

$$\mathbf{y}(t) = e^{Bt}\mathbf{c} = e^{B(t-t_0)}e^{At_0}\mathbf{y}(0). \tag{138}$$

From (138) we can, for instance, read off the behavior of the solution as $t \to \infty$.

Next, we make only a few sketchy remarks about the application of matrix methods to the approximate solution of partial differential equations.

Theorem 30. *Let A be a square, diagonalizable matrix, with eigenvalues $\lambda_1, \lambda_2, \ldots, \lambda_n$. In order that*

$$\lim_{\nu \to \infty} A^\nu = 0 \tag{139}$$

it is necessary and sufficient that

$$|\lambda_i| < 1 \qquad (i = 1, 2, \ldots, n). \tag{140}$$

Proof. Since

$$A^\nu = P\Lambda^\nu P^{-1}$$

it is clear that what we need is $\lambda_i^\nu \to 0$ for each i, which happens precisely when (140) holds. The hypothesis that A is diagonalizable is actually superfluous.

An immediate corollary is that

$$(I - A)^{-1} = I + A + A^2 + \cdots \tag{141}$$

if all the eigenvalues of A lie in the unit circle in the complex plane. Indeed, in that case we have

$$\begin{aligned} I + A + \cdots + A^\nu &= P\{I + \Lambda + \cdots + \Lambda^\nu\}P^{-1} \\ &\to P(1 - \Lambda)^{-1}P^{-1} \\ &= (I - A)^{-1} \end{aligned}$$

Let us now consider, say, Poisson's equation in a rectangle $\mathscr{R}$,

$$\nabla^2\phi(x, y) = S(x, y) \tag{142}$$

with values of ϕ given on the boundary of $\mathscr{R}$. If we overlay the rectangle $\mathscr{R}$ with a lattice formed by the lines $x_m = mh$ ($m = 0, 1, \ldots, M$), $y_n = nh$ ($n = 0, 1, \ldots, N$), then we may approximate

$$\left.\frac{\partial^2\phi}{\partial x^2}\right|_{(x_m, y_n)} = \frac{\phi(x_{m+1}, y_n) - 2\phi(x_m, y_n) + \phi(x_{m-1}, y_n)}{h^2}$$

$$\left.\frac{\partial^2\phi}{\partial y^2}\right|_{(x_m, y_n)} = \frac{\phi(x_m, y_{n+1}) - 2\phi(x_m, y_n) + \phi(x_m, y_{n-1})}{h^2}$$

If these approximations are substituted in (142) the result is a system of $(M + 1)(N + 1)$ simultaneous linear algebraic equations in the unknown numbers $\phi(x_m, y_n)$, of the form

$$\mathbf{\Phi} = A\mathbf{\Phi} + \mathbf{S} \tag{143}$$

The solution is manifestly $\mathbf{\Phi} = (I - A)^{-1}\mathbf{S}$, but because of the large size of A $[(M + 1)(N + 1) \times (M + 1)(N + 1)]$ and the large number of zeros it contains, a direct inversion of $I - A$ is quite wasteful of effort. If, however, we take an initial guess vector $\mathbf{\Phi}_0$ and define recursively

$$\mathbf{\Phi}_{\nu+1} = A\mathbf{\Phi}_\nu + \mathbf{S} \qquad (\nu = 0, 1, 2, \ldots) \tag{144}$$

then we have

$$\begin{aligned} \mathbf{\Phi}_1 &= A\mathbf{\Phi}_0 + \mathbf{S} \\ \mathbf{\Phi}_2 &= A^2\mathbf{\Phi}_0 + (I + A)\mathbf{S} \\ &\vdots \\ \mathbf{\Phi}_\nu &= A^\nu\mathbf{\Phi}_0 + (I + A + \cdots + A^{\nu-1})\mathbf{S} \end{aligned} \tag{145}$$

From (139) and (141) we see that if the eigenvalues of A lie in the unit circle (as is generally the case in practice) the first term in (145) tends to zero, and the second to the solution of the problem. The recursive process (144) is called *relaxation*, and converges rapidly or slowly depending on whether the largest eigenvalue of A is considerably or slightly less than unity in modulus.

1.22 BOUNDS FOR THE EIGENVALUES

The eigenvalues of a diagonal matrix are its diagonal elements. It is reasonable to suppose, therefore, that if the off-diagonal elements of a matrix are "small," then the eigenvalues are not "too far" from the diagonal elements. Our purpose here is to quantify this assertion.

Theorem 31. *Let $A = (a_{ij})_1^n$ be an arbitrary square matrix. If*

$$|a_{ii}| > \sum_{j \neq i} |a_{ij}| \qquad (i = 1, 2, \ldots, n) \tag{146}$$

then A is nonsingular.†

Proof. Suppose A is singular and let $\mathbf{x} = (x_1, \ldots, x_n)$ satisfy $A\mathbf{x} = \mathbf{0}, \mathbf{x} \neq \mathbf{0}$. Let $|x_s| = \max_i |x_i|$, then

$$\begin{aligned} |a_{ss}|\,|x_s| = |a_{ss}x_s| &= \left|\sum_{j \neq s} a_{sj}x_j\right| \\ &\leqq \sum_{j \neq s} |a_{sj}|\,|x_j| \leqq |x_s| \sum_{j \neq s} |a_{sj}| \\ &< |a_{ss}|\,|x_s| \end{aligned}$$

which is impossible.

Let us now apply this theorem to the matrix $A - \lambda I$. It tells us that if

$$|\lambda - a_{ii}| > \sum_{j \neq i} |a_{ij}| \qquad (i = 1, 2, \ldots, n)$$

then $A - \lambda I$ is nonsingular, that is, λ is not an eigenvalue of A. We have proved

Theorem 32.‡ *Every eigenvalue of A lies in at least one of the circles*

$$|\lambda - a_{ii}| \leqq \sum_{j \neq i} |a_{ij}| \qquad (i = 1, 2, \ldots, n) \tag{147}$$

in the complex plane.

Naturally, if A is Hermitian, we may interpret (147) as describing intervals rather than as circles.

This theorem can actually be considerably refined, though we shall not prove it here, to the statement that if p of the circles (147) overlap to form a connected region C which is disjoint from the rest of the circles, then exactly p eigenvalues of A lie in C. Hence, in particular, if the circles are disjoint, there is one eigenvalue in each.

1.23 MATRICES WITH NONNEGATIVE ELEMENTS

Let $A = (a_{ij})_1^n$ be a square matrix with real elements. We will write

$$A \geqq 0$$

if $a_{ij} \geqq 0\,(i, j = 1, 2, \ldots, n)$, and $A \leqq B$ if $B - A \geqq 0$. Furthermore, if B is a matrix with complex elements $(b_{ij})_1^n$, we write $B^+ = (|b_{ij}|)_1^n$. Plainly, $B^+ \geqq 0$, for any B.

† Hadamard [1].

‡ Lévy [1], Hadamard [1]. See also Gerşchgorin [1], Brauer [1].

We say that the matrix $A = (a_{ij})_1{}^n$ is *reducible* if there exist two sets of integers I, J having the properties

(i) $I \cup J$ is the set $\{1, 2, \ldots, n\}$.
(ii) $I \cap J$ is empty.
(iii) Neither I nor J is empty.
(iv) If i is in I and j is in J, then $a_{ij} = 0$.

Otherwise, if no such sets exist, A is said to be *irreducible*.

Example 1. The matrix

$$A = \begin{pmatrix} 0 & 1 & 1 \\ 0 & 0 & 0 \\ 1 & 0 & 0 \end{pmatrix}$$

is reducible, since we may take $I = \{2\}$, $J = \{1, 3\}$, and check that (i)–(iv) are satisfied.

Example 2. The companion matrix of an algebraic equation whose constant term is not zero is irreducible.

We postpone the proof of this important fact to a later chapter (exercise 1, Chapter 3).

Naturally a matrix with no zero elements is irreducible. On the other hand—as in Example 2—a matrix can have many zero entries and still be irreducible. The point is that the zeros must be strategically placed so that the sets I, J cannot be constructed. The main theorem of the subject of non-negative matrices gives certain information about the eigenvalues and eigenvectors of such matrices. This theorem is rather easy to prove if one confines attention to matrices with no zero entries. Most of the conclusions of this theorem, however, remain true if zero entries are permitted, provided the matrix remains irreducible. Since we wish to use the full generality of this theorem in later applications, the notion of irreducibility is necessary to our discussion.

For any vector $\mathbf{y}$, let $Z(\mathbf{y})$ denote the number of zero components of $\mathbf{y}$.

Lemma 1. Let $A \geqq 0$ be an $n \times n$ irreducible matrix, and let $\mathbf{y} \geqq 0$, $\mathbf{y} \neq \mathbf{0}$ be a non-negative n-vector. Then

$$Z((I + A)\mathbf{y}) < Z(\mathbf{y})$$

unless both sides are zero.

Proof. Write $\mathbf{u} = (I + A)\mathbf{y}$. Since $I + A \geqq 0$, clearly $Z(\mathbf{u}) \leqq Z(\mathbf{y})$. Suppose $Z(\mathbf{u}) = Z(\mathbf{y})$. The zero components of $\mathbf{u}$ and $\mathbf{y}$ occur in the same places in each vector because $u_i \geqq y_i$. Hence we may suppose, by relabeling the rows of $\mathbf{u}$ and $\mathbf{y}$, that

$$\mathbf{y} = \begin{pmatrix} \mathbf{x} \\ \mathbf{0} \end{pmatrix} \qquad \mathbf{u} = \begin{pmatrix} \mathbf{w} \\ \mathbf{0} \end{pmatrix}$$

where $\mathbf{x} > 0$, $\mathbf{w} > 0$ are vectors of length $n - Z(\mathbf{u})$. We then relabel the rows and the columns of A in accordance with the same permutation of the integers $1, 2, \ldots, n$ that was used in renumbering the rows of $\mathbf{u}$ and $\mathbf{y}$. The irreducibility of A is not affected by this process (see exercise 40). Next we partition the matrix A into

$$A = \begin{pmatrix} A_{11} & A_{12} \\ A_{21} & A_{22} \end{pmatrix}$$

where the submatrix A "matches" $\mathbf{x}$, etc. Then

$$\mathbf{u} = \begin{pmatrix} \mathbf{w} \\ \mathbf{0} \end{pmatrix} = (I + A)\mathbf{y} = \begin{pmatrix} I + A_{11} & A_{12} \\ A_{21} & I + A_{22} \end{pmatrix} \begin{pmatrix} \mathbf{x} \\ \mathbf{0} \end{pmatrix} = \begin{pmatrix} \mathbf{x} + A_{11}\mathbf{x} \\ A_{21}\mathbf{x} \end{pmatrix}$$

and it follows that $A_{21}\mathbf{x} = \mathbf{0}$. But since $\mathbf{x} > 0$, $A_{21} = 0$, and the matrix A has an $(n - p) \times p$ block of zeros in its lower left corner. It is easy to see that this contradicts the irreducibility of A, and the lemma is proved (see exercise 39).

Lemma 2. Let $A \geqq 0$ be $n \times n$ and irreducible. Then

$$(I + A)^{n-1} > 0.$$

Proof. Let $\mathbf{y} \geqq 0$, $\mathbf{y} \neq \mathbf{0}$ be a non-negative vector. By repeated application of Lemma 1, $(I + A)^{n-1}\mathbf{y}$ has no zero components. Since $\mathbf{y}$ was arbitrary, the lemma is proved.

Now let $A \geqq 0$ be a fixed, non-negative, irreducible matrix. For any vector $\mathbf{x} \geqq 0$, $\mathbf{x} \neq \mathbf{0}$, $\mathbf{x} = (\mathbf{x}_1, \mathbf{x}_2, \ldots, \mathbf{x}_n)$, define a number

$$r(\mathbf{x}) = \min_{1 \leqq i \leqq n} \frac{(A\mathbf{x})_i}{x_i}. \tag{148}$$

Lemma 3. For the given vector $\mathbf{x}$, $r(\mathbf{x})$ is the largest number r such that

$$r\mathbf{x} \leqq A\mathbf{x} \tag{149}$$

Proof. First, $r(\mathbf{x})$ has the property (149), since

$$\begin{aligned} (A\mathbf{x} - r(\mathbf{x})\mathbf{x})_j &= (A\mathbf{x})_j - r(\mathbf{x})x_j \\ &\geqq (A\mathbf{x})_j - \frac{(A\mathbf{x})_j}{x_j} x_j = 0. \end{aligned} \tag{150}$$

Further, there is a value of j in (150) for which the sign of equality holds throughout; hence no number larger than $r(\mathbf{x})$ satisfies (149).

The notation

$$\sup_{\mathbf{x} \text{ in } M} r(\mathbf{x})$$

denotes the least upper bound of all the values taken by $r(\mathbf{x})$ as $\mathbf{x}$ ranges through a set M of vectors.

Consider the following sets of vectors:

M'': The set of all vectors $\mathbf{x}$ such that $\mathbf{x} \geqq 0$, $\mathbf{x} \neq \mathbf{0}$.
M': The set of vectors $\mathbf{y} = \mathbf{x}/\|\mathbf{x}\|$, where $\mathbf{x}$ is in M''.
M: The set of vectors $\mathbf{y} = (I + A)^{n-1}\mathbf{x}$, where $\mathbf{x}$ is in M'.

Lemma 4. We have

$$\sup_{\mathbf{x} \text{ in } M''} r(\mathbf{x}) = \sup_{\mathbf{x} \text{ in } M'} r(\mathbf{x}) = \sup_{\mathbf{x} \text{ in } M} r(\mathbf{x}) = \max_{\mathbf{x} \text{ in } M} r(\mathbf{x}) \tag{151}$$

Proof. What this lemma asserts is that in order to find the largest value that $r(\mathbf{x})$ takes on any non-negative vector $\mathbf{x}$ it is enough to restrict attention to the special class M.

First, from the definition (148) of $r(\mathbf{x})$ we see that the value of $r(\mathbf{x})$ is unaltered if the vector $\mathbf{x}$ is multiplied by a scale factor. Hence we may restrict attention to normalized vectors, which proves the first of the equalities (151).

Next, by (149),

$$r(\mathbf{x})\mathbf{x} \leqq A\mathbf{x}$$

and multiplying both sides by the positive matrix $(I + A)^{n-1}$,

$$r(\mathbf{x})\mathbf{y} \leqq (I + A)^{n-1}A\mathbf{x} = A(I + A)^{n-1}\mathbf{x} = A\mathbf{y} \tag{152}$$

where $\mathbf{y} = (I + A)^{n-1}\mathbf{x}$. But $r(\mathbf{y})$ is the *largest* number with the property (152). Hence $r(\mathbf{x}) \leqq r(\mathbf{y})$, and

$$\sup_{\mathbf{x} \text{ in } M'} r(\mathbf{x}) \leqq \sup_{\mathbf{x} \text{ in } M} r(\mathbf{x})$$

The reverse inequality is obvious,

$$\sup_{\mathbf{x} \text{ in } M'} r(\mathbf{x}) = \sup_{\mathbf{x} \text{ in } M''} r(\mathbf{x}) \geqq \sup_{\mathbf{x} \text{ in } M} r(\mathbf{y})$$

since every vector in M is also in M''. Hence the second equality in (151) is proved. The third equality there states only that the least upper bound is actually attained on some vector $\mathbf{y}$ in the set M. But the set M' is just the surface of the unit sphere in Euclidean n-space, and is therefore compact. The set M is the image of this compact set under the continuous function $\mathbf{x} \to (I + A)^{n-1}\mathbf{x}$, and is also compact. Finally, since M consists only of *positive* vectors, $r(\mathbf{y})$ is a continuous function on M, which is compact, and therefore $r(\mathbf{y})$ attains its maximum value in M.

We now state the fundamental theorem of the subject of matrices with non-negative elements.

Theorem 33.† (*Perron-Frobenius*) *Let* $A \geqq 0$ be *irreducible. Then*

(i) *A has a positive eigenvalue r which is not exceeded in modulus by any other eigenvalue of A.*

† Perron [1], [2], Frobenius [1], [2].

(ii) *The eigenvector of A corresponding to the eigenvalue r has positive components and is essentially unique.*

(iii) *The number r is given by*

$$r = \max_{x \geqq 0} \min_{1 \leqq i \leqq n} \frac{(Ax)_i}{x_i} = \min_{x \geqq 0} \max_{1 \leqq i \leqq n} \frac{(Ax)_i}{x_i} \tag{153}$$

Proof. First, the number r defined by (153) is plainly non-negative. We show that it cannot be zero, for consider the vector $\mathbf{u} = (1, 1, \ldots, 1)$. We have

$$r(\mathbf{u}) = \min_{1 \leqq i \leqq n} \frac{(A\mathbf{u})_i}{u_i} = \min_{1 \leqq i \leqq n} (A\mathbf{u})_i$$

$$= \min_{1 \leqq i \leqq n} \sum_{j=1}^{n} a_{ij}$$

Since $A \geqq 0$, $\sum_{j=1}^{n} a_{ij} \neq 0$ for any i, for otherwise a whole row of A would consist of zeros, contradicting the irreducibility of A. Hence $r(\mathbf{u})$ is positive, and r, the maximum of $r(\mathbf{x})$ over all vectors $\mathbf{x} \geqq 0$ is, in particular, $\geqq r(\mathbf{u})$, and thus is positive also.

Next, let $\mathbf{v}$ be a vector on which $r(\mathbf{x})$ attains its maximum value, $r(\mathbf{v}) = r$, and set $\mathbf{y} = (I + A)^{n-1}\mathbf{v}$. By (149) we know that

$$A\mathbf{v} - r\mathbf{v} \geqq 0 \tag{154}$$

and we wish to show that the sign of equality actually holds here. Suppose $A\mathbf{v} - r\mathbf{v} \neq \mathbf{0}$. Multiplying (154) by $(I + A)^{n-1}$, we find

$$A\mathbf{y} - r\mathbf{y} > 0. \tag{155}$$

But $r(\mathbf{y})$ is the largest number such that $Ay - ry \geqq 0$, and (155) shows that $r(\mathbf{y}) > r$, which contradicts the maximum property of r. Hence we have shown that

$$A\mathbf{v} = r\mathbf{v} \tag{156}$$

and therefore the number r defined by (153) is an eigenvalue of A. This number is called the *Perron root* of A.

Next, we see that the eigenvector $\mathbf{v}$ is > 0, since

$$(I + A)^{n-1}\mathbf{v} = (1 + r)^{n-1}\mathbf{v} > 0.$$

Now let λ be any eigenvalue of A. We will show that $|\lambda| \leqq r$. Suppose

$$A\mathbf{u} = \lambda\mathbf{u}$$

then

$$|\lambda|\, \mathbf{u}^+ \leqq A\mathbf{u}^+$$

or

$$A\mathbf{u}^+ - |\lambda|\, \mathbf{u}^+ \geqq 0.$$

Thus, in view of Lemma 3,

$$|\lambda| \leqq r(\mathbf{u}^+) \leqq r.$$

Finally, we will prove that the eigenvector $\mathbf{v}$ corresponding to the Perron root r is unique (aside from a scale factor). Indeed, suppose $\mathbf{v}^{(1)}$ and $\mathbf{v}^{(2)}$ are two linearly independent eigenvectors corresponding to the eigenvalue r. We already know that we can take $\mathbf{v}^{(1)} > 0$, $\mathbf{v}^{(2)} > 0$. Defining

$$\sigma = -\min_{1 \leqq i \leqq n} \frac{v_i^{(2)}}{v_i^{(1)}}$$

the vector

$$\sigma \mathbf{v}^{(1)} + \mathbf{v}^{(2)} \geqq 0$$

is a non-negative eigenvector of A corresponding to the Perron root r, with one vanishing component, at least, which is impossible, completing the proof of the theorem.

Corollary 33. Let $A \geqq 0$ be irreducible, and let $\mathbf{x} \geqq 0$ be arbitrary. Then the Perron root of A satisfies

$$\min_{1 \leqq i \leqq n} \frac{(A\mathbf{x})_i}{x_i} \leqq r \leqq \max_{1 \leqq i \leqq n} \frac{(A\mathbf{x})_i}{x_i}. \tag{157}$$

Theorem 34.† *Let A be an irreducible $n \times n$ matrix with arbitrary complex elements. Let λ be any eigenvalue of A, and let r denote the Perron root of the matrix A^+. Then*

$$|\lambda| \leqq r. \tag{158}$$

Proof. Suppose

$$A\mathbf{y} = \lambda \mathbf{y}$$

Then

$$|\lambda|\, \mathbf{y}^+ \leqq A^+\mathbf{y}^+$$

or

$$A^+\mathbf{y}^+ - |\lambda|\, \mathbf{y}^+ \geqq 0$$

By Lemma 3,

$$|\lambda| \leqq r(\mathbf{y}^+) \leqq r$$

which was to be shown.

This theorem, coupled with (157), can often provide useful upper bounds for the eigenvalues of an irreducible complex matrix.

Bibliography

1. F. Riesz and B. v. sz. Nagy, *Leçons d'Analyse Fonctionelle*, Akadémiai Kiadó, Budapest, 1952.

provides a thorough and rigorous account of the theory of abstract Hilbert spaces and linear operators, as does

2. P. R. Halmos, *Introduction to Hilbert Space*, Chelsea Publishing Co., New York, 1957.

† Wielandt [1].

3. P. R. Halmos, *Finite Dimensional Vector Spaces*, Princeton University Press, 1948.
handles the subject of its title in a fully rigorous way.

Both volumes of

4. F. R. Gantmacher, *The Theory of Matrices*, Chelsea Publishing Co., New York, 1960.

are highly recommended and, in particular, are valuable references for the treatment of multiple characteristic numbers and elementary divisors, which we have slighted here, and for a complete discussion of matrices with non-negative elements.

The discussion in Chapter 1 of

5. F. B. Hildebrand, *Methods of Applied Mathematics*, Prentice-Hall, New York, 1952.

is close, in spirit, to the presentation given here although different in detail.

For numerical methods of inversion, calculation of eigenvalues, etc. one should first look in

6. A. Householder, *Principles of Numerical Analysis*, McGraw-Hill Book Co., New York, 1953.

Several individual methods of numerical matrix analysis are discussed in

7. A. Ralston and H. S. Wilf, *Mathematical Methods for Digital Computers*, John Wiley and Sons, New York, 1960.

For a rather complete list of theorems and references concerning matrices, see

8. C. C. MacDuffie, *Theory of Matrices*, Chelsea Publishing Co., New York, 1946.

Exercises

1. The familiar "triangle inequality" (the sum of the lengths of two sides of a triangle is not less than the length of the third side) has the following form in a vector space V:

$$\|\mathbf{x} + \mathbf{y}\| \leqq \|\mathbf{x}\| + \|\mathbf{y}\|.$$

   Prove the truth of this inequality, using only our system of axioms and the Schwarz inequality.
2. An orthogonal set of vectors is independent.
3. In the vector space V_2 of all polynomials of degree $\leqq 2$, with the inner product $(f, g) = \int_{-1}^{1} \overline{f(x)}g(x)\,dx$, the three vectors 1, x, x^2 are independent. Apply the Gram-Schmidt process to orthogonalize these vectors. The resulting polynomials are called Legendre polynomials.
4. (a) If T possesses an adjoint and an inverse, then $(T^*)^{-1} = (T^{-1})^*$.
   (b) $(\alpha T)^* = \bar{\alpha}T^*$; $(T_1 + T_2)^* = T_1^* + T_2^*$.
5. An operator T is skew-Hermitian if $T^* = -T$.
   (a) The eigenvalues of a skew-Hermitian operator are purely imaginary numbers.
   (b) Let S be an arbitrary operator possessing an adjoint S^*. Then S can be written in the form

$$S = H + K$$

   where H is Hermitian and K is skew-Hermitian.
6. If $AB = AC$, does $B = C$? (A, B, C are square matrices of order n.) Prove or construct a counterexample.

7. Let A be an arbitrary $m \times n$ matrix. Then $B = AA^*$ is square and Hermitian.
8. Discuss the eigenvalues and eigenvectors of $A = \begin{pmatrix} 0 & 1 \\ 0 & 0 \end{pmatrix}$. Is A diagonalizable?
9. The matrix $A = \begin{pmatrix} a & b \\ c & d \end{pmatrix}$ is a real projection operator on E_2. Find a, b, c, d, and discuss your answer geometrically. What are the spaces $\mathscr{M}$, $\mathscr{M}_\perp$ in terms of a, b, c, d?
10. The matrix $A = \begin{pmatrix} a & b \\ c & d \end{pmatrix}$ is a real unitary operator on E_2. Find a, b, c, d and discuss your answer geometrically.
11. The relation $A \sim B$ (A is similar to B) is an equivalence relation among square matrices of the same order. That is,
    (a) $A \sim A$.
    (b) If $A \sim B$, then $B \sim A$.
    (c) If $A \sim B$ and $B \sim C$, then $A \sim C$.
12. Using the result of (11), show that diagonalizable matrices with the same eigenvalues are similar.
13. $\mathrm{Tr}\,[A, B] = 0$, for arbitrary square matrices A, B.
14. Find the inverse of the matrix A of (98), using (97).
15. Find a 3×3 matrix whose eigenvalues are the squares of the roots of the equation $\lambda^3 + 11\lambda^2 - 2\lambda + 1 = 0$.
16. Let $\lambda_1, \lambda_2, \ldots, \lambda_n$ be the eigenvalues of A. Then the eigenvalues of $f(A)$ are $f(\lambda_1), \ldots, f(\lambda_n)$, where $f(\lambda)$ is a polynomial or entire function.
17. Let A be Hermitian and positive definite. Suppose A is bordered as in (102). What are the necessary and sufficient conditions (on $\mathbf{u}$ and α) that $\tilde{A}$ should also be positive definite?
18. Let A and B be Hermitian and positive definite. Is $A + B$? A^2? $f(A)$ (where $f(\lambda)$ is a polynomial)?
19. If A is $m \times n$, then AA^* is non-negative definite (compare (7)).
20. What is the analogue of (114) for negative definite matrices?
21. Show that each of the discriminants $\Delta_1, \Delta_2, \ldots, \Delta_n$ (114) is a real number.
22. Show that a real, symmetric, positive definite matrix A has a real, symmetric, positive definite square root, i.e., a matrix B such that $B^2 = A$.
23. Consider the matrix $A = (a_{mn})$ $(m, n = 1, 2, \ldots, N)$ where

$$a_{mn} = \cos\left(t \log \frac{m}{n}\right) \qquad (m, n = 1, 2, \ldots, N)$$

t being a real parameter.

(a) Show that, if the complex variable $s = \sigma + it$, then

$$\left| \sum_{n=1}^{N} n^{-s} \right|^2 = \sum_{m,n=1}^{N} m^{-\sigma} a_{mn} n^{-\sigma}$$

(b) Show that A is of rank 2.
(c) Show that A is non-negative definite and symmetric.
(d) Find the eigenvalues and eigenvectors of A.

24. Describe a method of inverting an arbitrary nonsingular matrix based on repeated application of (118). Count the multiplications needed for an $n \times n$ matrix.
25. Show that the inverse of an arbitrary nonsingular matrix can be found by inverting a non-negative definite matrix and two matrix multiplications.
26. Prove that ν linearly independent vectors $\mathbf{e}_1, \mathbf{e}_2, \ldots, \mathbf{e}_\nu$ can always be extended to a basis $\mathbf{e}_1, \ldots, \mathbf{e}_n$ for E_n.
27. Let A, B be Hermitian, positive definite matrices. Find necessary and sufficient conditions for AB to be also Hermitian and positive definite.
28. Is the relation $A \smile B$ (A commutes with B) an equivalence relation (see exercise 11)?
29. Let the matrix A be diagonalizable. Show that the Cauchy integral formula

$$f(A) = \frac{1}{2\pi i} \oint_C \frac{f(z)\,dz}{zI - A}$$

holds if the closed contour C encloses all the eigenvalues of A, and if $f(z)$ is regular in C. The formula is interpreted as

$$(f(A))_{ij} = \frac{1}{2\pi i} \oint_C (zI - A)_{ij}^{-1} f(z)\,dz \qquad (i, j = 1, \ldots, n).$$

30. Use the result of problem (30) to find e^{At} where

$$A = \begin{pmatrix} 1 & 2 \\ 0 & 2 \end{pmatrix}$$

and t is a real number. Thus solve the system of two differential equations $\mathbf{y}'(t) = A\mathbf{y}(t)$, with $\mathbf{y}(0)$ a given vector.

31. For Hermitian matrices A, B, let $A \geqq 0$ mean that A is non-negative definite, and $A \geqq B$ mean that $A - B \geqq 0$. If $A \geqq B$ and $B \geqq C$, is $A \geqq C$?
32. The rank of a matrix is a similarity invariant, i.e., if $A \sim B$, then $r(A) = r(B)$, and hence also, $\nu(A) = \nu(B)$ (see exercise 11).
33. Is the nullity of a matrix equal to the multiplicity of the eigenvalue zero? Is it for diagonalizable matrices?
34. A matrix which commutes with all diagonal matrices is diagonal.
35. Describe all Hermitian matrices of rank one.
36. Find $(I + uv^*)^n$.
37. Show that $r(AB) \geqq \min(r(A), r(B))$.
38. Is $e^{(A+B)} = e^A e^B$? When is this true?
39. Let the $n \times n$ matrix A have an $(n - p) \times p$ block of zeros in its lower left-hand corner. Then A is reducible.
40. Let A be a given $n \times n$ matrix. Let the rows of A be renumbered in any manner, and let the columns of A be renumbered exactly as the rows were. Denote the matrix which results by A_1. Then A_1 is reducible if and only if A was.

chapter 2

Orthogonal functions

2.1 INTRODUCTION

In the previous chapter we investigated certain general properties of vector spaces and linear operators on vector spaces and then specialized these results to Euclidean n-dimensional space. We noticed, in passing, that certain sets of functions also qualified for the title "vector space" and gave a few examples of these.

In this chapter some of these function spaces are discussed in detail. More specifically, we treat here orthogonal polynomials, Fourier series, and the convergence theory of orthogonal expansions—of which the Fourier series is the prototype. The concept of linear operators moves into the background, in this context, and is superseded by the notion of orthogonality itself.

Our presentation of the theory of orthogonal polynomials proceeds from the general to the specific, in the belief that the unity of the subject is best recognized in this manner. The degree of generality that we attain in this way is only very slightly less than optimum, the latter requiring a knowledge of the Lebesgue-Stieltjes integral which we do not presume. In practice, the slight loss of generality that we incur by using the Riemann integral means only that our discussion does not include the case of orthogonality with respect to summation over discrete point sets rather than with respect to integration over an interval. The reader who is familiar with the Stieltjes integral will find that our proofs carry over, virtually unchanged, to that case.

2.2 ORTHOGONAL POLYNOMIALS

Let $[a, b]$ be an interval (finite or infinite) of the real axis. Let $w(x)$ be a function defined on $[a, b]$ and satisfying the following hypotheses:

$H1$: $w(x) \geqq 0$ on $[a, b]$.
$H2$: The integrals

$$\int_a^b x^n w(x)\,dx \equiv \mu_n \qquad (n = 0, 1, 2, \ldots) \tag{1}$$

exist and are finite for each $n = 0, 1, 2, \ldots$, and $\mu_0 > 0$.

A function $w(x)$ satisfying $H1$, $H2$ will be called a *weight function for* $[a, b]$, or if the context is clear, a *weight function*. The numbers μ_n defined by (1) are the *moments* of $w(x)$.

Example. The function $w(x) = 1/x$ is a weight function for any finite interval $[a, b]$ where $b > a > 0$ and not for any other kind of interval.

If $f(x)$ and $g(x)$ are any real polynomials and $w(x)$ is a weight function for $[a, b]$, the number

$$(f, g) \equiv \int_a^b f(x)g(x)w(x)\,dx \tag{2}$$

being a linear combination of the μ_n, is finite, and is called the inner product of f and g. It should be verified that this inner product satisfies axioms (9) through (14) of Chapter 1 (exercise 1), so that the class of all real polynomials forms a vector space over the real numbers (i.e., the constants in the axioms (1) through (14) of Chapter 1 are all real).

That being so, we can remark that the vectors

$$1, x, x^2, x^3, \ldots$$

are linearly independent in this space; therefore we may apply the Gram-Schmidt process to orthogonalize them. The result of this procedure will be a sequence of polynomials

$$\phi_0(x), \phi_1(x), \phi_2(x), \ldots \tag{3}$$

where $\phi_j(x)$ is of degree j ($j = 0, 1, 2, \ldots$) having the property that

$$(\phi_i(x), \phi_j(x)) = \int_a^b \phi_i(x)\phi_j(x)w(x)\,dx = \delta_{ij} \qquad (i, j = 0, 1, 2, \ldots). \tag{4}$$

We call the sequence $\{\phi_n(x)\}_0^\infty$ the sequence of *orthogonal polynomials* associated with the weight function $w(x)$ and the interval $[a, b]$.

We give also the following, more explicit construction of the sequence (3). Assume that, for some fixed n, the polynomial $\phi_n(x)$ has the form

$$\phi_n(x) = \sum_{\nu=0}^{n} \alpha_\nu x^\nu. \tag{5}$$

We wish to determine the coefficients α_ν so that $\phi_n(x)$ is orthogonal to each power of x less than the nth. These conditions are

$$(6) \qquad (x^i, \phi_n(x)) = 0 \qquad (i = 0, 1, \ldots, n-1)$$

Using (5), (6) becomes

$$(7) \qquad \sum_{\nu=0}^{n} (x^i, x^\nu)\alpha_\nu = 0 \qquad (i = 0, 1, \ldots, n-1).$$

Now

$$(x^i, x^\nu) = \int_a^b x^i x^\nu w(x)\, dx = \int_a^b x^{i+\nu} w(x)\, dx$$
$$= \mu_{i+\nu}$$

Further, equations (7) are n linear equations in $n + 1$ unknowns; so we choose $\alpha_n = 1$, for example, and then must solve

$$(8) \qquad \sum_{\nu=0}^{n-1} \mu_{i+\nu}\alpha_\nu = -\mu_{n+i} \qquad (i = 0, 1, \ldots, n-1).$$

This will be possible if and only if the matrix

$$(9) \qquad M_{ij} = \mu_{i+j} \qquad (i, j = 0, 1, \ldots, n-1)$$

is nonsingular. This matrix is called the *moments matrix* of the weight function $w(x)$, and we have

Theorem 1. *The moments matrix of a weight function is strictly positive definite, and* a fortiori, *nonsingular.*

Proof. Let $\boldsymbol{\xi} = (\xi_0, \xi_1, \ldots, \xi_{n-1})$ be an n-vector. Then

$$(\boldsymbol{\xi}, M\boldsymbol{\xi}) = \sum_{i,j=0}^{n-1} \xi_i M_{ij} \xi_j = \sum_{i,j=0}^{n-1} \xi_i \mu_{i+j} \xi_j$$
$$= \left(\sum_{i=0}^{n-1} \xi_i x^i, \sum_{j=0}^{n-1} \xi_j x^j \right) > 0$$

unless $\boldsymbol{\xi} = \mathbf{0}$, which proves the theorem.

As an illustration, we take $w(x) = 1$, $[a, b] = [-1, 1]$. Then

$$(10) \qquad \mu_n = \int_{-1}^{1} x^n\, dx = \begin{cases} 2/n+1 & n \text{ even} \\ 0 & n \text{ odd} \end{cases}$$

The moments matrix is

$$(11) \qquad M = \begin{pmatrix} 2 & 0 & \frac{2}{3} & 0 & \cdots \\ 0 & \frac{2}{3} & 0 & \frac{2}{5} & \cdots \\ \frac{2}{3} & 0 & \frac{2}{5} & 0 & \cdots \\ 0 & \frac{2}{5} & 0 & \frac{2}{7} & \cdots \\ \cdot & \cdot & \cdot & \cdot & \\ \cdot & \cdot & \cdot & \cdot & \\ \cdot & \cdot & \cdot & \cdot & \end{pmatrix}$$

Taking, for example, $n = 2$, equations (8) are

$$\begin{pmatrix} 2 & 0 \\ 0 & \frac{2}{3} \end{pmatrix} \begin{pmatrix} \alpha_0 \\ \alpha_1 \end{pmatrix} = \begin{pmatrix} -\frac{2}{3} \\ 0 \end{pmatrix}$$

with solution

$$\begin{pmatrix} \alpha_0 \\ \alpha_1 \end{pmatrix} = \begin{pmatrix} -\frac{1}{3} \\ 0 \end{pmatrix}.$$

Hence $P_2(x) = \alpha_0 + \alpha_1 x + x^2 = -\frac{1}{3} + x^2$ is the second member of the sequence of orthogonal polynomials belonging to the weight function $w(x) = 1$ and interval $[-1, 1]$. This particular sequence is the sequence of Legendre's polynomials.

Clearly each member of a sequence of orthogonal polynomials can be multiplied by a constant without altering the orthogonality of the sequence. If this is done so that

$$(12) \qquad (\phi_i(x), \phi_i(x)) = 1 \qquad (i = 0, 1, 2, \ldots)$$

the sequence is said to be normalized. There is still a choice of sign to be made even though the sequence is normalized. When speaking of normalized sequences, we will assume that the sign has been chosen so that the highest power of x in $\phi_n(x)$ has a positive coefficient, for each $n = 0, 1, 2, \ldots$

We summarize with

Theorem 2. *Let $w(x)$ be a weight function for the interval $[a, b]$. Then there exists a unique sequence $\{\phi_n(x)\}_0^\infty$ of polynomials satisfying*

$$(13) \qquad (\phi_m(x), \phi_n(x)) = \delta_{mn} \qquad (m, n = 0, 1, \ldots)$$

$$(14) \qquad \phi_n(x) \text{ is of exact degree } n \qquad (n = 0, 1, \ldots)$$

$$(15) \qquad k_n = \text{highest coefficient of } \phi_n(x) \text{ is} > 0 \qquad (n = 0, 1, \ldots)$$

2.3 ZEROS

Let $w(x)$ be a fixed weight function on a given interval $[a, b]$. We wish to discuss some of the properties of the sequence of orthogonal polynomials associated with $w(x)$, $[a, b]$.

By (14), of course, $\phi_n(x)$ has n zeros (roots of the equation $\phi_n(x) = 0$) somewhere in the complex plane. Much more can be said, however, the following theorem being of fundamental importance.

Theorem 3. *The zeros of the polynomial $\phi_n(x)$ are*

(i) *real*

(ii) *distinct*

(iii) *contained in the interval (a, b).*

Proof. First, for $n \geqq 1$,

$$0 = (\phi_n(x), \phi_0(x)) = \phi_0 \int_a^b \phi_n(x) w(x)\, dx$$

where ϕ_0 is a nonzero constant. Thus $\phi_n(x)$ surely changes sign at least once in (a, b). Let $x_1, \ldots, x_r$ denote the zeros of odd multiplicity, each counted just once, of $\phi_n(x)$ which lie in (a, b). Since

$$\phi_n(x) = (x - x_1)^{\alpha_1} \cdots (x - x_r)^{\alpha_r} \psi(x)$$

where $\psi(x)$ is a polynomial of degree $n - \alpha_1 - \cdots - \alpha_r$, and does not change sign on (a, b), it follows that

$$\phi_n(x)(x - x_1) \cdots (x - x_r) = (x - x_1)^{\alpha_1 + 1} \cdots (x - x_r)^{\alpha_r + 1} \psi(x)$$

does not change sign in (a, b). On the other hand,

$$\int_a^b \phi_n(x)(x - x_1) \cdots (x - x_r) w(x)\, dx = 0$$

if $r < n$, which is impossible. Thus $r = n$ and the theorem is proved.

2.4 THE RECURRENCE FORMULA

Let k_n denote the highest coefficient of $\phi_n(x)$, as in (15). We remark first that any polynomial of degree n can be written as a linear combination of $\phi_0(x), \ldots, \phi_n(x)$, since, starting with

$$f(x) = c_0 + c_1 x + \cdots + c_n x^n \tag{16}$$

we may write x^n as a linear combination of $\phi_n(x)$ and lower powers of x and continue this process until the remainder is constant, i.e., a multiple of $\phi_0(x)$. More precisely, if

$$f(x) = \sum_{\nu=0}^{n} \alpha_\nu \phi_\nu(x) \tag{17}$$

is the expansion in question we have

$$(\phi_\nu(x), f(x)) = \alpha_\nu \qquad (\nu = 0, 1, \ldots, n) \tag{18}$$

as usual.

Now the polynomial

$$\phi_n(x) - \frac{k_n}{k_{n-1}} x \phi_{n-1}(x)$$

is clearly of degree $n - 1$, and therefore

$$\phi_n(x) - \frac{k_n}{k_{n-1}} x \phi_{n-1}(x) = \sum_{\nu=0}^{n-1} \alpha_\nu \phi_\nu(x). \tag{19}$$

Taking the inner product of both sides with $\phi_j(x)$,

$$\left(\phi_n - \frac{k_n}{k_{n-1}} x\phi_{n-1}, \phi_j\right) = \alpha_j(\phi_j, \phi_j) \tag{20}$$

$$= (\phi_n, \phi_j) - \frac{k_n}{k_{n-1}} (x\phi_{n-1}, \phi_j)$$

$$= (\phi_n, \phi_j) - \frac{k_n}{k_{n-1}} (\phi_{n-1}, x\phi_j)$$

If $j \leqq n - 3$, $x\phi_j(x)$ is of degree $\leqq n - 2$, and the last two terms in (20) vanish. Thus $\alpha_0 = \alpha_1 = \cdots = \alpha_{n-3} = 0$, and using (19),

$$\phi_n(x) - \frac{k_n}{k_{n-1}} x\phi_{n-1}(x) = \alpha_{n-2}\phi_{n-2}(x) + \alpha_{n-1}\phi_{n-1}(x)$$

or recalling that the numbers α_ν depend on n,

$$\phi_n(x) = \left(\frac{k_n}{k_{n-1}} x + B_n\right)\phi_{n-1}(x) - C_n\phi_{n-2}(x). \tag{21}$$

Theorem 4. *A sequence $\{\phi_n(x)\}_0^\infty$ of orthogonal polynomials satisfies a three term recurrence relation of the form* (21).

Concerning the coefficients C_n, we have

$$0 = (\phi_n, \phi_{n-2}) = \left(\phi_{n-2}, \left(\frac{k_n}{k_{n-1}} x + B_n\right)\phi_{n-1} - C_n\phi_{n-2}\right)$$

$$= \frac{k_n}{k_{n-1}} (x\phi_{n-1}, \phi_{n-2}) - C_n(\phi_{n-2}, \phi_{n-2})$$

$$= \frac{k_n}{k_{n-1}} (\phi_{n-1}, k_{n-2}x^{n-1} + \cdots) - C_n(\phi_{n-2}, \phi_{n-2})$$

$$= \frac{k_n k_{n-2}}{k_{n-1}^2} (\phi_{n-1}, k_{n-1}x^{n-1}) - C_n(\phi_{n-2}, \phi_{n-2})$$

$$= \frac{k_n k_{n-2}}{k_{n-1}^2} (\phi_{n-1}, \phi_{n-1}) - C_n(\phi_{n-2}, \phi_{n-2})$$

Hence if we define

$$\gamma_\nu = (\phi_\nu, \phi_\nu) \qquad (\nu = 0, 1, 2, \ldots) \tag{22}$$

then

$$C_n = \frac{k_n k_{n-2}}{k_{n-1}^2} \frac{\gamma_{n-1}}{\gamma_{n-2}}. \tag{23}$$

When the polynomials are normalized, $\gamma_\nu = 1$ ($\nu = 0, 1, 2, \ldots$), and the recurrence takes the form

$$\phi_n(x) = \left(\frac{k_n}{k_{n-1}}\, x + B_n\right)\phi_{n-1}(x) - \frac{k_n k_{n-2}}{k_{n-1}^2}\,\phi_{n-2}(x)$$

or, after some rearrangement,

$$x\phi_{n-1}(x) = \frac{k_{n-1}}{k_n}\,\phi_n(x) + \frac{k_{n-2}}{k_{n-1}}\,\phi_{n-2}(x) + \beta_{n-1}\phi_{n-1}(x). \tag{24}$$

Inspection of (24) shows how one can recognize a *normalized* sequence by its recurrence relation, for if n is replaced by $n - 1$ in the coefficient of $\phi_n(x)$, the coefficient of $\phi_{n-2}(x)$ must result.

Example. A certain sequence of orthogonal polynomials satisfies

$$\phi_{n+1}(x) = [(n + 1)x + 1]\phi_n(x) - 3(n + 1)\phi_{n-1}(x).$$

Is the sequence normalized? If not, find the normalization constants.

Solution. Solving for $x\phi_n(x)$,

$$x\phi_n(x) = \frac{1}{n + 1}\,\phi_{n+1}(x) + 3\phi_{n-1}(x) - \frac{1}{n + 1}\,\phi_n(x). \tag{25}$$

Thus the sequence is not normalized. Let $\psi_n(x) = \lambda_n\phi_n(x)$ denote the normalized sequence, where the constants $\lambda_0, \lambda_1, \ldots$ are to be found. Substituting in (25),

$$x\psi_n(x) = \left[\frac{\lambda_n}{(n + 1)\lambda_{n+1}}\right]\psi_{n+1}(x) + 3\left[\frac{\lambda_n}{\lambda_{n-1}}\right]\psi_{n-1}(x) - \frac{1}{n + 1}\,\psi_n(x).$$

The condition for normalization is

$$\frac{\lambda_{n-1}}{n\lambda_n} = 3\,\frac{\lambda_n}{\lambda_{n-1}}$$

or

$$\lambda_n = \frac{1}{\sqrt{3n}}\,\lambda_{n-1}.$$

Hence

$$\lambda_n = 3^{-n/2}(n!)^{1/2}\lambda_0 \tag{26}$$

are the required normalization constants. In other words, for this sequence we have found that

$$\gamma_n = (\phi_n, \phi_n) = 3^n n!\,\gamma_0 \qquad (n = 0, 1, 2, \ldots). \tag{27}$$

It is also possible to give a remarkable proof of the reality of the zeros of

$\phi_n(x)$ based on the recurrence relation. Indeed, let us write (24) for $n = 1, 2, \ldots, N$, in the form

$$x\begin{pmatrix} \phi_0(x) \\ \phi_1(x) \\ \phi_2(x) \\ \cdot \\ \cdot \\ \cdot \\ \cdot \\ \phi_{N-1}(x) \end{pmatrix} = \begin{pmatrix} \beta_0 & \frac{k_0}{k_1} & 0 & 0 & \cdots & 0 \\ \frac{k_0}{k_1} & \beta_1 & \frac{k_1}{k_2} & 0 & \cdots & 0 \\ 0 & \frac{k_1}{k_2} & \beta_2 & \frac{k_2}{k_3} & \cdots & 0 \\ \cdot & \cdot & \cdot & \cdot & & 0 \\ \cdot & \cdot & \cdot & \cdot & & \cdot \\ \cdot & \cdot & \cdot & \cdot & & \cdot \\ \cdot & \cdot & \cdot & \cdot & & \cdot \\ 0 & 0 & 0 & 0 & \cdots & \beta_{N-1} \end{pmatrix} \begin{pmatrix} \phi_0(x) \\ \phi_1(x) \\ \phi_2(x) \\ \cdot \\ \cdot \\ \cdot \\ \cdot \\ \phi_{N-1}(x) \end{pmatrix} + \begin{pmatrix} 0 \\ 0 \\ 0 \\ \cdot \\ \cdot \\ \cdot \\ 0 \\ \frac{k_{N-1}}{k_N}\phi_N(x) \end{pmatrix} \tag{28}$$

Now, suppose we choose the number x to be a zero of $\phi_N(x)$, say $x = x_i$. Then (28) reads, with obvious notation,

$$x_i\mathbf{\Phi}(x_i) = J\mathbf{\Phi}(x_i) \tag{29}$$

that is, the eigenvalues of the $N \times N$ matrix J are the zeros of $\phi_N(x.)$ Since J is symmetric, these zeros are real. The matrix J is called the *Jacobi matrix* associated with the sequence $\{\phi_n(x)\}$. We have proved

Theorem 5. *The eigenvalues* $x_1, \ldots, x_N$ *of the Jacobi matrix in* (28) *are the zeros of* $\phi_N(x)$. *The eigenvector corresponding to* x_i *is* $(\phi_0(x_i), \phi_1(x_i), \ldots, \phi_{N-1}(x_i))$.

2.5 THE CHRISTOFFEL-DARBOUX IDENTITY

Write (28) in the vector form

$$x\mathbf{\Phi}(x) = J\mathbf{\Phi}(x) + \frac{k_{N-1}}{k_N}\phi_N(x)\mathbf{e}_N \tag{30}$$

where $\mathbf{e}_N = (0, 0, 0, \ldots, 0, 1)$. Replacing x by y,

$$(31) \qquad y\mathbf{\Phi}(y) = J\mathbf{\Phi}(y) + \frac{k_{N-1}}{k_N}\phi_N(y)e_N.$$

Taking the inner product of (30) with $\mathbf{\Phi}(y)$, (31) with $\mathbf{\Phi}(x)$ and subtracting, we get, using the symmetry of J,

$$(32) \quad (x - y)(\mathbf{\Phi}(x), \mathbf{\Phi}(y)) = [\phi_{N-1}(y)\phi_N(x) - \phi_{N-1}(x)\phi_N(y)]\frac{k_{N-1}}{k_N}.$$

This is the Christoffel-Darboux identity.† In scalar form it reads

$$(33) \qquad \sum_{\nu=0}^{N-1}\phi_\nu(x)\phi_\nu(y) = \frac{k_{N-1}}{k_N}\frac{\phi_{N-1}(y)\phi_N(x) - \phi_{N-1}(x)\phi_N(y)}{x - y}$$

and is valid, as written, for normalized polynomials only. If the $\{\phi_\nu(x)\}$ are not normalized, then $\phi_\nu(x)$ must be replaced by $\gamma_\nu^{-1/2}\phi_\nu(x)$ on both sides of (33) (note that this changes k_ν also).

If we let $y \to x$ in (33), the right-hand side becomes

$$\frac{k_{N-1}}{k_N}\lim_{y\to x}\left\{\phi_N(x)\left[\frac{\phi_{N-1}(y) - \phi_{N-1}(x)}{x - y}\right] + \phi_{N-1}(x)\left[\frac{\phi_N(x) - \phi_N(y)}{x - y}\right]\right\}$$
$$= \frac{k_{N-1}}{k_N}\{-\phi_N(x)\phi'_{N-1}(x) + \phi_{N-1}(x)\phi'_N(x)\}$$

and therefore

$$(34) \qquad \sum_{\nu=0}^{N-1}[\phi_\nu(x)]^2 = \frac{k_{N-1}}{k_N}\{-\phi_N(x)\phi'_{N-1}(x) + \phi_{N-1}(x)\phi'_N(x)\}.$$

If we take $x = x_0$, a zero of $\phi_N(x)$, in (34),

$$(35) \qquad \sum_{\nu=0}^{N-1}[\phi_\nu(x_0)]^2 = \frac{k_{N-1}}{k_N}\phi_{N-1}(x_0)\phi'_N(x_0).$$

On the other hand, setting $n = N + 1$ and $x = x_0$ in (24), we find

$$(36) \qquad \phi_{N-1}(x_0) = -\frac{k_N^{\,2}}{k_{N+1}k_{N-1}}\phi_{N+1}(x_0)$$

and substituting in (35), we obtain finally

$$(37) \qquad \sum_{\nu=0}^{N-1}[\phi_\nu(x_0)]^2 = -\frac{k_N}{k_{N+1}}\phi_{N+1}(x_0)\phi_N'(x_0)$$

a formula which will be of use presently.

† Christoffel [1], Darboux [1].

2.6 MODIFYING THE WEIGHT FUNCTION

Suppose $\{\phi_\nu(x)\}_0^\infty$ is the sequence of orthogonal polynomials associated with the weight function $w(x)$ on the interval $[a, b]$. Let $p(x)$ be a polynomial of degree r which is non-negative on $[a, b]$. Then $w(x)p(x)$ is again a weight function for $[a, b]$, and we may seek the sequence $\{\psi_n(x)\}_0^\infty$ of orthogonal polynomials associated with this modified weight function on the same interval $[a, b]$. Christoffel has given the following ingenious construction for the $\psi_n(x)$.

Consider the determinant

$$D_n(x) = \begin{vmatrix} \phi_n(x) & \phi_{n+1}(x) & \cdots & \phi_{n+r}(x) \\ \phi_n(x_1) & \phi_{n+1}(x_1) & \cdots & \phi_{n+r}(x_1) \\ \cdot & \cdot & & \cdot \\ \cdot & \cdot & & \cdot \\ \cdot & \cdot & & \cdot \\ \phi_n(x_r) & \phi_{n+1}(x_r) & \cdots & \phi_{n+r}(x_r) \end{vmatrix} \tag{38}$$

where $x_1, x_2, \ldots, x_r$ are the zeros of $p(x)$ which we suppose, for the moment, to be distinct.

Clearly $D_n(x)$ is a polynomial of degree $n + r$ and

$$D_n(x_i) = 0 \qquad (i = 1, 2, \ldots, r) \tag{39}$$

since, for these values of x, two rows of $D_n(x)$ are the same. Thus $D_n(x)$ is divisible by $p(x)$, so that

$$D_n(x) = p(x)\psi_n(x) \tag{40}$$

where $\psi_n(x)$ is a polynomial of degree n. Next, if we expand $D_n(x)$ across the first row,

$$D_n(x) = A_0\phi_n(x) + A_1\phi_{n+1}(x) + \cdots + A_r\phi_{n+r}(x)$$

where the A_j are certain constants. Thus, if $f(x)$ is any polynomial of degree $\leq n - 1$,

$$\begin{aligned} \int_a^b \psi_n(x)f(x)p(x)w(x)\,dx &= \int_a^b D_n(x)f(x)w(x)\,dx \\ &= A_0\int_a^b \phi_n(x)f(x)w(x)\,dx + \cdots + A_r\int_a^b \phi_{n+r}(x)f(x)w(x)\,dx \\ &= 0 \end{aligned}$$

by definition of the $\{\phi_n(x)\}$. Thus $\psi_n(x)$ is orthogonal to every polynomial of lower degree with respect to the weight function $p(x)w(x)$, as required.

Theorem 6.† *The orthogonal sequence* $\{\psi_n(x)\}$ *associated with the weight function* $p(x)w(x)$ *on the interval* $[a, b]$ *is*

$$\psi_n(x) = \frac{1}{p(x)} D_n(x) \tag{41}$$

where $D_n(x)$ *is given by* (38), *if the zeros of* $p(x)$ *are distinct.*

It is not hard to see that if a zero of $p(x)$ has multiplicity m, the corresponding rows of $D_n(x)$ must be replaced by the derivatives of order $0, 1, \ldots, m - 1$ of the polynomials $\phi_n(x), \ldots, \phi_{n+r}(x)$ evaluated at this repeated root. The proof is identical with that already given.

Example. Find the sequence belonging to the weight function $1 - x$ on $[-1, 1]$.

The Legendre polynomials $P_n(x)$ belong to $w(x) = 1$ on this interval. Taking $p(x) = 1 - x$, $x_1 = 1$, $r = 1$ in (38),

$$D_n(x) = \begin{vmatrix} P_n(x) & P_{n+1}(x) \\ P_n(1) & P_{n+1}(1) \end{vmatrix}$$

and the required polynomials are

$$\begin{aligned} \psi_n(x) &= (1 - x)^{-1} D_n(x) \\ &= (1 - x)^{-1}[P_n(x)P_{n+1}(1) - P_{n+1}(x)P_n(1)] \qquad (n = 0, 1, 2, \ldots). \end{aligned}$$

2.7 RODRIGUES' FORMULA

Suppose that our weight function $w(x)$, in addition to satisfying the usual hypotheses for weight functions, is also infinitely often differentiable at each point of the interval (a, b). In that case, we can give yet another method of constructing the sequence of orthogonal polynomials.

We define

$$\phi_n(x) = \frac{1}{w(x)} \frac{d^n G_n(x)}{dx^n} \qquad (n = 0, 1, 2, \ldots) \tag{42}$$

and will now determine $G_n(x)$ so that the $\phi_n(x)$ so defined are the required sequence. First, $\phi_n(x)$ is to be a polynomial of degree n, and hence we must have

$$\frac{d^{n+1}}{dx^{n+1}}\left(\frac{1}{w(x)} \frac{d^n G_n(x)}{dx^n}\right) = 0. \tag{43}$$

† Christoffel [1].

Since this is a differential equation of order $2n + 1$, we may also impose the end conditions

(44) $$G_n(a) = G_n'(a) = \cdots = G_n^{(n-1)}(a) = 0$$

(45) $$G_n(b) = G_n'(b) = \cdots = G_n^{(n-1)}(b) = 0$$

and we state

Theorem 7. *Let $G_n(x)$ satisfy* (43–45), *$w(x)$ being an infinitely differentiable weight function. Then the polynomials $\phi_n(x)$ of* (42) *are the orthogonal sequence associated with $w(x)$, $[a, b]$.*

Proof. Let $q(x)$ be any polynomial of degree $\leqq n - 1$, then

$$\int_a^b \phi_n(x)q(x)w(x)\,dx = \int_a^b \frac{d^nG(x)}{dx^n}\,q(x)\,dx.$$

This can be integrated n times by parts. At each stage the integrated part will vanish, by (44) and (45), and at the last step we will have an integral involving the nth derivative of $q(x)$, which vanishes because the degree of $q(x)$ is $<n$. Thus the integral is zero, and $\phi_n(x)$ is orthogonal to every polynomial of degree $<n$, as required.

Equation (42) is known as Rodrigues' formula.

Example. $w(x) = 1$, $[a, b] = [-1, 1]$. Here (43)–(45) are

$$\frac{d^{(2n+1)}}{dx^{2n+1}}\,G_n(x) = 0$$

$$G_n(-1) = \cdots = G_n^{(n-1)}(-1) = G_n(1) = \cdots = G_n^{(n-1)}(1) = 0.$$

The solution is clearly

$$G_n(x) = (1 - x^2)^n.$$

Therefore Legendre's polynomials are given by

(46) $$P_n(x) = \frac{d^n}{dx^n}\,(1 - x^2)^n$$

aside from a multiplicative constant.

2.8 LOCATION OF THE ZEROS

The matrix J of (28) is a ready source of precise information concerning the zeros of $\phi_N(x)$. Indeed, by direct application of Theorem 32 of Chapter 1, we find at once

Theorem 8. *Each zero of $\phi_N(x)$ lies in one of the intervals*

$$(47)\qquad \begin{cases} |x - \beta_0| \leqq \dfrac{k_0}{k_1} \\ |x - \beta_i| \leqq \dfrac{k_{i-1}}{k_i} + \dfrac{k_i}{k_{i+1}} \qquad (i = 1, 2, \ldots, N-2) \\ |x - \beta_{N-1}| \leqq \dfrac{k_{N-2}}{k_{N-1}} \end{cases}$$

The utility of this result is illustrated by the example of the Hermite polynomials, which belong to $(-\infty, \infty)$, e^{-x^2}. They satisfy the recurrence

$$xH_n(x) = \tfrac{1}{2}H_{n+1}(x) + nH_{n-1}(x).$$

To find the recurrence relation for the normalized polynomials $\tilde{H}_n(x)$, write, as usual, $H_n(x) = \lambda_n \tilde{H}_n(x)$. Then, substituting,

$$(48)\qquad x\tilde{H}_n(x) = \frac{1}{2}\frac{\lambda_{n+1}}{\lambda_n}\tilde{H}_{n+1}(x) + \frac{n\lambda_{n-1}}{\lambda_n}\tilde{H}_{n-1}(x)$$

and the normalization condition is

$$\frac{1}{2}\frac{\lambda_n}{\lambda_{n-1}} = n\frac{\lambda_{n-1}}{\lambda_n}.$$

From this we find $\lambda_n = \sqrt{2^n n!}$, and from (48),

$$(49)\qquad x\tilde{H}_n(x) = \sqrt{(n+1)/2}\,\tilde{H}_{n+1}(x) + \sqrt{n/2}\,\tilde{H}_{n-1}(x).$$

is the recurrence formula satisfied by the normalized Hermite polynomials. Theorem 8 gives

Theorem 9. *The zeros of $H_n(x)$ lie in the interval*

$$(50)\qquad |x| \leqq \sqrt{(n-1)/2} + \sqrt{(n-2)/2}.$$

Notice how much more exact this is than Theorem 5, which, in this case, tells us only that the zeros are somewhere on the real axis. When n is large, the right side of (50) is about $\sqrt{2n}$, which happens to give exactly the correct asymptotic rate of growth of the largest zero of $H_n(x)$, as can be shown by deeper methods of analysis.

Another method—based on the Perron-Frobenius theorem—gives more precise information about the largest zero. First we need

Lemma 1. If the numbers $\beta_n \geqq 0$ $(n = 0, 1, \ldots, N-1)$, then the matrix J is non-negative and irreducible.

Proof. The non-negativity is clear, for we may always suppose the leading coefficients k_n to be positive. For the irreducibility, notice that $J_{i+1,i} \neq 0$

$(i = 1, \ldots, N-1)$ and $J_{i,i+1} \neq 0$ $(i = 1, \ldots, N-1)$. Now suppose J is reducible, and let I, K be the two sets of integers which show this. Let i_0 be the largest integer in I. Then $i_0 - 1$ is not in K because that would imply $J_{i_0, i_0-1} = 0$, which is false. Hence $i_0 - 1$ is in I. Continuing in this way, all the integers $i_0, i_0 - 1, \ldots, 1$ are in I. Thus K contains $i_0 + 1, \ldots, N$. But this means $J_{i_0, i_0+1} = 0$—which is also impossible—and we have the desired contradiction. Now from (157) of Chapter 1, the Perron root r of J being the largest zero x_{NN} of $\phi_N(x)$, we find

$$\min_{1 \leq i \leq N} \frac{(J\mathbf{x})_i}{x_i} \leqq x_{NN} \leqq \max_{1 \leq i \leq N} \frac{(J\mathbf{x})_i}{x_i} \tag{51}$$

as bounds for the largest zero of $\phi_N(x)$, the vector $\mathbf{x}$ being arbitrary, aside from non-negativity. Equation (50), for example, is a special case of (51) in which $\mathbf{x} = (1, 1, \ldots, 1)$, but considerably more freedom exists in (51) because the components of $\mathbf{x}$ can be chosen for optimum results.

2.9 GAUSS QUADRATURE

One of the most interesting and useful applications of the theory of orthogonal polynomials occurs in the theory of numerical integration. The reader is no doubt familiar with the simple numerical integration formulas, such as the Trapezoidal rule

$$\int_0^1 f(x)\,dx = \tfrac{1}{2}f(0) + \tfrac{1}{2}f(1) + \text{error} \tag{52}$$

and Simpson's rule

$$\int_0^1 f(x)\,dx = \tfrac{1}{6}f(0) + \tfrac{2}{3}f(\tfrac{1}{2}) + \tfrac{1}{6}f(1) + \text{error}. \tag{53}$$

The first of these is exact if $f(x)$ is linear on $(0, 1)$, the second if $f(x)$ is quadratic on $(0, 1)$. In general, the sizes of the error terms involved depend on the second and third derivatives, respectively, of $f(x)$. Both of these formulas are of Newton-Cotes type, which means that equally spaced abscissas are used. The most general Newton-Cotes formula has the form

$$\int_0^1 f(x)\,dx = \sum_{\nu=0}^{N} H_\nu f\left(\frac{\nu}{N}\right) + \text{error}. \tag{54}$$

In such a formula there are $N + 1$ free parameters, the "weights" H_0, $H_1, \ldots, H_N$. These can be chosen so as to satisfy any $N + 1$ compatible conditions, the ones usually used being the requirement that (54) shall be exact for polynomials of degree $\leqq N$. These conditions are

$$\int_0^1 x^n\,dx = \frac{1}{n+1} = \sum_{\nu=0}^{N} H_\nu \left(\frac{\nu}{N}\right)^n \qquad (n = 0, 1, \ldots, N) \tag{55}$$

which are clearly $N + 1$ linear equations that determine the H uniquely.

It was Gauss who first pointed out that the form (54) is unnecessarily restrictive in that there is no real reason for using equidistant points $x_\nu = \nu/N$ at which to evaluate the function. He suggested that we adopt the more general

$$\int_0^1 f(x)\,dx = \sum_{\nu=1}^{N} H_\nu f(x_\nu) + \text{error} \tag{56}$$

in which we have $2N$ parameters $H_1, \ldots, H_N, x_1, \ldots, x_n$ free, and hence may hope to satisfy $2N$ exactness conditions of the type

$$\int_0^1 x^n\,dx = \frac{1}{n+1} = \sum_{\nu=1}^{N} H_\nu x_\nu^{\,n} \qquad (n = 0, 1, \ldots, 2N-1). \tag{57}$$

The word "hope" was used in the previous sentence because these last equations, while linear in the H_ν, are nonlinear in the x_ν. Hence these might (*a priori*) be complex numbers or even real numbers outside of (0, 1). In either case the formula (56) would be useless.

As an example, consider the formula

$$\int_{-1}^{1} f(x)\,dx = H_1 f(x_1) + H_2 f(x_2) \tag{58}$$

which we require to be exact if $f(x)$ is a polynomial of degree ≤ 3. The conditions are

$$\begin{cases} H_1 + H_2 = 2 \\ H_1 x_1 + H_2 x_2 = 0 \\ H_1 x_1^{\,2} + H_2 x_2^{\,2} = \frac{2}{3} \\ H_1 x_1^{\,3} + H_2 x_2^{\,3} = 0 \end{cases} \tag{59}$$

The solution here is readily found to be $H_1 = H_2 = 1$, $x_1 = -x_2 = 1/\sqrt{3}$. We notice that the weights H_1, H_2 are positive, and the abscissas x_1, x_2 lie inside the interval $(-1, 1)$, so that the resulting approximate integration formula

$$\int_{-1}^{1} f(x)\,dx = f\left(\frac{1}{\sqrt{3}}\right) + f\left(-\frac{1}{\sqrt{3}}\right) + \text{error} \tag{60}$$

is actually useful. We propose now to show that this benign state of affairs persists in general. The principal theorem of the subject is

Theorem 10.† *Let $w(x)$ be a weight function for the interval $[a, b]$. Then there exist real numbers $x_1, x_2, \ldots, x_N$; $H_1, \ldots, H_N$ having the properties*

(i) $a < x_1 < x_2 < \cdots < x_N < b$
(ii) $H_\nu > 0 \qquad (\nu = 1, 2, \ldots, N)$

† Gauss [2], Jacobi [1].

(iii) *The formula*

$$\int_a^b f(x)w(x)\,dx = \sum_{\nu=1}^{N} H_\nu f(x_\nu) \tag{61}$$

is true for every polynomial $f(x)$ *of degree* $\leqq 2N - 1$.

Proof. Let $\phi_N(x)$ be the Nth member of the sequence of orthonormal polynomials belonging to $w(x)$, $[a, b]$, and let $x_1, x_2, \ldots, x_N$ be the zeros of $\phi_N(x)$. Define the polynomials

$$l_\nu(x) = \frac{\phi_N(x)}{(x - x_\nu)\phi_N'(x_\nu)} \qquad (\nu = 1, 2, \ldots, N). \tag{62}$$

Clearly

$$l_\nu(x_\mu) = \delta_{\mu\nu} \qquad (\mu, \nu = 1, \ldots, N). \tag{63}$$

Hence if we write

$$L_{N-1}(x) = \sum_{\nu=1}^{N} l_\nu(x) f(x_\nu)$$

then $L_{N-1}(x)$ is a polynomial of degree $N - 1$, which passes through the points $(x_\nu, f(x_\nu))$, $\nu = 1, 2, \ldots, N$, i.e.,

$$L_{N-1}(x_\nu) = f(x_\nu) \qquad (\nu = 1, 2, \ldots, N). \tag{64}$$

But this says that the zeros of $\phi_N(x)$ are also zeros of the polynomial $L_{N-1}(x) - f(x)$, which is of degree $\leqq 2N - 1$. Hence there is a polynomial $r(x)$ of degree $\leqq N - 1$ such that

$$f(x) - L_{N-1}(x) = \phi_N(x)r(x). \tag{65}$$

Thus

$$\begin{aligned}\int_a^b f(x)w(x)\,dx &= \int_a^b L_{N-1}(x)w(x)\,dx + \int_a^b \phi_N(x)r(x)w(x)\,dx \\ &= \int_a^b L_{N-1}(x)w(x)\,dx\end{aligned}$$

since $\phi_N(x)$ is orthogonal to $r(x)$. Therefore

$$\int_a^b f(x)w(x)\,dx = \sum_{\nu=1}^{N} f(x_\nu) \int_a^b l_\nu(x)w(x)\,dx \tag{66}$$

which is precisely equation (61) with

$$H_\nu = \int_a^b \frac{\phi_N(x)w(x)\,dx}{\phi_N'(x_\nu)(x - x_\nu)} \qquad (\nu = 1, 2, \ldots, N). \tag{67}$$

Conclusion (i) is clearly true, for the abscissas $x_1, \ldots, x_N$ of the Gauss quadrature formula have been revealed as the zeros of $\phi_N(x)$. Finally, let us take $f(x) = [l_\mu(x)]^2$ in (61), remembering (63). We get

$$\int_a^b [l_\mu(x)]^2 w(x)\, dx = H_\mu \qquad (\mu = 1, 2, \ldots, N)$$

which shows clearly that $H_\mu > 0$, completing the proof.

To find a more convenient expression for the weights H_ν, let us put $y = x_\mu$ in (33), where x_μ is a zero of $\phi_N(x)$. Then

$$\sum_{\nu=0}^{N-1} \phi_\nu(x)\phi_\nu(x_\mu) = \frac{k_{N-1}}{k_N} \frac{\phi_N(x)\phi_{N-1}(x_\mu)}{x - x_\mu}.$$

Multiplying by $w(x)$ and integrating over $[a, b]$,

$$1 = \frac{k_{N-1}}{k_N} \phi_{N-1}(x_\mu) \int_a^b \frac{\phi_N(x)w(x)\, dx}{x - x_\mu}.$$

Comparing with (67),

$$H_\mu = \frac{k_N}{k_{N-1}} [\phi_{N-1}(x_\mu)\phi_N'(x_\mu)]^{-1} \qquad (\mu = 1, 2, \ldots, N) \tag{68}$$

which is perhaps the most useful form. Another interesting relation results from comparing (68) with (35), namely,

$$\frac{1}{H_\nu} = \sum_{n=0}^{N-1} [\phi_n(x_\nu)]^2 \qquad (\nu = 1, 2, \ldots, N) \tag{69}$$

which again clearly shows the positivity of the weights. We emphasize that the $\phi_N(x)$ are here assumed *normalized*, and must be replaced by $\phi_n(x)\gamma_n^{-1/2}$ if this is not the case.

2.10 THE CLASSICAL POLYNOMIALS

The classical polynomials are those obtained from Rodrigues' formula (equation (42)) where $G_n(x)$ has the special form

$$G_n(x) = w(x)[G(x)]^n \tag{70}$$

and $G(x)$ is a quadratic polynomial

$$G(x) = \alpha + \beta x + \gamma x^2. \tag{71}$$

To satisfy the end conditions (44), (45), we require

$$\frac{d^k}{dx^k} \{w(x)[G(x)]^n\}_{x=a,b} = 0 \qquad (k = 0, 1, \ldots, n - 1). \tag{72}$$

Now, consider

$$(73)\quad \frac{d^{n+1}}{dx^{n+1}}\left(G(x)\frac{d}{dx}(w(x)G^n(x))\right) = G(x)\frac{d^2}{dx^2}\frac{d^n}{dx^n}(w(x)G^n(x)) + (n+1)G'(x)\frac{d}{dx}\left(\frac{d^n}{dx^n}(w(x)G^n(x))\right) + \frac{n(n+1)}{2}G''(x)\frac{d^n}{dx^n}(w(x)G^n(x))$$

where we have used Leibniz' rule

$$(74)\quad \frac{d^k}{dx^k}(f(x)g(x)) = \sum_{\nu=0}^{k}\binom{k}{\nu}f^{(k-\nu)}(x)g^{(\nu)}(x)$$

and the form (71) of $G(x)$.

Now, using the definition (42) of $\phi_N(x)$, (73) becomes

$$(75)\quad \frac{d^{n+1}}{dx^{n+1}}\left(G(x)\frac{d}{dx}(w(x)G^n(x))\right) = G(x)(w(x)\phi_n(x))'' + (n+1)G'(x)(w(x)\phi_n(x))' + \frac{n(n+1)}{2}G''(x)(w(x)\phi_n(x)).$$

Now, set $n = 1$ in (42), using (70), and get

$$(76)\quad (\phi_1(x) - G'(x))w(x) = G(x)w'(x).$$

Thus

$$(77)\quad G(x)\frac{d}{dx}(w(x)G^n(x)) = G(x)\{nw(x)G^{n-1}(x)G'(x) + G^n(x)w'(x)\} = nw(x)G^n(x)G'(x) + G^n(x)\{\phi_1(x) - G'(x)\}w(x) = G^n(x)w(x)\{(n-1)G'(x) + \phi_1(x)\}.$$

Now, apply the operator d^{n+1}/dx^{n+1} to (77), using (74) again,

$$(78)\quad \frac{d^{n+1}}{dx^{n+1}}\left(G(x)\frac{d}{dx}(w(x)G^n(x))\right) = \frac{d^{n+1}}{dx^{n+1}}\{G^n(x)w(x)[(n-1)G'(x) + \phi_1(x)]\}$$
$$= [(n-1)G'(x) + \phi_1(x)]\frac{d}{dx}\frac{d^n}{dx^n}(G^n(x)w(x)) + (n+1)[(n-1)G''(x) + \phi_1'(x)]\frac{d^n}{dx^n}(w(x)G^n(x))$$
$$= [(n-1)G'(x) + \phi_1(x)](w\phi_n)' + (n+1)[(n-1)G''(x) + \phi_1'(x)](w\phi_n).$$

Comparing (75) with (78), we find

$$\begin{aligned}(79)\quad G(x)(w(x)\phi_n(x))'' &+ (n+1)G'(x)(w(x)\phi_n(x))' \\ &+ \frac{n(n+1)}{2} G''(x)(w(x)\phi_n(x)) \\ &= [(n-1)G'(x) + \phi_1(x)](w(x)\phi_n(x))' \\ &\quad + (n+1)[(n-1)G''(x) + \phi_1'(x)](w(x)\phi_n(x))\end{aligned}$$

and finally

Theorem 11. *Let $\{\phi_n(x)\}_0^\infty$ be a sequence of classical orthogonal polynomials, with weight function $w(x)$. Then the function $y_n(x) = w(x)\phi_n(x)$ satisfies the differential equation of the second order*

$$(80)\quad G(x)y'' + \{2G'(x) - \phi_1(x)\}y' - \left\{\frac{n^2 - n - 2}{2} G'' + (n+1)\phi_1'(x)\right\}y = 0$$

in which the coefficient of $y''(x)$ is quadratic, of $y'(x)$ is linear, and of $y(x)$ is constant.

Next, consider the polynomial

$$h(x) = G(x)\phi_n'(x) - \tfrac{1}{2}nG''(x)x\phi_n(x)$$

On the surface, this would appear to be of degree $n + 1$. Referring to (71), however, we see that the coefficient of x^{n+1} is

$$nk_n\gamma - \tfrac{1}{2}n \cdot 2\gamma k_n = 0$$

and therefore $h(x)$ is of degree $\leqq n$. Hence

$$(81)\quad h(x) = G(x)\phi_n'(x) - \tfrac{1}{2}nG''(x)x\phi_n(x) = \sum_{\nu=0}^{n} \alpha_\nu \phi_\nu(x).$$

Multiplying by $w(x)\phi_k(x)$ $(0 \leqq k \leqq n)$,

$$(82)\quad \alpha_k\gamma_k = (G\phi_n', \phi_k) - \frac{n}{2}(\phi_n, xG''\phi_k).$$

The second term clearly vanishes for $k \leqq n - 2$. For the first,

$$\begin{aligned}\int_a^b \phi_n'(x)G(x)\phi_k(x)w(x)\,dx &= -\int_a^b \phi_n(x)\frac{d}{dx}\{G(x)\phi_k(x)w(x)\}\,dx \\ &= -\int_a^b \phi_n(x)\{w(x)[G\phi_k]' + G(x)\phi_k(x)w'(x)\}\,dx \\ &= -\int_a^b \phi_n(x)\{w(x)[G(x)\phi_k(x)]' + \phi_k(x)[\phi_1(x) - G'(x)]w(x)\}\,dx\end{aligned}$$

where (76) has been used. Now the integral of the first term vanishes when the degree of $\{(G\phi_k)'\}$ is $\leqq n - 1$, or, when $k + 1 \leqq n - 1$, i.e., $k \leqq n - 2$. The integral of the second term vanishes when $k + 1 \leqq n - 1$ also; hence we have shown that $\alpha_\nu = 0$ $(0 \leqq \nu \leqq n - 2)$ by an argument very much like that in the proof of Theorem 4. Thus, by (81),

$$G(x)\phi_n'(x) = \left(\frac{n}{2}\gamma x + \sigma_n\right)\phi_n(x) + \tau_n\phi_{n-1}(x). \tag{83}$$

Theorem 12. *The classical polynomials satisfy a differential recurrence relation of the form* (83).

It should be emphasized that the classical polynomials are a very special case of orthogonal polynomials. This can be seen clearly in (76), which shows that if $w(x)$ is a classical weight function, the $w'(x)/w(x)$ is the ratio of a linear and a quadratic polynomial.

2.11 SPECIAL POLYNOMIALS

In this section we tabulate, for reference purposes, the developments of this chapter as they apply to four special orthogonal sequences: the Legendre (P), Tschebycheff (T), Laguerre (L), and Hermite (H) polynomials.

(α) *Weight Function*

P: $w(x) = 1$

T: $w(x) = (1 - x^2)^{-1/2}$

L: $w(x) = e^{-x}$

H: $w(x) = e^{-x^2}$

(β) *Interval:*

P: $[-1, 1]$

T: $[-1, 1]$

L: $[0, \infty)$

H: $(-\infty, \infty)$

(γ) *Conventional Normalization:*

P: $\gamma_\nu = 2/(2\nu + 1) \qquad (\nu = 0, 1, 2, \ldots)$

T: $\gamma_0 = \pi, \quad \gamma_\nu = \pi/2 \qquad (\nu = 1, 2, \ldots)$

L: $\gamma_\nu = (\nu!)^2 \qquad (\nu = 0, 1, 2, \ldots)$

H: $\gamma_\nu = 2^\nu \nu! \sqrt{\pi} \qquad (\nu = 0, 1, \ldots)$

(δ) *Recurrence Formulae:*

P: $(l+1)P_{l+1}(x) = (2l+1)xP_l(x) - lP_{l-1}(x)$

T: $T_{n+1}(x) = 2xT_n(x) - T_{n-1}(x)$

L: $L_{n+1}(x) = (2n+1-x)L_n(x) - n^2L_{n-1}(x)$

H: $H_{n+1}(x) = 2xH_n(x) - 2nH_{n-1}(x)$

(ε) *First Two Polynomials*

P: $P_0(x) = 1 \quad P_1(x) = x$

T: $T_0(x) = 1 \quad T_1(x) = x$

L: $L_0(x) = 1 \quad L_1(x) = 1 - x$

H: $H_0(x) = 1 \quad H_1(x) = 2x$

(ζ) *Highest Coefficient*

P: $k_n = 1 \cdot 3 \cdot 5 \cdots (2n-1)/n!$

T: $k_n = 2^{n-1} \qquad (n > 0)$

L: $k_n = (-1)^n$

H: $k_n = 2^n$

(η) *Rodrigues' Formula*

$$P\colon\ P_n(x) = (2^n n!)^{-1}\left(\frac{d}{dx}\right)^n (x^2-1)^n$$

$$T\colon\ T_n(x) = (-1)^n[2^n\Gamma(n+\tfrac{1}{2})]^{-1}\sqrt{1-x^2}\,\Gamma\left(\frac{1}{2}\right)\left(\frac{d}{dx}\right)^n (1-x^2)^{n-\frac{1}{2}}$$

$$L\colon\ L_n(x) = e^x\left(\frac{d}{dx}\right)^n (x^n e^{-x})$$

$$H\colon\ H_n(x) = (-1)^n e^{x^2}\left(\frac{d}{dx}\right)^n e^{-x^2}$$

(θ) *Christoffel-Darboux Identity*

$$P\colon\ (x-y)\sum_{n=0}^{N}(2n+1)P_n(x)P_n(y)$$
$$= (N+1)[P_{N+1}(x)P_N(y) - P_N(x)P_{N+1}(y)]$$

$$T\colon\ \frac{1}{2} + \sum_{n=1}^{N} T_n(x)T_n(y)$$
$$= (x-y)^{-1}[T_{N+1}(x)T_N(y) - T_N(x)T_{N+1}(y)]$$

$$L\colon\ \sum_{n=1}^{N}\frac{L_n(x)L_n(y)}{(n!)^2}$$
$$= [(N!)^2(x-y)]^{-1}[L_{N+1}(x)L_N(y) - L_{N+1}(y)L_N(x)]$$

$$H\colon\ \sum_{n=0}^{N}\frac{H_n(x)H_n(y)}{2^n n!}$$
$$= [2^{N+1}N!\,(x-y)]^{-1}[H_{N+1}(x)H_N(y) - H_N(x)H_{N+1}(y)]$$

(ι) *Gauss Quadrature*

$$P\colon \int_{-1}^{1} f(x)\,dx = \sum_{\nu=1}^{N} H_\nu f(x_\nu)$$

where $P_N(x_\nu) = 0 \qquad (\nu = 1, 2, \ldots, N)$

$$H_\nu = \frac{2}{1 - x_\nu^{\,2}}[P_N'(x_\nu)]^{-2} \qquad (\nu = 1, 2, \ldots, N)$$

$$T\colon \int_{-1}^{1} f(x)\frac{dx}{\sqrt{1-x^2}} = \frac{\pi}{N}\sum_{\nu=1}^{N} f\left(\cos\frac{2\nu-1}{2N}\pi\right)$$

$$L\colon \int_{0}^{\infty} e^{-x} f(x)\,dx = \sum_{\nu=1}^{N} H_\nu f(x_\nu)$$

where $L_N(x_\nu) = 0 \qquad (\nu = 1, 2, \ldots, N)$

$$H_\nu = \frac{(N!)^2}{x_\nu}[L_N'(x_\nu)]^{-2} \qquad (\nu = 1, 2, \ldots, N)$$

$$H\colon \int_{-\infty}^{\infty} f(x) e^{-x^2}\,dx = \sum_{\nu=1}^{N} H_\nu f(x_\nu)$$

where $H_N(x_\nu) = 0 \qquad (\nu = 1, 2, \ldots, N)$

$$H_\nu = 2^{N+1} N!\,\sqrt{\pi}[H_N'(x_\nu)]^{-2} \qquad (\nu = 1, 2, \ldots, N)$$

(κ) *Differential Equation:*

$$P\colon (1 - x^2)P_n''(x) - 2xP_n'(x) + n(n+1)P_n(x) = 0$$
$$T\colon (1 - x^2)T_n''(x) - xT_n'(x) + n^2T_n(x) = 0$$
$$L\colon xL_n''(x) + (1 - x)L_n'(x) + nL_n(x) = 0$$
$$H\colon H_n''(x) - 2xH_n'(x) + 2nH_n(x) = 0$$

(λ) *Differential Recurrence:*

$$P\colon (1 - x^2)P_n'(x) = -nxP_n(x) + nP_{n-1}(x)$$
$$T\colon (1 - x^2)T_n'(x) = -nxT_n(x) + nT_{n-1}(x)$$
$$L\colon xL_n'(x) = nL_n(x) - n^2L_{n-1}(x)$$
$$H\colon H_n'(x) = 2nH_{n-1}(x)$$

2.12 THE CONVERGENCE OF ORTHOGONAL EXPANSIONS

Let $w(x)$ be a weight function for $[a, b]$. Let $f(x)$ be defined on $[a, b]$, and suppose that the integrals

$$\int_a^b x^n f(x) w(x)\,dx = \mu_n[f] \qquad (n = 0, 1, 2, \ldots) \tag{84}$$

exist. If $\{\phi_n(x)\}_0^\infty$ is the normalized sequence of orthogonal polynomials belonging to $w(x)$ on $[a, b]$, we may calculate the Fourier coefficients of $f(x)$ with respect to the $\{\phi_n(x)\}$,

$$(85)\qquad c_n = \int_a^b \phi_n(x)f(x)w(x)\,dx$$
$$= (\phi_n, f) \qquad (n = 0, 1, 2, \ldots).$$

The Fourier series of $f(x)$ is

$$(86)\qquad f(x) \sim \sum_{n=0}^{\infty} c_n\phi_n(x).$$

The relation (86) simply means that the constants c_n have been calculated in the manner (85) and not that the series converges at all or, in particular, to $f(x)$.

We wish to inquire as to what restrictions on the function $f(x)$, in addition to (84), will insure that the series actually does converge—and to the sum $f(x)$. Before proceeding, we remark that there exist functions $f(x)$ for which the integrals in (85) exist, such that any of the following possibilities may occur: (a) the series diverges everywhere in (a, b); (b) the series converges at some points of (a, b) and not at others; (c) the series converges everywhere on (a, b) but to $f(x)$ nowhere; (d) the series converges everywhere on (a, b) at some points to $f(x)$ and at others not to $f(x)$; (e) the series converges everywhere on (a, b) to $f(x)$.

Even if $f(x)$ is *continuous* on $[a, b]$, possibility (e) *need not hold.*

Now let us consider the Nth partial sum of the series (86) evaluated at a fixed point x lying interior to (a, b).

$$(87)\qquad S_N(x) = \sum_{n=0}^{N} c_n\phi_n(x)$$

We wish to determine conditions under which $S_N(x) \to f(x)$ as $N \to \infty$. Now

$$(88)\qquad S_N(x) = \sum_{n=0}^{N} (\phi_n, f)\phi_n(x)$$
$$= \sum_{n=0}^{N} \int_a^b \phi_n(y)f(y)w(y)\,dy\ \phi_n(x)$$
$$= \int_a^b f(y)w(y)\left\{\sum_{n=0}^{N} \phi_n(x)\phi_n(y)\right\} dy.$$

Next, from the Christoffel-Darboux formula,

$$\sum_{\nu=0}^{N} \phi_\nu(x)\phi_\nu(y) = \frac{k_N}{k_{N+1}} \frac{\phi_N(y)\phi_{N+1}(x) - \phi_N(x)\phi_{N+1}(y)}{x - y}$$

we deduce, by multiplying by $w(y)\,dy$ and integrating over (a, b), that

$$1 = \frac{k_N}{k_{N+1}} \int_a^b \frac{\phi_N(y)\phi_{N+1}(x) - \phi_N(x)\phi_{N+1}(y)}{x - y}\, w(y)\, dy.$$

Multiplying this by $f(x)$,

$$f(x) = \frac{k_N}{k_{N+1}} \int_a^b f(x)\left\{\frac{\phi_N(y)\phi_{N+1}(x) - \phi_N(x)\phi_{N+1}(y)}{x - y}\right\} w(y)\, dy$$

and subtracting from (88),

$$\text{(89)} \quad S_N(x) - f(x)$$

$$= \frac{k_N}{k_{N+1}} \int_a^b [f(y) - f(x)]\left\{\frac{\phi_N(y)\phi_{N+1}(x) - \phi_N(x)\phi_{N+1}(y)}{x - y}\right\} w(y)\, dy.$$

Theorem 13. *In order that the Fourier series of $f(x)$ should converge to $f(x)$, it is necessary and sufficient that, at the given point x, the right-hand side of* (89) *should tend to zero as $N \to \infty$.*

Next, we propose to show that on a finite interval, the factor k_N/k_{N+1} can be ignored in the sense that the right side of (89) converges to zero if the integral does by itself. We suppose that the sequence $\{\phi_n(x)\}$ has been normalized so that $k_n > 0$, $\beta_n \geqq 0$ $(n = 0, 1, 2, \ldots)$, the β_n being defined by (24). Now, by (113) of Chapter 1, we know that x_{NN}, the largest zero of $\phi_N(x)$, being the largest eigenvalue of the Jacobi matrix J, satisfies

$$x_{NN} \geqq \frac{(\mathbf{x}, J\mathbf{x})}{(\mathbf{x}, \mathbf{x})}$$

for any N-vector $\mathbf{x}$. Taking, in particular, $\mathbf{x} = (0, 0, \ldots, 0, 1, 1)$,

$$\text{(90)} \qquad x_{NN} \geqq \tfrac{1}{2}(\beta_{N-1} + \beta_{N-2}) + \frac{k_{N-2}}{k_{N-1}}.$$

$$\geqq \frac{k_{N-2}}{k_{N-1}}.$$

If the sequence k_{N-2}/k_{N-1} were unbounded (as $N \to \infty$), we would find arbitrarily large zeros of the sequence $\{\phi_n(x)\}$, contradicting Theorem 3, which asserts that all zeros lie in the finite interval (a, b). In the same way, from (90),

$$x_{NN} \geqq \tfrac{1}{2}\beta_{N-1}$$

and therefore the sequence $\{\beta_n\}$ is likewise bounded.

Theorem 14. *Let the sequence $\{\phi_n(x)\}_0^\infty$ of orthogonal polynomials belong to a finite interval (a, b). Then the coefficients k_{n-1}/k_n, β_n appearing in its recurrence formula* (24) *are bounded functions of n.*

Referring back to (89), we see that in order to prove convergence of a Fourier series on a finite interval it is enough to show that the integral tends to zero as $N \to \infty$.

2.13 TRIGONOMETRIC SERIES

We wish now to push our analysis further in the case of the classical trigonometric series. This could be done for cosine series by specialization of our general result, Theorem 13, to the case of Tschebycheff polynomials, using the result of exercise 7(a). For full generality, however, we start from the beginning, supposing that a function $f(x)$ is given on $(0, 2\pi)$, and defined by periodicity elsewhere; that its Fourier coefficients

$$a_n = \frac{1}{\pi}\int_0^{2\pi} f(x)\cos nx\,dx \tag{91}$$

$$b_n = \frac{1}{\pi}\int_0^{2\pi} f(x)\sin nx\,dx \tag{92}$$

exist and have been calculated, and that we are to study the convergence of the sequence of partial sums

$$S_N(x) = \tfrac{1}{2}a_0 + \sum_{n=1}^{N}(a_n \sin nx + b_n \cos nx) \tag{93}$$

to the function $f(x)$. As before,

(94)

$$S_N(x) = \frac{1}{2\pi}\int_0^{2\pi} f(t)\,dt + \frac{1}{\pi}\sum_{n=1}^{N}\int_0^{2\pi} f(t)[\cos nt \sin nx + \sin nt \cos nx]\,dt$$

$$= \frac{1}{\pi}\int_0^{2\pi} f(t)[\tfrac{1}{2} + \cos(t - x) + \cos 2(t - x) + \cdots + \cos N(t - x)]\,dt.$$

Now, for any fixed θ, consider the sum

$$C_m = \cos\theta + \cos 2\theta + \cdots + \cos m\theta.$$

We have

$$2\sin\frac{\theta}{2}C_m = \sum_{\nu=1}^{m} 2\sin\frac{\theta}{2}\cos\nu\theta$$

$$= \sum_{\nu=1}^{m}\left[-\sin\frac{2\nu - 1}{2}\theta + \sin\frac{2\nu + 1}{2}\theta\right]$$

$$= -\sin\frac{\theta}{2} + \sin\frac{2m + 1}{2}\theta = 2\sin\frac{m\theta}{2}\cos\frac{m + 1}{2}\theta.$$

Hence

$$\sum_{\nu=1}^{m} \cos \nu\theta = \frac{\sin m(\theta/2) \cos ((m+1)/2)\theta}{\sin \theta/2}$$

and

$$\frac{1}{2} + \sum_{\nu=1}^{m} \cos \nu\theta = \frac{\sin (2m+1)\theta/2}{2 \sin \theta/2}. \tag{95}$$

Substituting (95) in (94),

$$S_N(x) = \frac{1}{2\pi} \int_0^{2\pi} f(t) \frac{\sin (N + \frac{1}{2})(t-x)}{\sin \frac{1}{2}(t-x)} \, dt.$$

Replacing $t - x$ by t, and using the periodicity of $f(t)$,

$$S_N(x) = \frac{1}{2\pi} \int_0^{2\pi} f(x+t) \frac{\sin (2N+1)(t/2)}{\sin t/2} \, dt. \tag{96}$$

If we break up this integral into $\int_0^{\pi}$ and $\int_{\pi}^{2\pi}$, replace t by $-t$ in the second and use the periodicity of $f(t)$ again, we find

$$S_N(x) = \frac{1}{\pi} \int_0^{\pi} \frac{f(x+t) + f(x-t)}{2} \frac{\sin (2N+1)(t/2)}{\sin t/2} \, dt$$

and finally, replacing t by $2t$,

$$S_N(x) = \frac{2}{\pi} \int_0^{\pi/2} \frac{f(x+2t) + f(x-2t)}{2} \frac{\sin (2N+1)t}{\sin t} \, dt. \tag{97}$$

This is Dirichlet's integral. It is analogous to (88) in the general case. Next, as before, from (95) we find by integration,

$$\int_0^{2\pi} \frac{\sin (2N+1)(t/2)}{2 \sin t/2} \, dt = \pi$$

and repeating the chain of transformations following (96),

$$\frac{2}{\pi} \int_0^{\pi/2} \frac{\sin (2N+1)t}{\sin t} \, dt = 1.$$

Hence

$$f(x) = \frac{2}{\pi} \int_0^{\pi/2} f(x) \frac{\sin (2N+1)t}{\sin t} \, dt$$

and subtracting from (97)

$$S_N(x) - f(x) = \frac{2}{\pi} \int_0^{\pi/2} \left[\frac{f(x+2t) + f(x-2t)}{2} - f(x) \right] \frac{\sin (2N+1)t}{\sin t} \, dt. \tag{98}$$

Theorem 15. *In order that the Fourier series of $f(x)$ converge to $f(x)$ at the point x it is necessary and sufficient that the integral in* (98) *tend to zero as $N \to \infty$.*

Now let $\delta > 0$ be an arbitrarily small, fixed positive number. The function

$$\psi(t) = \left[\frac{f(x+2t)+f(x-2t)}{2} - f(x)\right]\frac{1}{\sin t}$$

is bounded for $\delta \leqq t \leqq \pi/2$, and hence integrable over $[\delta, \pi/2]$. Thus, by the Riemann-Lebesgue lemma (equation (41) of Chapter 1),

$$\int_{\delta}^{\pi/2}\left[\frac{f(x+2t)+f(x-2t)}{2} - f(x)\right]\frac{\sin(2n+1)t}{\sin t}\,dt \to 0$$

as $n \to \infty$.

Thus we see that the question of convergence of the Fourier series at a point x depends only on the behavior of the function $f(x)$ in a small neighborhood of x. More precisely, what we need is

$$\int_{0}^{\delta}\left[\frac{f(x+2t)+f(x-2t)}{2} - f(x)\right]\frac{\sin(2n+1)t}{\sin t}\,dt \to 0 \tag{99}$$

as $n \to \infty$, for some fixed $\delta > 0$.

Next, writing

$$\frac{1}{\sin t} = \frac{1}{t} + \left[\frac{1}{\sin t} - \frac{1}{t}\right] \tag{100}$$

we see that the second term is bounded as $t \to 0$; hence, by the same argument, we may replace (99) by

(101)

$$\int_{0}^{\delta}\left[\frac{f(x+2t)+f(x-2t)}{2} - f(x)\right]\sin(2n+1)t\,\frac{dt}{t} \to 0 \qquad (n \to \infty).$$

We notice that this condition cannot be met unless

$$\lim_{t\to 0}\left\{\frac{f(x+2t)+f(x+2t)}{2} - f(x)\right\} = 0$$

(assuming the limit exists at all), i.e., unless

$$f(x) = \frac{f(x+0)+f(x-0)}{2} \tag{102}$$

which is no restraint if x is a point of continuity of $f(x)$, but shows that at any point of discontinuity of $f(x)$, the function must be standardized by defining it as the average of its right-hand and left-hand limits.

Theorem 16. *Suppose there are constants $A > 0$, $\beta > 0$ such that*

$$\left|\frac{f(x+2t)+f(x-2t)}{2} - f(x)\right| \leqq At^{\beta} \tag{103}$$

for $|t| \leqq \delta$. Then the Fourier series of $f(x)$ converges to $f(x)$ at the point x.

Proof. If Lipschitz' condition (103) holds, then

$$\left|\int_0^{\delta}\left[\frac{f(x+2t)+f(x-2t)}{2} - f(x)\right]\frac{dt}{t}\right| \leqq \int_0^{\delta} At^{\beta-1}\,dt$$
$$= \frac{A\delta^{\beta}}{\beta}.$$

Thus the coefficient of $\sin(2n+1)t$ in (101) is integrable, and by the Riemann-Lebesgue lemma, the integral tends to zero.

The Lipschitz condition (103) is perhaps the easiest and most general condition to use in practice. It obviously holds if $f'(x)$ exists but is less severe than differentiability.

2.14 FEJÉR SUMMABILITY

The developments of the preceding section, while immensely interesting from a mathematical point of view, must be regarded as disturbing in the sense that the Fourier series of a function which is merely continuous need not converge to the function. This difficulty can be ameliorated by an elegent device, known as *summability*, whose importance was first recognized by Fejér. Before proceeding with the discussion of summability of Fourier series, we illustrate the ideas involved with a simple example drawn from the theory of ordinary power series. Consider the relation

$$\frac{1}{1+x} = 1 - x + x^2 - x^3 + \cdots \tag{104}$$

valid for $x < 1$, in the usual sense of convergence. If we formally replace x by 1 on both sides, we find that

$$\tfrac{1}{2} = 1 - 1 + 1 + \cdots$$

which is, of course, nonsense, since the partial sums of the series on the right are

$$S_n = \begin{cases} 1 & n \text{ odd} \\ 0 & n \text{ even} \end{cases} \tag{105}$$

and the sequence $S_1, S_2, S_3, \ldots$, naturally, does not converge. Consider, however, the average of the first n partial sums,

$$\sigma_n = \frac{1}{n}\sum_{\nu=1}^{n} S_{\nu}. \tag{106}$$

A short calculation shows that

$$\sigma_n = \begin{cases} \frac{1}{2} & n \text{ even} \\ \dfrac{1}{2}\left(1 + \dfrac{1}{n}\right) & n \text{ odd} \end{cases} \tag{107}$$

This sequence is

$$1, \tfrac{1}{2}, \tfrac{2}{3}, \tfrac{1}{2}, \tfrac{3}{5}, \tfrac{1}{2}, \tfrac{4}{7}, \ldots$$

and it is clear that the sequence converges to the value $\frac{1}{2}$. Hence we may say that if we understand the word "convergence" in the sense of convergence of the σ_n, rather than of the s_n, the relation (104) is actually valid for $0 \leq x \leq 1$.

Definition. Let $\sum_{n=1}^{\infty} a_n$ be a formal series. Defining

$$S_n = \sum_{\nu=1}^{n} a_\nu \tag{108}$$

$$\sigma_n = \frac{1}{n} \sum_{\nu=1}^{n} S_\nu \tag{109}$$

we say that the given series is *summable to the value A* if

$$\lim_{n\to\infty} \sigma_n = A. \tag{110}$$

It is a simple matter to verify that if a series converges in the ordinary sense to a value A, then it is summable to the same value, so the notion of summability actually broadens the class of series to which sums can be assigned in a natural manner.

Theorem 17. [*Fejér*].† *Let $f(x)$ be integrable on* $[0, 2\pi]$ *and periodic with period* 2π. *Let x_0 be a point of the interval* $[0, 2\pi]$ *at which* $f(x_0 + 0)$, $f(x_0 - 0)$ *exist. Then the Fourier series of $f(x)$ is summable at x_0 to the value* $\frac{1}{2}[f(x_0 + 0) + f(x_0 - 0)]$. *In particular, the Fourier series of a continuous function $f(x)$ is everywhere summable to $f(x)$.*

Proof. If

$$\frac{a_0}{2} + \sum_{n=1}^{\infty} (a_n \cos nx_0 + b_n \sin nx_0)$$

is the given series, by (97), we have

$$S_n(x_0) = \frac{2}{\pi} \int_0^{\pi/2} \tfrac{1}{2}[f(x_0 + 2t) + f(x_0 - 2t)] \frac{\sin (2n + 1)t}{\sin t}\, dt. \tag{111}$$

† Fejér [1].

Forming the Fejér means (109),

$$
\begin{aligned}
(112)\qquad n\sigma_{n-1} &= S_0(x_0) + S_1(x_0) + \cdots + S_{n-1}(x_0) \\
&= \frac{2}{\pi}\int_0^{\pi/2} \tfrac{1}{2}[f(x_0 + 2t) + f(x_0 - 2t)] \\
&\quad \times \frac{\sin t + \sin 3t + \cdots + \sin (2n-1)t}{\sin t}\, dt \\
&= \frac{2}{\pi}\int_0^{\pi/2} \tfrac{1}{2}[f(x_0 + 2t) + f(x_0 - 2t)]\left(\frac{\sin nt}{\sin t}\right)^2 dt.
\end{aligned}
$$

We have used without proof the identity

$$
(113)\qquad \sum_{\nu=0}^{n-1} \sin (2\nu + 1)t = \frac{\sin^2 nt}{\sin t}
$$

whose proof is similar to (95). Hence

$$
(114)\qquad \sigma_{n-1}(x_0) = \frac{2}{n\pi}\int_0^{\pi/2} \tfrac{1}{2}[f(x_0 + 2t) + f(x_0 - 2t)]\left(\frac{\sin nt}{\sin t}\right)^2 dt.
$$

Comparing this with (111), we note the essential difference that "Fejér's kernel" $(\sin nt/\sin t)^2$ is everywhere positive.

Now from (113) we observe that

$$
\begin{aligned}
(115)\qquad \int_0^{\pi/2}\left(\frac{\sin nt}{\sin t}\right)^2 dt &= \sum_{\nu=0}^{n-1}\int_0^{\pi/2} \frac{\sin (2\nu + 1)t}{\sin t}\, dt \\
&= \sum_{\nu=0}^{n-1} \frac{\pi}{2} \\
&= n\frac{\pi}{2}
\end{aligned}
$$

where (95) has been used. Thus, as before,

$$
f(x_0) = \frac{2}{n\pi}\int_0^{\pi/2} f(x_0)\left(\frac{\sin nt}{\sin t}\right)^2 dt
$$

and

$$
\begin{aligned}
(116)\qquad & \sigma_{n-1}(x_0) - f(x_0) = \\
& \frac{2}{n\pi}\int_0^{\pi/2}\left[\frac{f(x_0 + 2t) + f(x_0 - 2t)}{2} - f(x_0)\right]\left(\frac{\sin nt}{\sin t}\right)^2 dt.
\end{aligned}
$$

Referring to the hypotheses of the theorem, we see that the expression in brackets approaches zero (assuming $f(x)$ to be standardized (102)). Let us abbreviate

$$
h(t) = \frac{f(x_0 + 2t) + f(x_0 - 2t)}{2} - f(x_0).
$$

Since $h(t) \to 0$ as $t \to 0$, there is a number $\delta > 0$ such that, $\varepsilon > 0$ being given, $|h(t)| < \varepsilon/2$ for $0 \leqq t \leqq \delta$. Fixing δ, we observe that $h(t)$ is bounded on $[\delta, \pi/2]$, say $|h(t)| \leqq M$ there. Then

$$\frac{2}{n\pi}\left|\int_0^\delta h(t)\left(\frac{\sin nt}{\sin t}\right)^2 dt\right| \leqq \frac{\varepsilon}{2}\frac{2}{n\pi}\int_0^\delta \left(\frac{\sin nt}{\sin t}\right)^2 dt < \frac{\varepsilon}{2}$$

and

$$\begin{aligned}\frac{2}{n\pi}\left|\int_\delta^{\pi/2} h(t)\left(\frac{\sin nt}{\sin t}\right)^2 dt\right| &\leqq \frac{2}{n\pi} M \int_\delta^{\pi/2} \left(\frac{\sin nt}{\sin t}\right)^2 dt \\ &\leqq \frac{2}{n\pi} M \frac{1}{\sin^2 \delta}\int_\delta^{\pi/2} (\sin nt)^2\, dt \\ &\leqq \frac{2}{n\pi} M \frac{1}{\sin^2 \delta}\int_0^{\pi/2} (\sin nt)^2\, dt \\ &= \frac{1}{n}\frac{M}{\sin^2 \delta}.\end{aligned}$$

Thus, for the given ε, we can choose n_0 so large that this last integral does not exceed $\varepsilon/2$. For all $n \geqq n_0$, then,

$$|f(x_0) - \sigma_{n-1}(x_0)| < \varepsilon$$

and the theorem is proved.

One of the most important applications of the convergence theory of Fourier series concerns "equiconvergence."

Two series

$$\sum a_n f_n(x) \qquad \sum b_n g_n(x)$$

are called *equiconvergent* if the convergence of one of them at a point x_0 implies that of the other. There is a wide class of theorems to the effect that if $\{\phi_n(x)\}$ is an orthogonal sequence of functions, then the series expansion of $f(x)$,

$$f(x) = \sum a_n \phi_n(x)$$

converges to $f(x)$ if and only if the trigonometric series expansion of $f(x)$

$$f(x) = \sum (b_n \cos nx + c_n \sin nx) \tag{117}$$

converges to $f(x)$. This is true, for example, for expansions in the classical orthogonal polynomials and for expansions in the eigenfunctions of certain linear differential operators, particularly those of Sturm-Liouville type

$$\frac{d}{dx}\left(p(x)\frac{dy}{dx}\right) + (\lambda q(x) + r(x))y = 0$$

with homogeneous end conditions. One invariably finds, for instance, that expansions in series for which equiconvergence with (117) is true are always

Fejér summable to $f(x)$ if $f(x)$ is continuous or to (102) otherwise. Thus the range and significance of the Fejér summability theorem is greatly amplified to the class of all orthogonal expansions equiconvergent with trigonometric series.

Bibliography

The standard reference for the theory of orthogonal polynomials is

1. G. Szegö, *Orthogonal Polynomials*, American Mathematical Society Colloquium Publications, vol. 23, 1939.

A complete bibliography of virtually every paper ever published on the subject of orthogonal polynomials before 1939, with coded abstracts, is contained in

2. J. Shohat et al. *A Bibliography of Orthogonal Polynomials*, National Research Council, 1940.

For Gauss quadrature and its variants see

3. F. B. Hildebrand, *An Introduction to Numerical Analysis*, McGraw-Hill Book Co., New York, 1956.
4. Z. Kopal, *Numerical Analysis*, John Wiley and Sons, New York, 1955.

which has also an interesting account of Tschebycheff quadrature.

A table of weights and abscissas for Gauss quadrature with the common weight functions is in

5. *Tables of Functions and of Zeros of Functions*, National Bureau of Standards, *Appl. Math. Series*, no. 37, 1954.

The theory of Fourier Series is completely discussed in the treatise

6. A. Zygmund, *Trigonometric Series*, Cambridge University Press, 1959,

and on a more elementary level, in

7. K. Knopp, *Theory and Application of Infinite Series*, Blackie and Son, London and Glasgow, 1928.

Equiconvergence of Fourier series and Sturm-Liouville expansions is treated in

8. E. L. Ince, *Ordinary Differential Equations*, Dover Publications, New York, 1944,

while equiconvergence with orthogonal polynomial expansions appears in

9. F. G. Tricomi, *Vorlesungen über Orthogonalreihen*, Springer-Verlag, Berlin, 1955.

Exercises

1. For the weight function $w(x) = e^{-x}$ on the interval $(0, \infty)$,
   (a) find the moments matrix M.
   (b) construct the first three members of the sequence $\{\phi_n(x)\}$ given by (8), in this case.
   (c) verify Theorem 3 for these three polynomials.
2. Let $x_1, \ldots, x_N$ be the zeros of $\phi_N(x)$ and let $H_1, \ldots, H_N$ be the weights for Gauss quadrature. The $N \times N$ matrix

$$U_{ij} = \sqrt{H_i}\,\phi_j(x_i) \qquad (i = 1, 2, \ldots, N;\ j = 0, 1, \ldots, N-1)$$

   is unitary.
3. If J is the matrix of (28), then
   (a) $\phi_N(x) = k_N \det(xI - J)$.
   (b) $\phi_N(J) = 0$.

4. If $\phi_n(x)$ is a sequence of orthogonal polynomials, then the zeros of $\phi_n(x)$ separate the zeros of $\phi_{n+1}(x)$.
5. At a zero, x_i, of $\phi_n(x)$, we have $\phi_{n-1}(x_i)\phi_{n+1}(x_i) < 0$.
6. The Laguerre polynomials $L_n(x)$ satisfy the recurrence

$$L_{n+1}(x) = (2n + 1 - x)L_n(x) - n^2L_{n-1}(x).$$

(a) Find the recurrence relation satisfied by the normalized Laguerre polynomials.
(b) Write down the 2×2 matrix J in this case, and verify directly that its eigenvalues are the zeros of $L_2(x)$.
(c) Find an upper bound for the largest zero of $L_n(x)$, using Theorem 8.
(d) Find the weights and abscissas for the Gauss quadrature

$$\int_0^\infty f(x)e^{-x}\,dx = H_1f(x_1) + H_2f(x_2).$$

(e) Evaluate $\int_0^\infty x^4e^{-x}\,dx$ both exactly and by the formula of part (d).

7. (a) Using the recurrence relation for the Tschebycheff polynomials $T_n(x)$, prove that $T_n(x) = \cos(n\cos^{-1}x)$ $(n = 0, 1, 2, \ldots)$.
(b) Thus find an explicit expression for the zeros

$$x_{N\nu}(\nu = 1, \ldots, N)$$

of $T_N(x)$.
(c) Using the results of parts (a) and (b), verify, from the general formula (68), the relation $H_\nu = \pi/N$ $(\nu = 1, \ldots, N)$ for the weights of Gauss-Tschebycheff quadrature, which is stated without proof in the text.
(d) Prove the result of part (a) by direct integration, showing that the correct orthogonality condition is indeed satisfied.
(e) Using the result of (a), translate the Christoffel-Darboux formula for $T_n(x)$ into a trigonometric identity.

8. The remarkable result of exercise 7(c), that the weights for Gauss-Tschebycheff quadrature are all *equal*, leads one to investigate the possibility of finding other formulas of the type

$$\int_a^b f(x)w(x)\,dx = H_0[f(x_1) + f(x_2) + \cdots + f(x_N)].$$

(a) If $[a, b] = [0, 1]$, $w(x) = 1$, $N = 2$, is there such a formula with $H_0 > 0$, $0 \leq x_1 < x_2 \leq 1$, exact for polynomials of degree ≤ 2?
(b) If $[a, b] = [0, \infty]$, $w(x) = e^{-x}$, investigate the same question for $N = 2$ and $N = 3$ separately.

The general question posed here is that of Tschebycheff-Bernstein quadrature. The question of characterizing the weight functions and intervals for which this formula exists is still unsettled. (See Bernstein [1], Wilf [2], Ullman [1].

9. $1/H_\nu$ is the square of the length of the νth eigenvector of J if the vector is normalized so that its first component is $\mu_0^{-1/2}$, where $\mu_0 = \int_a^b w(x)\,dx$.

10. Consider the formal series

$$\delta(x) = \frac{1}{2} + \sum_{n=1}^{\infty} \cos nx.$$

(a) Prove that this series converges nowhere.
(b) Calculate the nth Fejér mean of the series, and hence find all points at which the series is summable and the values to which it is summable.
(c) Let a, $0 < a < \pi$ be given, and let $f(x)$ be a given continuous function on $(-\pi, \pi)$. Show that

$$\frac{2}{\pi} \lim_{n\to\infty} \int_{-a}^{a} f(t)\sigma_n(t)\, dt = f(0)$$

where $\sigma_n(t)$ denotes the Fejér mean formed in (b).

The moral of this story is that we may regard the formal power series as "representing" the Dirac δ-function and formally write

$$\frac{2}{\pi} \int_{-a}^{a} f(t)\, \delta(t)\, dt = f(0)$$

provided that this last equation is precisely understood in the sense of the result of part (c).

chapter 3

The roots of polynomial equations

3.1 INTRODUCTION

A function $f(z)$ of the form

$$f(z) = a_0 + a_1 z + \cdots + a_n z^n \tag{1}$$

of the complex variable z, with complex coefficients $a_0, a_1, \ldots, a_n$ is a polynomial of degree n. A complex number z_0, having the property that $f(z_0) = 0$ is called a *root* of the equation $f(z) = 0$, or a *zero* of the polynomial $f(z)$. We will assume throughout, for simplicity, that in (1) $a_0 \neq 0$ and $a_n \neq 0$, which can always be achieved in a trivial manner.

We further assume that the reader is familiar with the fact that $f(z)$ of (1) always has exactly n zeros $z_1, z_2, \ldots, z_n$ in the complex plane and may be factored in the form

$$f(z) = a_n(z - z_1)(z - z_2) \cdots (z - z_n). \tag{2}$$

Our concern in this chapter is almost entirely with the analytic (as opposed to the algebraic) theory of polynomial equations. Roughly speaking, this theory is concerned with describing the position of the zeros in the complex plane, without actually solving the equation, as accurately as possible in terms of easily calculated functions of the coefficients.

Specifically, we list the following questions, all of which are answered more or less completely in the following sections:

1. Suppose we know the zeros of $f(z)$. What can be said about the zeros $f'(z)$?
2. What circle $|z| \leqq R$, in the complex plane, surely contains all the zeros of $f(z)$?
3. How many zeros does $f(z)$ have in the left (right) half plane? In the unit circle? On the real axis? On the real interval $[a, b]$? In the sector $\alpha \leqq \arg z \leqq \beta$?
4. How can we efficiently calculate the zeros of $f(z)$?

3.2 THE GAUSS-LUCAS THEOREM

Let us recall—from elementary calculus—the theorem of Rolle, which asserts that if $f(a) = f(b) = 0$, then $f'(x) = 0$ somewhere between a and b, $f(x)$ being continuously differentiable in (a, b). Viewed otherwise, this theorem states that if z_1, z_2 are two *real* zeros of $f(z)$, then $f'(z)$ has a zero somewhere between z_1, z_2. We propose to generalize this result to the case of arbitrary complex zeros, $z_1, z_2, \ldots, z_n$. We need first

Lemma 1. Let the complex numbers $\zeta_1, \zeta_2, \ldots, \zeta_n$ all lie on the same side of some straight line through the origin, in the complex plane. Suppose, further, that at least one of the points is not on this line. Then

$$\sum_{\nu=1}^{n} \zeta_\nu \neq 0. \tag{3}$$

Proof. This result is geometrically obvious, by repeated use of the "parallelogram law" of adding complex numbers. For a proof, however, our hypotheses state that

$$\theta_0 \leqq \arg \zeta_\nu \leqq \theta_0 + \pi \qquad (\nu = 1, 2, \ldots, n)$$

with each equality sign excluded for at least one value of γ. But then,

$$-\frac{\pi}{2} \leqq \arg \{e^{-i[(\pi/2)+\theta_0]}\zeta_\nu\} \leqq \frac{\pi}{2} \qquad (\nu = 1, 2, \ldots, n)$$

i.e.,

$$\mathbf{Re}\, \{e^{-i[(\pi/2)+\theta_0]}\zeta_\nu\} \geqq 0 \qquad (\nu = 1, 2, \ldots, n) \tag{4}$$

with equality excluded for some ν. Hence

$$\mathbf{Re} \sum_{\nu=1}^{n} \{e^{-i[(\pi/2)+\theta_0]}\zeta_\nu\} > 0$$

and

$$\sum_{\nu=1}^{n} e^{-i[(\pi/2)+\theta_0]}\zeta_\nu = e^{-i[(\pi/2)+\theta_0]}\sum_{\nu=1}^{n}\zeta_\nu \neq 0$$

completing the proof.

Next, let us recall that a set of points is *convex* if it contains, with any two points P, Q in the set, the line segment joining P and Q.

Theorem 1.† *(Gauss-Lucas.) Let K be any convex polygon enclosing all the zeros of the polynomial $f(z)$. Then the zeros of $f'(z)$ lie in K.*

Proof. First, let us write the factorization (2) in the form

$$f(z) = a_n(z - z_1)^{m_1}(z - z_2)^{m_2} \cdots (z - z_p)^{m_p} \tag{5}$$

where $z_1, \ldots, z_p$ are the *distinct* zeros of $f(z)$ and m_j is the multiplicity of z_j. Of course,

$$\sum_{j=1}^{p} m_j = n. \tag{6}$$

Now

$$\frac{f'(z)}{f(z)} = \frac{d}{dz}\log f(z) = \sum_{j=1}^{p}\frac{m_j}{z - z_j} \tag{7}$$

and therefore

$$f'(z) = a_n(z - z_1)^{m_1} \cdots (z - z_p)^{m_p}\sum_{j=1}^{p}\frac{m_j}{z - z_j}. \tag{8}$$

From this relation, it is clear that if z_j is a zero of $f(z)$ of multiplicity m_j, then z_j is a zero of $f'(z)$ of multiplicity $m_j - 1$. In this way we account for

$$\sum_{j=1}^{p}(m_j - 1) = n - p \tag{9}$$

of the $n - 1$ zeros of $f'(z)$. The remaining $p - 1$ zeros of $f'(z)$ are the zeros of

$$F(z) = \sum_{j=1}^{p}\frac{m_j}{z - z_j}. \tag{10}$$

Now let ζ denote any zero of $f'(z)$. If ζ is one of the zeros accounted for in (9), ζ is identical with a zero of $f(z)$ and plainly lies in the polygon K. It remains to show the same if ζ is a zero of $F(z)$. Suppose ζ lies outside the polygon K. Since K is convex, it subtends an angle $\theta_0 < \pi$ when viewed from ζ. Now the vectors $\zeta - z_j$ join the point ζ to each of the z_j, and hence all lie in the angle subtended by K at ζ. Thus the "spread" in the arguments of the numbers $\zeta - z_j$ $(j = 1, \ldots, p)$ is less than π. The same is true of the vectors $1/\zeta - z_j$, and therefore also of the vectors $m_j/\zeta - z_j$. Thus the

† Gauss [1], Lucas [1].

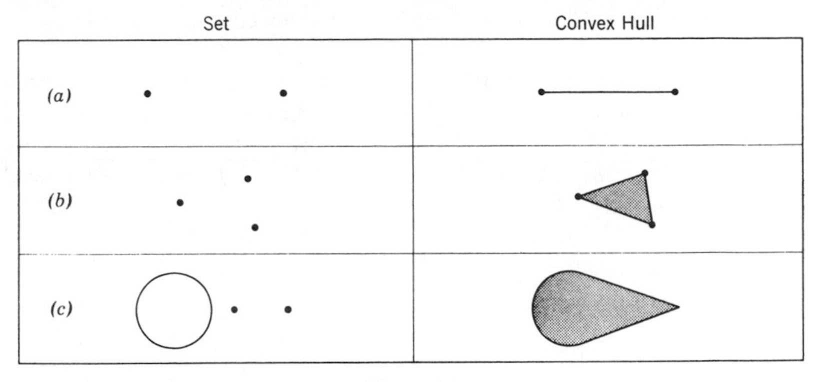

Figure 3.1

points $m_j/\zeta - z_j$ all lie on the same side of some line through the origin, and hence, referring to (10),

$$F(\zeta) = \sum_{j=1}^{p} m_j/\zeta - z_j$$

is not zero, and ζ is not a zero of $F(z)$, which was to be shown.

The "smallest" convex set containing the points $z_1, \ldots, z_n$ is called the *convex hull* of the points $z_1, \ldots, z_n$. It is the set K having the properties (i) K is convex, (ii) K contains $z_1, z_2, \ldots, z_n$, (iii) If $\tilde{K}$ is any other set satisfying (i), (ii), then $K \subseteq \tilde{K}$.

The theorem of Gauss-Lucas then says, in particular, that the zeros of $f'(z)$ lie in the convex hull of the zeros of $f(z)$. The figure above shows three point sets and their convex hulls.

Example. The polynomial

$$f(z) = z^4 + 4$$

has zeros at $\pm(1 \pm i)$, i.e., at the corners of the square of side 2 centered at the origin. Hence the zeros of $f'(z)$ must lie in this square, as they obviously do.

3.3 BOUNDS FOR THE MODULI OF THE ZEROS

Again referring to the polynomial

$$f(z) = a_0 + a_1 z + \cdots + a_n z^n \tag{11}$$

where the a_i are arbitrary complex numbers except that $a_0 a_n \neq 0$, we would like to make statements of the type that all zeros of $f(z)$ surely lie in the circle $|z| \leqq R$, where R is some (more or less) easily computable function of the coefficients.

Let C be the companion matrix [Chapter 1, equation (100)] of $f(z)$, and let $z_1, \ldots, z_n$ denote the zeros of $f(z)$ arranged in nondecreasing order of magnitude $|z_1| \leqq |z_2| \leqq \cdots \leqq |z_n|$. The eigenvalues of C, of course, are $z_1, \ldots, z_n$. Let C^+ denote, as usual, the matrix C with its elements replaced by their moduli:

$$C^+ = \begin{pmatrix} \left|\frac{a_{n-1}}{a_n}\right| & \left|\frac{a_{n-2}}{a_n}\right| & \cdots & & \left|\frac{a_0}{a_n}\right| \\ 1 & 0 & \cdots & & 0 \\ 0 & 1 & \cdots & & 0 \\ \cdot & \cdot & & & \cdot \\ \cdot & \cdot & & & \cdot \\ \cdot & \cdot & & & \cdot \\ 0 & 0 & \cdots & 1 & 0 \end{pmatrix} \tag{12}$$

We leave the proof of the irreducibility of C^+ as an exercise. Assuming this, by Theorem 34 of Chapter 1 we know that the eigenvalues of C, i.e., the numbers $z_1, z_2, \ldots, z_n$, do not exceed in modulus the Perron root r of C^+.

But C^+ is itself a companion matrix, namely of the polynomial

$$\tilde{f}(z) = |a_0| + |a_1|\, z + \cdots + |a_{n-1}|\, z^{n-1} - |a_n|\, z^n. \tag{13}$$

Hence the Perron root r of C^+ is the largest positive zero of $\tilde{f}(z)$. We claim that $\tilde{f}(z)$ has exactly one positive zero. Indeed, the function,

$$\frac{|a_0|}{x^n} + \frac{|a_1|}{x^{n-1}} + \cdots + \frac{|a_{n-1}|}{x}$$

is clearly monotone, decreasing from $+\infty$ to zero as x goes from zero to $+\infty$. Hence this function attains the value $|a_n|$ exactly once, which was to be shown. We have proved

Theorem 2.† *All the zeros of the polynomial* $f(z)$ *of* (11) *lie in the circle* $|z| \leqq r$, *where* r *is the unique positive real root of* (13).

Now, for this number r, we get the inequality

$$r \leqq \max_{1 \leqq i \leqq n} \frac{((C^+)^T x)_i}{x_i} \qquad (\mathbf{x} > 0, \mathbf{x} \neq \mathbf{0}) \tag{14}$$

immediately from equation (157) of Chapter 1, where we have replaced C^+ by its transpose, which, of course, does not alter its eigenvalues. Now, by

† Cauchy [1]. See also Pellet [1].

direct multiplication, using (12),

$$(C^+)^T x = (C^+)^T \begin{pmatrix} x_1 \\ x_2 \\ \cdot \\ \cdot \\ \cdot \\ x_n \end{pmatrix} = \begin{pmatrix} \left|\dfrac{a_{n-1}}{a_n}\right| x_1 + x_2 \\ \left|\dfrac{a_{n-2}}{a_n}\right| x_1 + x_3 \\ \cdot \\ \cdot \\ \cdot \\ \left|\dfrac{a_0}{a_n}\right| \end{pmatrix}$$

and from (14), deduce at once

Theorem 3.† *Let $x_1, x_2, \ldots, x_n$ be arbitrary positive numbers, and let $x_{n+1} = 0$. Then all the zeros of the polynomial $f(z)$ of* (11) *lie in the circle*

$$|z| \leqq \max_{1 \leqq i \leqq n} \left\{ \left| \frac{a_{n-i}}{a_n} \right| \frac{x_1}{x_i} + \frac{x_{i+1}}{x_i} \right\}. \tag{15}$$

Furthermore, there exists a choice of $x_1, \ldots, x_n$ for which the right side of (15) *is the positive root of* (13).

From this general theorem we may get several results as special cases. First, take all $x_i = 1$.

Theorem 4.‡ *All the roots of the polynomial $f(z)$ of* (11) *lie in the circle*

$$|z| \leqq \max \left\{ \left| \frac{a_{n-1}}{a_n} \right| + 1, \left| \frac{a_{n-2}}{a_n} \right| + 1, \ldots, \left| \frac{a_1}{a_n} \right| + 1, \left| \frac{a_0}{a_n} \right| \right\}. \tag{16}$$

Next, take $x_i = \rho^i$ $(i = 1, 2, \ldots, n)$, for some $\rho > 0$. Then all the roots lie in

$$|z| \leqq \max_{1 \leqq i \leqq n} \left\{ \left| \frac{a_{n-i}}{a_n} \right| \rho^{1-i} \right\} + \rho \tag{17}$$

for example (actually (15) gives a slightly better inequality than this). Now suppose the "max" in (17) is attained when $i = p$, say. Then, denoting the maximum value by M,

$$M = \left| \frac{a_{n-p}}{a_n} \right| \rho^{1-p}.$$

Solving for ρ,

$$\rho = M^{-(1/p-1)} \left| \frac{a_{n-p}}{a_n} \right|^{1/p-1} \tag{18}$$

† Wilf [1].
‡ Cauchy [1].

Suppose we choose ρ so that $\rho = M$. Setting (18) equal to M,

$$M = \left| \frac{a_{n-p}}{a_n} \right|^{1/p}. \tag{19}$$

Since $\rho = M$, the two terms on the right side of (17) are equal, and therefore all roots lie in the circle

$$|z| \leqq 2\rho = 2M = 2\left| \frac{a_{n-p}}{a_n} \right|^{1/p}.$$

Now we do not know the integer p but can state with assurance

Theorem 5.† *All the zeros of the polynomial* $f(z)$ *of* (11) *lie in the circle*

$$|z| \leqq 2 \max_{1 \leqq i \leqq n} \left| \frac{a_{n-i}}{a_n} \right|^{1/i} \tag{20}$$

since one of the values of i on the right is p.

As a final example, let us choose the x_i so that the two terms in the braces in (15) are equal, for each $1 = 1, 2, \ldots, n - 1$, i.e., so that

$$x_{i+1} = \left| \frac{a_{n-i}}{a_n} \right| x_1 \qquad (i = 1, 2, \ldots, n - 1).$$

Then,

$$\left| \frac{a_{n-i}}{a_n} \right| \frac{x_1}{x_i} + \frac{x_{i+1}}{x_i} = \left| \frac{a_{n-i}}{a_n} \right| \frac{x_1 |a_n|}{|a_{n-i+1}| x_1} + \left| \frac{a_{n-i}}{a_{n-i+1}} \right|$$
$$= 2\left| \frac{a_{n-i}}{a_{n-i+1}} \right|$$

and we have

Theorem 6.‡ *All the zeros of the polynomial* $f(z)$ *of* (11) *lie in the circle*

$$|z| \leqq \max \left\{ \left| \frac{a_0}{a_1} \right|, 2\left| \frac{a_1}{a_2} \right|, \ldots, 2\left| \frac{a_{n-2}}{a_{n-1}} \right|, 2\left| \frac{a_{n-1}}{a_n} \right| \right\}. \tag{21}$$

To illustrate these theorems, consider the polynomial

$$f(z) = 1 + z + \frac{z^2}{2!} + \frac{z^3}{3!} + \cdots + \frac{z^n}{n!} \tag{22}$$

which is the nth partial sum of the Taylor's series for e^z. We have

$$a_\nu = \frac{1}{\nu!} \qquad (\nu = 0, 1, \ldots, n). \tag{23}$$

† Fujiwara [1].
‡ Kojima [1].

From (16) we get the circle

$$|z| \leqq \max\{n+1, n(n-1)+1, \ldots, n!+1, n!\} = 1 + n! \tag{24}$$

while from (20),

$$\begin{aligned} |z| &\leqq 2 \max_{1 \leqq i \leqq n} \left| \frac{n!}{(n-i)!} \right|^{1/i} \\ &= 2 \max\{n, \sqrt{n(n-1)}, \sqrt[3]{n(n-1)(n-2)}, \ldots, n!^{1/n}\} \\ &= 2n. \end{aligned} \tag{25}$$

Finally, (21) gives the circle

$$\begin{aligned} |z| &\leqq \max\{2n, 2(n-1), \ldots, 2, 1\} \\ &= 2n. \end{aligned} \tag{26}$$

Thus the best result we get from any of them is that the zeros of (22) are in the circle $|z| \leqq 2n$.

To find a lower bound for the zero of smallest modulus of a polynomial, see exercise 2. To find a lower bound for the zero of largest modulus, we may use the Gauss-Lucas theorem repeatedly, as follows. Let $f^{(\nu)}(z)$ denote the νth derivative of $f(z)$, and let $z_{1\nu}, z_{2\nu}, \ldots, z_{n-\nu,\nu}$ be the zeros of $f^{(\nu)}(z)$, arranged in nondecreasing order of modulus. Now

$$\begin{aligned} f^{(\nu)}(z) &= \sum_{k=0}^{n} a_k \frac{d^\nu}{dz^\nu} z^k = \sum_{k=\nu}^{n} a_k k(k-1)\cdots(k-\nu+1) z^{k-\nu} \\ &= \sum_{k=\nu}^{n} a_k \frac{k!}{(k-\nu)!} z^{k-\nu}. \end{aligned} \tag{27}$$

The ratio of the constant term of $f^{(\nu)}(z)$ to the coefficient of the highest power of z in $f^{(\nu)}(z)$ is the product of the zeros of $f^{(\nu)}(z)$, aside from a sign. Hence

$$\begin{aligned} |z_{1\nu} z_{2\nu} \cdots z_{n-\nu,\nu}| &= \left| \frac{a_\nu}{a_n} \frac{\nu!\,(n-\nu)!}{n!} \right| \\ &= \binom{n}{\nu}^{-1} \left| \frac{a_\nu}{a_n} \right| \\ &\leqq |z_{n-\nu,\nu}|^{n-\nu} \end{aligned}$$

and therefore

$$|z_{n-\nu,\nu}| \geqq \left\{ \left| \frac{a_\nu}{a_n} \right| \Big/ \binom{n}{\nu} \right\}^{1/n-\nu}$$

On the other hand, the Gauss-Lucas theorem tells us that

$$|z_{n-\nu,\nu}| \leqq |z_{n-\nu+1,\nu-1}| \leqq \cdots \leqq |z_{n0}|$$

thus

$$|z_{n0}| \geqq \left\{ \binom{n}{\nu}^{-1} \left| \frac{a_\nu}{a_n} \right| \right\}^{1/n-\nu}$$

This being true for each $\nu = 0, 1, \ldots, n-1$, we have proved

Theorem 7.† *The modulus of the zero of $f(z)$ of largest modulus is at least*

$$\max_{0 \leqq \nu \leqq n-1} \left\{ \binom{n}{\nu}^{-1} \left| \frac{a_\nu}{a_n} \right| \right\}^{1/n-\nu} \tag{28}$$

Using (22), again, as an illustration, (28) becomes

$$\begin{aligned} \max_{0 \leqq \nu \leqq n-1} \left\{ \frac{\nu!\,(n-\nu)!}{n!} \frac{n!}{\nu!} \right\}^{1/n-\nu} &= \max_{0 \leqq \nu \leqq n-1} \{(n-\nu)!\}^{1/n-\nu}. \\ &= \max_{1 \leqq r \leqq n} \{r!\}^{1/r} \\ &\geqq \max_{1 \leqq r \leqq n} \left\{ \frac{r}{e} \right\} \\ &= \frac{n}{e}. \end{aligned} \tag{29}$$

Thus the largest zero of the nth partial sum of e^z lies in the ring

$$\frac{n}{e} \leqq |z| \leqq 2n.$$

(In (29) we used the relation $(n!)^{1/n} \geqq n/e$, which may be proved easily, for

$$\begin{aligned} n! &= \int_0^\infty x^n e^{-x}\, dx \geqq \int_y^\infty x^n e^{-x}\, dx \qquad (y > 0) \\ &\geqq y^n \int_0^\infty e^{-x}\, dx = y^n e^{-y}. \end{aligned}$$

Taking $y = n$, the result follows.)

3.4 STURM SEQUENCES

We turn next to the location of zeros on the real axis, considering only the case where the coefficients of $f(z)$ are real.

Let (a, b) be a finite or infinite interval of the real axis, and let $f_1(x), \ldots, f_p(x)$ be p continuous functions defined on (a, b). We say that $f_1(x), \ldots, f_p(x)$ are a *Sturm sequence* for (a, b) if

(i) at a zero x_0 of $f_k(x)$, $f_{k+1}(x_0)$ and $f_{k-1}(x_0)$ have opposite signs and are not zero ($k = 2, 3, \ldots, p-1$; $a < x_0 < b$).

(ii) the function $f_p(x)$ is never zero in (a, b).

† L. A. Rubel (unpublished); compare Throumolopoulos [1].

Now, suppose $f_1(x), \ldots, f_p(x)$ is a given Sturm sequence for (a, b). Let x_0 be a fixed point of (a, b), and suppose first that none of the $f_k(x)$ vanishes at x_0 $(k = 1, \ldots, p)$. We define $V(x_0)$, the number of variations of sign in the sequence $f_1(x_0), f_2(x_0), \ldots, f_p(x_0)$, by writing down the vector

$$(\operatorname{sgn} f_1(x_0), \operatorname{sgn} f_2(x_0), \ldots, \operatorname{sgn} f_p(x_0))$$

and counting the number of times the sign changes from $+$ to $-$ or from $-$ to $+$ as we pass from left to right along the vector. Next, if one of the functions is zero at x_0, say $f_k(x_0) = 0$, then by axiom (i), the functions $f_{k-1}(x), f_{k+1}(x)$ have opposite signs at x_0, and it is clear that in determining $V(x_0)$ we may give either sign to $f_k(x_0)$ without affecting the answer. Finally, at the endpoints a, b, $V(a)$ means $V(a + 0)$, and $V(b)$ means $V(b - 0)$.

Theorem 8. *Let $f(x)$ be defined and continuously differentiable on (a, b), and suppose*

$$f(x), f'(x), f_3(x), \ldots, f_n(x) \tag{30}$$

is a Sturm sequence on (a, b). Then the number of zeros of $f(x)$ in the interval (a, b) is precisely $V(a) - V(b)$.

Proof. Notice, first, that the theorem asserts that if we can form a Sturm sequence which begins with $f(x)$, $f'(x)$, then we can find the number of distinct zeros of $f(x)$ in (a, b) by examining $V(a)$ and $V(b)$.

To prove this, let us trace the behavior of the function $V(x)$ as x moves from $a + 0$ to $b - 0$. Initially $V(x)$ has the value $V(a)$. Clearly $V(x)$ can change only at a point where one of the functions $f_k(x)$ changes sign, i.e., vanishes. Let x_0 be such a point and suppose $f_k(x_0) = 0$. Now $k \neq n$ by axiom (ii); thus $1 \leqq k \leqq n - 1$.

Suppose that $2 \leqq k \leqq n - 1$. By axiom (i) for Sturm sequences there are exactly the following possibilities:

Left of x_0			Right of x_0		
$f_{k-1}(x)$	$f_k(x)$	$f_{k+1}(x)$	$f_{k-1}(x)$	$f_k(x)$	$f_{k+1}(x)$
$+$	$+$	$-$	$+$	$-$	$-$
$+$	$-$	$-$	$+$	$+$	$-$
$-$	$+$	$+$	$-$	$-$	$+$
$-$	$-$	$+$	$-$	$+$	$+$

In each of these four cases, the number of variations of sign in the sequence is unchanged as we pass through x_0. In other words, at a zero x_0, of $f_k(x_0)$, where $2 \leqq k \leqq n - 1$, we have

$$V(x_0 + 0) = V(x_0 - 0)$$

Hence the only points at which $V(x)$ can change are those at which $f_1(x) = f(x)$ is zero. Let x_0 be such a point. Since $f'(x)$ is the slope of $f(x)$, we have only the following possibilities:

Left of x_0		Right of x_0	
$f(x)$	$f'(x)$	$f(x)$	$f'(x)$
$+$	$-$	$-$	$-$
$-$	$+$	$+$	$+$

In each case the sequence loses exactly one sign variation passing through x_0, i.e.,

$$V(x_0 + 0) = V(x_0 - 0) - 1.$$

Therefore, as we move from a to b, the sequence loses as many variations of sign as there are zeros of $f(x)$, which was to be shown.

Notice that if $f(x)$ has multiple zeros in (a, b), then there are points at which $f(x)$ and $f'(x)$ simultaneously vanish, contradicting axiom (i). Then there is no Sturm sequence starting with $f(x), f'(x)$, although, as we shall see, the situation is not irretrievable.

Now let $f(x)$ be a polynomial of degree n with real coefficients. We propose to construct a Sturm sequence beginning with $f(x), f'(x)$, if this is possible (i.e., if $f(x)$ has no multiple zeros in (a, b)), and in any case to construct a Sturm sequence which will give the number of distinct zeros of $f(x)$ in (a, b). Define $f_1(x) = f(x), f_2(x) = f'(x)$. Now divide $f_1(x)$ by $f_2(x)$, getting a quotient $q_1(x)$ and a remainder $r_1(x)$. Take $f_3(x) = -r_1(x)$. Then

$$f_1(x) = q_1(x) \cdot f_2(x) - f_3(x). \tag{31}$$

Next, divide $f_2(x)$ by $f_3(x)$, taking $f_4(x)$ to be the negative of the remainder so obtained. In general, if $f_1(x), \ldots, f_k(x)$ have been found, write

$$f_{k-1}(x) = q_{k-1}(x) f_k(x) - f_{k+1}(x) \qquad (k = 2, 3, \ldots, m) \tag{32}$$

thereby determining $f_{k+1}(x)$. Since the degrees of the $f_k(x)$ are steadily decreasing, this process will terminate after $m \leqq n + 1$ steps with $f_{m-1}(x)$, $f_m(x)$, where $f_m(x)$ divides $f_{m-1}(x)$ with no remainder, i.e.,

$$f_{m-1}(x) = q_{m-1}(x) f_m(x). \tag{33}$$

If $m = n + 1$, $f_m(x)$ is a constant, and we claim that the sequence $f_1(x)$, $f_2(x), \ldots, f_{n+1}(x)$ so generated is a Sturm sequence. Indeed, axiom (ii) is clear since $f_{n+1}(x)$ is constant. Next, if x_0 is a zero of $f_k(x)$, then from (32) we see that $f_{k-1}(x_0)$ and $f_{k+1}(x_0)$ have opposite signs, unless one of them vanishes. But this cannot happen, since if two consecutive functions vanish at x_0, then by the recurrence (32) all succeeding functions vanish at x_0, hence the constant $f_{n+1}(x)$ vanishes at x_0, and therefore identically, hence working backward, all functions vanish identically, which is impossible.

It remains to consider the case where $m < n + 1$, i.e., where the process terminates after fewer then n steps. From (33), $f_m(x)$ divides $f_{m-1}(x)$. From (32) with $k = m - 1$, $f_m(x)$ divides $f_{m-2}(x)$, and so on; $f_m(x)$ divides each of $f_1(x), \ldots, f_{m-1}(x)$; and $f_m(x)$ is clearly the highest common factor of $f_1(x), \ldots, f_{m-1}(x)$. Then it is easy to see that the functions

$$\frac{f_1(x)}{f_m(x)}, \frac{f_2(x)}{f_m(x)}, \ldots, \frac{f_{m-1}(x)}{f_m(x)}, 1 \tag{34}$$

form a Sturm sequence. Since $f_2(x)/f_m(x)$ is not the derivative of $f_1(x)/f_m(x)$, Theorem 8 is not directly applicable, but examination of the proof of that theorem shows that its conclusion remains true in this case also. We summarize with

Theorem 9. *The sequence of polynomials formed from* (32) *with* $f_1(x) = f(x)$, $f_2(x) = f'(x)$ *is a Sturm sequence if* $m = n + 1$, *and* (34) *is such a sequence if* $m < n + 1$. *In either case, the number of distinct zeros of* $f(x)$ *in* (a, b) *is* $V(a) - V(b)$, *where* $V(x)$ *is defined on the appropriate sequence.*

As an illustration, take

$$f(x) = x^3 - 2x^2 - x + 2 = (x^2 - 1)(x - 2). \tag{35}$$

With $f_1(x) = f(x)$, $f_2(x) = f'(x) = 3x^2 - 4x - 1$, we divide $f_1(x)$ by $f_2(x)$ and get $f_3(x) = 7x - 8$, then $f_4(x) = +1$ (we have multiplied the $f_k(x)$ by positive constants, to clear of fractions). Thus $V(x)$ is the number of sign variations in the sequence

$$(x^3 - 2x^2 - x + 2, \quad 3x^2 - 4x - 1, \quad 7x - 8, \quad +1).$$

To find, for example, the number of zeros of (35) in (0, 5), we have

$$V(0) = 2, \quad V(5) = 0, \quad V(0) - V(5) = 2$$

thus there are exactly two zeros in this interval. To find the number of real zeros of (35) we take $a = -\infty, b + \infty$. When x is very large and negative, $V(x) = 3$, while, when x is very large and positive, $V(x) = 0$; therefore all of the zeros of (35) are real.

To illustrate a degenerate case, take

$$f(x) = x^3 - x^2 - x + 1 = (x - 1)^2(x + 1). \tag{36}$$

Here we find $f_2(x) = 3x^2 - 2x - 1$, $f_3(x) = x - 1$, and the process then terminates because $f_2(x)$ is divisible by $x - 1$. Thus we form, instead, the sequence (34), which is

$$x^2 - 1, \quad 3x + 1, \quad 1,$$

and use this sequence in the usual manner, interpreting the answers as the number of *distinct* zeros of $f(x)$ in (a, b). At this stage, however, one actually has complete information about the zeros, since the "extra" zeros are those of $f_m(x)$, in this case simply $x = 1$.

We state, without proof, two theorems which give less precise information than the above, although they are considerably easier to use.

Theorem 10. (*Descartes' Rule of Signs.*) *The number of positive zeros of the polynomial*

$$f(x) = a_0 + a_1x + \cdots + a_nx^n \tag{37}$$

is either equal to the number of variations of sign in the sequence

$$(a_0, \quad a_1, \ldots, a_n)$$

or less by an even number.

Theorem 11. (*Budan's Rule.*) *The number of zeros of* (37) *in* (a, b) *is either equal to* $V(a) - V(b)$ *or less by an even number, where* $V(x)$ *is the number of variations of sign in the sequence*

$$f(x), f'(x), f''(x), \ldots, f^{(n)}(x). \tag{38}$$

Budan's rule reduces to Descartes' when $a = 0$, $b = \infty$. It gives exact information when $V(a) - V(b)$ is zero or one, or if it is somehow known that the sequence of derivatives of $f(x)$ forms a Sturm sequence. The conditions under which this happens are in exercise 6.

Theorems 10 and 11 give upper bounds for the number of zeros of $f(x)$ in an interval. We propose next to find lower bounds for this number.

If $f(x)$ is given, define, recursively,

$$\begin{cases} f_0(x) = f(x) \\ f_{n+1}(x) = \int_0^x f_n(t)\,dt \qquad (n = 0, 1, \ldots, N-1). \end{cases} \tag{39}$$

Lemma 2.† The number of zeros of $f(x)$ in the interval $(0, a)$ is not less than the number of changes in sign of the sequence

$$f_0(a), f_1(a), \ldots, f_N(a). \tag{40}$$

Proof. We first remark that N is arbitrary here and is not related to the degree of $f(x)$; indeed, the result holds for any continuous function. Now the conclusion is obvious if $N = 0$. Suppose the result has been proved for $0, 1, \ldots, k-1$. Suppose the sequence $f_1(a), f_2(a), \ldots, f_k(a)$ has m variations of sign. By the inductive hypothesis applied to $f_1(x)$, $f_1(x)$ has at least m zeros. Considering the full sequence

$$f_0(a), \quad f_1(a), \ldots, f_k(a)$$

† Féjér [2].

what we have to prove is that if $f(a)$ and $f_1(a)$ have the same sign, then $f(x)$ has at least m zeros, and if $f(a)$ and $f_1(a)$ have opposite signs, then $f(x)$ has at least $m + 1$ zeros. Consider the first case, where $f(a)$, $f_1(a)$ have the same sign. Since $f_1(x)$ has m changes of sign in $(0, a)$, and vanishes at $x = 0$, by Rolle's theorem we know that between each pair of zeros of $f_1(x)$ lies a zero of $f_1'(x) = f(x)$. Hence $f(x)$ has at least m changes of sign in $(0, a)$. In the second case, suppose $f_1(a) > 0$, $f(a) = f_1'(a) < 0$. Then the curve $y = f_1(x)$ decreases at a to its positive value $f_1(a)$ at a. Hence the curve turns around between its rightmost zero and a. Therefore, as above, $f(x) = f_1'(x)$ has m zeros between the zeros of $f_1(x)$ and another between the rightmost zero of $f_1(x)$ and a, completing the proof.

Theorem 12.† *The polynomial (or continuous function) $f(x)$ has at least as many zeros in $(0, a)$ as the sequence*

$$f(0), \int_0^a f(t)\,dt, \ldots, \int_0^a t^n f(t)\,dt$$

has variations of sign, n being arbitrary.

Proof. Since

$$f_\nu(x) = \frac{1}{(\nu - 1)!} \int_0^x (x - t)^{\nu-1} f(t)\,dt$$

we have

$$f_\nu(a) = \frac{1}{(\nu - 1)!} \int_0^a (a - t)^{\nu-1} f(t)\,dt = \frac{1}{(\nu - 1)!} \int_0^a f(a - t) t^{\nu-1}\,dt.$$

Hence the sequence (40) can be replaced by

$$f(a), \int_0^a f(a - t)\,dt, \ldots, \int_0^a f(a - t) t^{N-1}\,dt.$$

Since $f(a - t)$ has just as many zeros as $f(t)$, we can replace $f(a - t)$ by $f(t)$, thereby proving the theorem.

3.5 ZEROS IN A HALF-PLANE

Now let $f(z)$ denote a polynomial with complex coefficients, of degree n. We are interested in determining the number of zeros of $f(z)$ in a half-plane, which we take to be the upper half-plane, although obvious changes of variable make our results applicable to any half-plane.

We suppose that $f(z)$ has no zeros on the real axis. Now, let z traverse the real axis from $-\infty$ to ∞, and let z_1 be any fixed point in the upper half-plane. Then, as z moves from $-\infty$ to ∞, the argument of the number

† Fekete [1].

$z_1 - z$, increases by π, as can be seen from the diagram. Similarly, if z_1 is fixed in the lower half-plane, the argument of $z - z_1$ *decreases* by π as z moves from $-\infty$ to ∞ along the real axis.

Now, consider the factored form of $f(z)$,

(41) $$f(z) = (z - z_1)(z - z_2) \cdots (z - z_n).$$

Clearly,

(42) $$\arg f(z) = \arg(z - z_1) + \arg(z - z_2) + \cdots + \arg(z - z_n).$$

Hence, as z moves from $-\infty$ to ∞, $\arg f(z)$ increases by π, for each zero in the upper half-plane and decreases by π for each zero in the lower half-

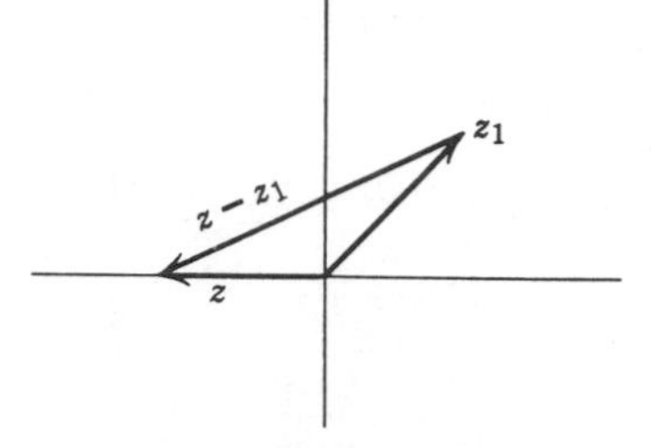

Figure 3.2

plane. Denoting by $\Delta_R \arg f(z)$ the net change in the argument of $f(z)$ as z traverses the x-axis from left to right, we have

(43) $$\Delta_R \arg f(z) = p\pi - q\pi = (p - q)\pi$$

where p, q are, respectively, the number of zeros of $f(z)$ in the upper and lower half-planes. Since also

(44) $$p + q = n$$

we may solve (43), (44) simultaneously for p and q, getting

(45) $$p = \frac{n}{2} + \frac{1}{2\pi} \Delta_R \arg f(z)$$

(46) $$q = \frac{n}{2} - \frac{1}{2\pi} \Delta_R \arg f(z).$$

Now let us write $f(z)$ in the form

(47) $$f(z) = a_0 + a_1 z + \cdots + a_{n-1} z^{n-1} + z^n$$

where

(48) $$a_\nu = \sigma_\nu + i\tau_\nu \qquad (\nu = 0, 1, \ldots, n - 1)$$

σ_ν, τ_ν real. Thus

(49) $$(x) = P(x) + iQ(x)$$

where

$$P(x) = \sigma_0 + \sigma_1 x + \cdots + \sigma_{n-1} x^{n-1} + x^n \tag{50}$$

$$Q(x) = \tau_0 + \tau_1 x + \cdots + \tau_{n-1} x^{n-1} \tag{51}$$

are both real polynomials. Now, when x is real, from (49) we see that

$$f(x) = |f(x)| \exp\left\{i \cot^{-1} \frac{P(x)}{Q(x)}\right\}$$

and therefore

$$\arg f(x) = \cot^{-1} \psi(x) \tag{52}$$

where

$$\psi(x) = \frac{P(x)}{Q(x)}. \tag{53}$$

We wish to calculate $\Delta_R \arg f(x)$, which from (52), (53) obviously depends on the real zeros of $P(x)$. Suppose the real zeros of $P(x)$ are

$$\xi_1 < \xi_2 < \cdots < \xi_m \qquad (m \leqq n). \tag{54}$$

None of the ξ_ν is also a zero of $Q(x)$, since that would imply, by (49), that $f(x) = 0$ on the real axis, contrary to our supposition.

Now consider the interval $(\xi_k + \varepsilon, \xi_{k+1} - \varepsilon)$ of the real axis. As x traverses this interval from left to right, if $P(x)/Q(x)$ changes from $-$ to $+$, $\cot^{-1} \psi(x)$, and therefore $\arg f(x)$ increases by π, while if $P(x)/Q(x)$ changes from $+$ to $-$, $\arg f(x)$ decreases by π. Combining these results, we see that $1/\pi \, \Delta_R \arg f(x)$ is the excess of the number of points of the real axis at which $P(x)/Q(x)$ changes from $+$ to $-$ over the number of points at which it changes from $-$ to $+$ (see exercise 8).

Now, examination of the proof of Theorem 8 shows that this excess is precisely measured by $V(\infty) - V(-\infty)$, where $V(x)$ is the number of changes of sign in any Sturm sequence beginning with $P(x)$, $Q(x)$. Summarizing, we have

Theorem 13.† *Let $f(z)$ of* (47) *be given, let $P(x)$, $Q(x)$ be defined by* (48), (50), (51), *and suppose $f(z)$ has no real zeros. If $P(x)$, $Q(x)$, $f_3(x)$, . . . is any Sturm sequence beginning with $P(x)$, $Q(x)$, then the number of zeros of $f(z)$ in the upper half-plane is*

$$p = \tfrac{1}{2}\{n + V(\infty) - V(-\infty)\}. \tag{55}$$

As an illustration, we ask for the number of zeros in the left half-plane of the polynomial

$$f(z) = z^3 - 3z^2 + 4z - 2 \tag{56}$$

† Routh [1], Hurwitz [1].

whose zeros are actually at the points $z = 1$, $z = 1 - i$, $z = 1 + i$. Now this number is the number of zeros in the upper half-plane of the polynomial

$$g(z) = if(iz) = z^3 + 3iz^2 - 4z - 2i. \tag{57}$$

Referring to (50), (51), we have

$$P(x) = x^3 - 4x \tag{58}$$

$$Q(x) = 3x^2 - 2. \tag{59}$$

By the usual division algorithm, the rest of the Sturm sequence is

$$f_3(x) = \tfrac{10}{3}x \tag{60}$$

$$f_4(x) = 2. \tag{61}$$

Hence, from (55),

$$\begin{aligned} p &= \tfrac{1}{2}\{3 + V(\infty) - V(-\infty)\} \\ &= \tfrac{3}{2} + \tfrac{1}{2}V(+, +, +, +) - \tfrac{1}{2}V(-, +, -, +) \\ &= {}_2 + 0 - \tfrac{3}{2} \\ &= 0 \end{aligned}$$

as is obviously true.

The process of forming the Sturm sequence and examining the number of sign variations can be mechanized, in a certain sense, by constructing certain matrices, known as the Routh and Hurwitz matrices. It is the author's opinion that in small problems the straightforward construction of Sturm sequences is not troublesome, while large problems require the use of computing machinery in any case. In the latter event, paradoxically enough, the best way to evaluate the determinants of Routh and Huritz turns out to be by essentially rederiving them from the algorithm above. In view of this circumstance it appears that in any event the best mode of approach may well be the direct construction of Sturm sequences. For these reasons we will not discuss the Routh-Hurwitz determinants here, but instead refer the reader to the bibliography at the end of the chapter.

3.6 ZEROS IN A SECTOR, ERDÖS-TURÁN'S THEOREM

By proceeding in the manner of the preceding section one can construct certain Sturm sequences which give exact information about the number of zeros of a polynomial in a sector. As these results are of rather limited utility in practice, we devote this section to a most remarkable theorem on this same subject, which, even though it does not give exact information, nonetheless gives some insight into those properties of a polynomial which tend to make it have a uniform or nonuniform distribution of zeros in angle.

Suppose we choose a polynomial

$$f(z) = a_0 + a_1 z + \cdots + a_n z^n \tag{62}$$

with complex coefficients "at random," in some sense. How many zeros would it be expected to have in the sector $\alpha \leqq \arg z \leqq \beta$? In the absence of further information about $f(z)$, the answer must clearly be $(\beta - \alpha)n/2\pi$. Let $V(\alpha, \beta, f)$ denote the number of zeros of $f(z)$ in this sector. Then if

$$V(\alpha, \beta, f) - \frac{\beta - \alpha}{2\pi} n \tag{63}$$

is never very large for any $0 \leqq \alpha \leqq \beta < 2\pi$, we may say that the zeros of $f(z)$ are rather uniformly distributed in angle.

The prototype of those polynomials whose zeros are so distributed is

$$f_0(z) = 1 + z^n \tag{64}$$

with zeros evenly spaced around $|z| = 1$. The salient feature of (64) is that its "middle coefficients" $a_1, a_2, \ldots, a_{n-1}$ are small in modulus compared to the end coefficients $a_0 = 1, a_n = 1$. One might reasonably expect, therefore, to see a theorem which states that if $a_1, a_2, \ldots, a_{n-1}$ are not too large, in modulus, compared with a_0, a_n, then the quantity (63), which measures the equidistribution of zeros in angle, will also never become too large. The precise result is

Theorem 14.† *If $f(z)$ is defined by* (62), *let*

$$P = \frac{|a_0| + |a_1| + \cdots + |a_n|}{\sqrt{|a_0 a_n|}}. \tag{65}$$

Then

$$\left| V(\alpha, \beta, f) - \frac{\beta - \alpha}{2\pi} n \right| < 16\sqrt{n \log P}. \tag{66}$$

We are unable to prove this theorem here because that would require a rather deep analysis of the relationship between the maximum modulus of a polynomial whose zeros are on the unit circle and the number of zeros of that polynomial on an arc of the circle.

Instead we shall attempt, by a naive argument, to make the inequality (66) appear reasonable. To do this, observe that the left-hand side of (66) is never larger than n and reaches n only when the zeros of $f(z)$ are on a ray $\arg z = \gamma$, and $\alpha = \beta = \gamma$. Let us suppose that this is so and that all the zeros of $f(z)$ are on the negative real axis. In this class of polynomials let us try to minimize the right-hand side of (66). We have the

Lemma. If $f(z)$ has negative real zeros, then

$$P \geqq 2^n$$

the sign of equality holding only for $f(z) = (1 + z)^n$.

† P. Erdös and P. Turán [1].

Proof. Such a polynomial can be written in two ways,

$$f(z) = a_n \prod_{i=1}^{n} (z + x_i) = a_0 + a_1 z + \cdots + a_n z^n$$

where $x_i > 0$, $a_i > 0$ $(i = 0, 1, \ldots, n)$. Now,

$$\begin{aligned} P &= \frac{|a_0| + \cdots + |a_n|}{\sqrt{|a_0 a_n|}} = \frac{a_0 + a_1 + \cdots + a_n}{\sqrt{a_0 a_n}} \\ &= \frac{f(1)}{\sqrt{a_0 a_n}} = \frac{a_n \prod(1 + x_i)}{\sqrt{a_0 a_n}} \\ &= \sqrt{(a_n/a_0)} \prod(1 + x_i) = \frac{\prod(1 + x_i)}{\sqrt{\prod(x_i)}} \\ &= \frac{\prod(1 + x_i)}{\prod(\sqrt{x_i})} = \prod_{i=1}^{n} \left(\frac{1}{\sqrt{x_i}} + \sqrt{x_i} \right) \geqq 2^n \end{aligned}$$

since $x > 0$ implies $x + 1/x \geqq 2$, with equality only if $x = 1$.

Thus the minimum value of $\log P$ in the class of polynomials with negative real zeros is $n \log 2$. Hence for all such polynomials we have

$$\left| V(\alpha, \beta, f) - \frac{\beta - \alpha}{2\pi} n \right| \leq \frac{1}{\sqrt{\log 2}} \sqrt{n \log P}$$

and the general theorem of Erdös-Turán asserts that this actually holds for all polynomials if $(\log 2)^{-1/2}$ is replaced by 16 (although this may not be necessary).

3.7 NEWTON'S SUMS

In this section we write $f(z)$ in the form

$$f(z) = z^n + c_1 z^{n-1} + \cdots + c_{n-1} z + c_n. \tag{67}$$

Let the zeros of $f(z)$ be $z_1, z_2, \ldots, z_n$. The power sums

$$S_k = \sum_{\nu=1}^{n} z_\nu^{\ k} \tag{68}$$

are called the Newton sums of $f(z)$. The first few are

$$\begin{aligned} S_0 &= n \\ S_1 &= z_1 + z_2 + \cdots + z_n \\ S_2 &= z_1^2 + z_2^2 + \cdots + z_n^2 \\ &\ \ \vdots \end{aligned} \tag{69}$$

It was first shown by Newton that the S_k can be found without solving the equation $f(z) = 0$, by a simple recurrence relation. Indeed, taking the logarithm of both sides of (2), and differentiating,

$$
\begin{aligned}
f'(z) &= f(z)\sum_{j=1}^{n} \frac{1}{z - z_j} \\
&= f(z)\sum_{j=1}^{n} \frac{1}{z}\frac{1}{1 - z_j/z} \\
&= f(z)\sum_{j=1}^{n} \frac{1}{z}\sum_{\nu=0}^{\infty} \frac{z_j^{\nu}}{z^{\nu}} \\
&= \frac{f(z)}{z}\sum_{\nu=0}^{\infty} z^{-\nu}\sum_{j=1}^{n} z_j^{\nu} \\
&= \frac{f(z)}{z}\sum_{\nu=0}^{\infty} \frac{S_\nu}{z^{\nu}}
\end{aligned}
\tag{70}
$$

the expansion being valid for $|z| > \max |z_j|$. Thus

$$
\begin{aligned}
zf'(z) &= f(z)\sum_{\nu=0}^{\infty} \frac{S_\nu}{z^{\nu}} \\
&= (z^n + c_1 z^{n-1} + \cdots + c_n)\sum_{\nu=0}^{\infty} \frac{S_\nu}{z^{\nu}} \\
&= nz^n + (n-1)c_1 z^{n-1} + \cdots + c_{n-1}z.
\end{aligned}
\tag{71}
$$

Let $1 \leqq p \leqq n$. The coefficient of z^{n-p} on the left is $(n - p)c_p$ and on the right is $nc_p + c_{p-1}S_1 + c_{p-2}S_2 + \cdots + c_1 S_{p-1} + S_p$. Hence

$$(n - p)c_p = S_p + c_1 S_{p-1} + \cdots + c_{p-1}S_1 + nc_p$$

or transposing,

$$S_p + c_1 S_{p-1} + c_2 S_{p-2} + \cdots + c_{p-1}S_1 + pc_p = 0 \qquad (p = 1, 2, \cdots, n). \tag{72}$$

For $p \geqq n$, the coefficient of z^{n-p} on the left of (71) is zero and on the right is $S_p + c_1 S_{p-1} + \cdots + c_n S_{p-n}$; hence

$$S_p + c_1 S_{p-1} + \cdots + c_n S_{p-n} = 0 \qquad (p = n + 1, n + 2, \ldots). \tag{73}$$

Theorem 15. *The Newton sums* (68) *of the polynomial* (67) *may be determined recursively from the coefficients by means of Newton's identities* (72), (73).

The first few sums are found to be

$$
\begin{aligned}
S_0 &= n = 1 + 1 + \cdots + 1 \\
S_1 &= -c_1 = z_1 + \cdots + z_n \\
S_2 &= c_1^2 - 2c_2 = z_1^2 + \cdots + z_n^2 \\
S_3 &= 3(c_1c_2 - c_3) - c_1^3 = z_1^3 + \cdots + z_n^3 \qquad (74) \\
&\;\;\vdots \qquad\qquad \vdots
\end{aligned}
$$

etc.

Now, suppose that one of the zeros of $f(z)$, say z_1, exceeds all others in modulus. Then

$$
\frac{S_{k+1}}{S_k} = \frac{z_1^{k+1} + \cdots + z_n^{k+1}}{z_1^{\,k} + \cdots + z_n^{\,k}} \qquad (75)
$$

$$
= \frac{z_1^{k+1}\{1 + (z_2/z_1)^{k+1} + \cdots + (z_n/z_1)^{k+1}\}}{z_1^{\,k}\{1 + (z_2/z_1)^k + \cdots + (z_n/z_1)^k\}}
$$

and making $k \to \infty$, clearly

$$
\lim_{k\to\infty} \frac{S_{k+1}}{S_k} = z_1. \qquad (76)
$$

If our assumption concerning z_1 is fulfilled, then equation (76) gives an elegant technique for calculating z_1 numerically. This procedure, known as Bernoulli's method, is as follows:

(a) Using Newton's identities, calculate $S_1, S_2, S_3, \ldots$ recursively.

(b) When the ratio S_{k+1}/S_k has converged sufficiently, take z_1 to be the last value of this ratio.

(c) Form the reduced polynomial

$$
\frac{f(z)}{z - z_1}
$$

and repeat the process until either all roots have been found or a root which is repeated in modulus is reached.

To deal with this last eventuality, suppose first that z_1, z_2 are complex conjugates of each other and that the remaining zeros have smaller modulus. If

$$
z_1 = re^{i\varphi}, \quad z_2 = re^{-i\varphi}
$$

then

$$
S_k = z_1^{\,k} + z_2^{\,k} + \cdots + z_n^{\,k} = 2r^k \cos k\varphi + o(r^k) \qquad (k \to \infty)
$$

where the term $o(r^k)$ refers to a function of k which, when divided by r^k, tends to zero (see Chapter 4 for exact definitions of these symbols) as $k \to \infty$. Hence

$$\frac{S_k}{2r^k} = \cos k\varphi + o(1) \tag{77}$$

$$\frac{S_{k+1}}{2r^{k+1}} = \cos (k + 1)\varphi + o(1) \tag{78}$$

$$\frac{S_{k+2}}{2r^{k+2}} = \cos (k + 2)\varphi + o(1). \tag{79}$$

Multiplying (77) by 1, (78) by $-2 \cos \varphi$, (79) by 1 and adding,

$$\frac{S_{k+2}}{2r^{k+2}} - 2 \cos \varphi \frac{S_{k+1}}{2r^{k+1}} + \frac{S_k}{2r^k} = o(1)$$

or

$$S_{k+2} - 2r \cos \varphi S_{k+1} + r^2 S_k = o(r^{k+2}) \tag{80}$$

and replacing k by $k - 1$,

$$S_{k+1} - 2r \cos \varphi S_k + r^2 S_{k-1} = o(r^{k+1}). \tag{81}$$

We may regard (80), (81) as two simultaneous equations in two unknowns r^2, $r \cos \varphi$, which are the squared modulus and real part of the root we seek. Solving,

$$2r \cos \varphi = \frac{S_{k-1}S_{k+2} - S_k S_{k+1}}{S_{k+1}S_{k-1} - S_k^{\,2}} + o(1)$$

$$r^2 = \frac{S_k S_{k+2} - S_{k+1}^2}{S_{k+1}S_{k-1} - S_k^2} + o(1)$$

or what is the same thing,

$$2r \cos \varphi = \lim_{k \to \infty} \frac{S_{k-1}S_{k+2} - S_k S_{k+1}}{S_{k+1}S_{k-1} - S_k^{\,2}} \tag{82}$$

$$r^2 = \lim_{k \to \infty} \frac{S_k S_{k+2} - S_{k+1}^2}{S_{k+1}S_{k-1} - S_k^{\,2}}. \tag{83}$$

This analysis shows first that the presence of a root of repeated modulus can be detected during the calculation by the oscillatory behavior of the S_k. It will be observed that the ratio in (76) is not tending to a limit. In this case the ratios in (82), (83) should be checked for smooth behavior. If those ratios tend to limits, then those limits are respectively twice the real part and the squared modulus of the conjugate pair being sought. If neither of these eventualities occurs, then a multiplicity of some order is present. Although these can be dealt with similarly, the method is probably unsuitable in such cases for multiplicities of order higher than the second.

3.8 OTHER NUMERICAL METHODS

The Newton-Raphson iteration for finding the roots of polynomial equations, which we now discuss, converges very rapidly, if at all. It is somewhat less reliable than the Bernoulli iteration described in the previous section in that it requires a moderately good estimate of the root to be available at the start of the process. The iteration is carried out by choosing an initial "guess" z_0, and calculating recursively

$$z_{\nu+1} = z_\nu - \frac{f(z_\nu)}{f'(z_\nu)} \qquad (\nu = 0, 1, 2, \ldots). \tag{84}$$

If the process converges at all, $z_\nu \to z$, and if $f'(z) \neq 0$, then (84) shows clearly that $f(z) = 0$, i.e., z is a zero of $f(z)$. The following theorem gives a sufficient condition for the convergence of the method.

Theorem 16. *If the initial guess, z_0, is contained in some circle C,*

$$|z - \zeta| \leqq \rho$$

about a zero, ζ, of $f(z)$ such that if z', z'' are any two points of C, we have

$$\left| z' - \frac{f(z')}{f'(z')} - z'' + \frac{f(z'')}{f'(z'')} \right| \leqq \gamma\, |z' - z''| \qquad \gamma < 1 \tag{85}$$

then the sequence $\{z_\nu\}$ generated by (84) *converges to ζ.*
Proof. First, we claim that all z_ν lie in C if z_0 does. Indeed,

$$\begin{aligned} |z_{\nu+1} - \zeta| &= \left| z_\nu - \frac{f(z_\nu)}{f'(z_\nu)} - \zeta \right| \\ &= \left| z_\nu - \frac{f(z_\nu)}{f'(z_\nu)} - \zeta + \frac{f(\zeta)}{f'(\zeta)} \right| \\ &\leqq \gamma\, |z_\nu - \zeta| < |z_\nu - \zeta| \end{aligned} \tag{86}$$

which proves the previous assertion. Next, from (86) we see that

$$|z_\nu - \zeta| \leqq \gamma^\nu\, |z_0 - \zeta| \qquad (\nu = 1, 2, \ldots)$$

whence z_ν converges to ζ with geometric rapidity.

Because of the slow-but-sure character of the Bernoulli iteration, as contrasted to the rapid but unsure behavior of Newton's iteration, a combination of the two is a reasonably good calculation scheme. Bernoulli's method is then used to provide the initial guess z_0 for (84).

Next, we propose to find a family of numerical methods by relating the Bernoulli process to matrix iterations. Indeed, referring to (99), (100) of

Chapter 1, the companion matrix of the polynomial

$$f(z) = z^n + c_1 z^{n-1} + \cdots + c_n \tag{87}$$

is

$$A = \begin{pmatrix} -c_1 & -c_2 & \cdots & -c_n \\ 1 & 0 & \cdots & 0 \\ 0 & 1 & \cdots & 0 \\ \cdot & & & \cdot \\ \cdot & & & \cdot \\ \cdot & & & \cdot \\ 0 & 0 & \cdots & 0 \end{pmatrix} \tag{88}$$

Any method of calculating the eigenvalues of A is a method of calculating the zeros of $f(z)$. One such method is to choose a starting vector $\mathbf{y}_0$, and form $A\mathbf{y}_0, A^2\mathbf{y}_0, \ldots$. If the eigenvalues of A (zeros of f) are distinct in modulus, the ratio of components of successive members of this sequence tend to the dominant zero of $f(z)$ (see Section 1.20; compare (75) of this chapter). This is essentially Bernoulli's method, as can be seen by writing down $\mathbf{y}_0$, $A\mathbf{y}_0$, $A^2\mathbf{y}_0$ and comparing with Newton's identities. The rate of convergence of this process is determined by $|z_2/z_1|$ where z_2 is the subdominant and z_1 the dominant root, as is clear from (75). Therefore, any transformation which diminishes $|z_2/z_1|$ will accelerate the convergence of the iteration.

One such transformation consists in squaring the companion matrix before beginning the iteration. The eigenvalues of A^2 are $z_1^2, z_2^2, \ldots, z_n^2$, and the convergence is now governed by

$$\left|\frac{z_2^2}{z_1^2}\right| = \left|\frac{z_2}{z_1}\right|^2 < \left|\frac{z_2}{z_1}\right|$$

since $|z_2/z_1| < 1$. Indeed, one can form the matrices $A^2, A^4, A^8, A^{16}, \ldots$ successively, by repeated squaring, the separation of the roots being enhanced by each successive matrix multiplication. This procedure is quite effective when the roots are close together in modulus and is known as Graeffe's process. It has the disadvantage of being unstable against buildup of roundoff error, and it is wise, for this reason, to take the roots as found and correct them once or twice with Newton's iteration.

As a final remark on numerical methods, let us recall, from the theory of functions, that the number of zeros of the function $f(z)$ inside a simple closed curve C which lies inside its domain of analyticity is equal to the change in the amplitude of $f(z)$ around C, divided by 2π.

Hence, let $f(z)$ be a polynomial, in particular, and C such a curve. As z goes around C in the counterclockwise direction, a curve $w = f(z)$ is traced out in the w-plane. The theorem just referred to states that the number of zeros of $f(z)$ in C is the number of times this image of C winds around the origin in the w-plane. This can be made into an effective numerical procedure by choosing several points around C, calculating $f(z)$ at those points, plotting the resulting curve, and counting the number of turns around the origin.

Bibliography

There are three general references on the theory of root location. In English there is

1. M. Marden, *The Geometry of the Zeros of a Polynomial in a Complex Variable*, American Mathematical Society, Survey III, New York, 1949.

Perhaps more suitable for the beginner is

2. J. Dieudonné, *La Theorie Analytique des Polynomes d'une Variable*, Mémorial des Sciences Mathématiques, vol. 93, Paris, 1938.

The reader will also find much reward in the elegant treatment of the subject in problems 16–27 of Chapter 3 and all of Chapter 5 in

3. G. Pólya and G. Szegö, *Aufgaben und Lehrsätze aus der Analysis*, Springer-Verlag, Berlin, 1954.

Surveys of numerical methods are in

4. F. B. Hildebrand, An Introduction to Numerical Analysis, McGraw-Hill Book Co., New York, 1955.

A complete treatment of the Routh-Hurwitz theory of zeros in the left half-plane may be found in

5. F. R. Gantmacher, *Applications of the Theory of Matrices*, Interscience Publishers, New York, 1959.

A proof of Budan's theorem is, among much interesting elementary material, in

6. L. E. Dickson, *First Course in the Theory of Equations*, John Wiley and Sons, New York, 1922.

Exercises

1. Prove that the companion matrix of (11) is irreducible.
2. (a) If $f(z) = a_0 + a_1z + \cdots + a_nz^n$, display the function $g(z) = z^nf(1/z)$.
   (b) If the zeros of $f(z)$ are $z_1, \ldots, z_n$, what are the zeros of $g(z)$?
   (c) Using the result of (b), find a *lower bound* for the modulus of the zero of $f(z)$ of *smallest* modulus, and hence find an annular ring containing all the zeros of $f(z)$.
   (d) What does your result say about the zeros of

$$f(z) = 1 + z + z^2 + \cdots + z^n?$$

   What are the zeros of this polynomial?
3. Find a circle which contains the zeros of the $(2n)$th partial sum of the Taylor's series for $\cos z$.
4. What is the convex hull of the set consisting of the interval [0, 1] of the x-axis and [0, 1] of the y-axis? Of the set consisting of the entire real axis and the entire imaginary axis?

5. If no zero of $f(z)$ exceeds R in modulus, then the convex hull of the zeros of $f(z)$ is contained in the circle $z \leq R$.
6. Let $f(x)$ be a polynomial of degree n, with real coefficients. Suppose that Budan's theorem gives exact information for every interval (a, b). For this to happen it is necessary and sufficient that all the zeros of $f(x)$ be real.
7. If $\varphi_0(x), \varphi_1(x), \ldots, \varphi_n(x)$ are the first $n + 1$ members of a sequence of orthogonal polynomials on (a, b), then they form a Sturm sequence on any interval of the real axis.
8. Prove the assertion made in the paragraph immediately preceding Theorem 13, and carry out the operation called "combining these results," in the paragraph preceding that.
9. Derive (17) directly from (16) by considering $f(\rho z)$.
10. If $f(z)$ has complex coefficients, how can Sturm's theorem be used to give the number of zeros in a real interval (a, b)?
11. It is desired to find precisely the number of zeros of the polynomial $f(z)$ in the circle $|z| < R$. How can this problem be transformed into that of Theorem 13? Exactly how many zeros has the polynomial

$$f(z) = z^3 - 3z^2 + z - 1$$

in the unit circle?

chapter 4

Asymptotic expansions

4.1 INTRODUCTION; THE O, o, $\sim$ SYMBOLS

Asymptotics is the art of finding a simple function which is a good approximation to a given complicated function, the accuracy of the approximation increasing as the argument of the given function behaves in a certain preassigned manner. It is a branch of mathematics in which intuition, experience, and even luck play an important role, since particular problems have a habit of being highly individual, and not special cases of any theorem. With these cautions, we proceed to summarize in this chapter a few of the rules which do exist.

Consider the function

$$f(x) = \frac{1 + x^2}{1 + x} \tag{1}$$

as $x \to \infty$. The crudest statement we could make is simply that $f(x) \to \infty$ as $x \to \infty$. The next question concerns the rate at which $f(x) \to \infty$. Does it, for example, grow like e^x? x^{23}? $\log \log x$? $\Gamma(x)$? The answer is quite clear here, even though we have not defined the word "rate" yet; this function grows like x as $x \to \infty$. Next we may ask about the growth of $f(x) - x$. To answer this we write

$$\begin{aligned} f(x) &= \frac{x^2 + 1}{x(1 + 1/x)} = \frac{x^2 + 1}{x}\left\{1 - \frac{1}{x} + \frac{1}{x^2} - \frac{1}{x^3} + \cdots\right\} \\ &= \left(x + \frac{1}{x}\right)\left(1 - \frac{1}{x} + \frac{1}{x^2} - \cdots\right) \\ &= x - 1 + \frac{2}{x} - \frac{2}{x^2} + \cdots \end{aligned} \tag{2}$$

From this expansion, which converges for $|x| > 1$, we see that $f(x) - x$ remains bounded as $x \to \infty$, and, actually, that $f(x) - x$ approaches -1 as a limit. Next we ask about the behavior of $f(x) - x + 1$ as $x \to \infty$. From (2) we can make either the crude statement that $f(x) - x + 1$ approaches zero as $x \to \infty$ or the more precise statement that $f(x) - x + 1$ "behaves like" $2/x$ when x is large. The process can be continued indefinitely, and we notice that all such questions will be answered by the expansion (2), which is therefore both a convergent development of $f(x)$ in a series for $|x| > 1$ and an *asymptotic expansion* of $f(x)$ for large x, which means roughly that the chain of questions asked above can be answered by inspection of the series (we give a precise definition below).

The relation (2) is quite useless for discovering the behavior of $f(x)$ as $x \to 0$, for it is neither convergent nor asymptotic, but writing

$$f(x) = \frac{1 + x^2}{1 + x} = (1 + x^2)(1 - x + x^2 + \cdots) \tag{3}$$

$$= 1 - x + 2x^2 - x^3 + \cdots$$

gives an expansion which is both convergent and asymptotic in a neighborhood of $x = 0$.

Passing, by way of contrast, to a more difficult situation, consider the function

$$f(N) = \sum_{n=1}^{N} \cos(\log n) \tag{4}$$

and let us ask about the growth of $f(N)$ as $N \to \infty$. The only obvious fact is that

$$|f(N)| \leq \sum_{n=1}^{N} |\cos(\log n)| \leq N \tag{5}$$

so that $|f(N)|$ grows no faster than N. The indiscriminate use of absolute value signs in (5) has, however, destroyed the entire delicacy of the problem (4), which arises from the cancellation between terms of (4) caused by the changes in sign of the cosine. It is by no means clear even that $|f(N)| \to \infty$ as $N \to \infty$, or—if it does—whether it does so at the rate of N, $N^{1/2}$, $N^{1/8} \log N$ etc. What we are saying is that not only do we not have an asymptotic expansion like (2) but also that we are completely in the dark about the first term in that expansion.

We now wish to give precise definitions of three symbols which are used to compare the rates of growth of functions.

Let $f(x)$, $g(x)$ be given functions, $g(x)$ continuous, and let x_0 be a given point. We say that $f(x) = O(g(x))$ as $x \to x_0$, written

$$f(x) = O(g(x)) \qquad (x \to x_0) \tag{6}$$

if there is a constant A such that

$$|f(x)| \leqq A\,|g(x)| \tag{7}$$

for all values of x in some neighborhood of x_0.

We say that $f(x) = o(g(x))$ as $x \to x_0$, written

$$f(x) = o(g(x)) \qquad (x \to x_0) \tag{8}$$

if

$$\lim_{x \to x_0} \left| \frac{f(x)}{g(x)} \right| = 0. \tag{9}$$

Finally, we say that $f(x) \sim g(x)$ as $x \to x_0$, written

$$f(x) \sim g(x) \qquad (x \to x_0) \tag{10}$$

if

$$\lim_{x \to x_0} \frac{f(x)}{g(x)} = 1. \tag{11}$$

In the definition of the O symbol, if $x_0 = \infty$, the phrase "in some neighborhood of x_0" means "for all sufficiently large x." In some cases we are interested in the rate of growth as $x \to x_0$ from one side only, say $x \to x_0$ from above. In such cases we write, for instance,

$$f(x) = o(g(x)) \qquad (x \to x_0^{+}) \tag{12}$$

with corresponding modifications in the other cases.

Roughly speaking, the symbols O, o, $\sim$ have the following meanings:

(*a*) $f(x) = O(g(x))$ means $f(x)$ does not grow faster than $g(x)$ as $x \to x_0$.
(*b*) $f(x) = o(g(x))$ means $f(x)$ grows slower than $g(x)$ as $x \to x_0$.
(*c*) $f(x) \sim g(x)$ means $f(x)$ and $g(x)$ grow at the same rate as $x \to x_0$.

Needless to say, the last three statements are mnemonic devices only, and the formal definitions given above must always be used.

The equations

$$f(x) = g(x) + O(h(x)) \qquad (x \to x_0) \tag{13}$$

$$f(x) = g(x) + o(h(x)) \qquad (x \to x_0) \tag{14}$$

mean, respectively,

$$f(x) - g(x) = O(h(x)) \qquad (x \to x_0) \tag{15}$$

$$f(x) - g(x) = o(h(x)) \qquad (x \to x_0). \tag{16}$$

The following examples should be carefully studied before proceeding further.

(17) $$\sin x = O(1) \qquad (x \to \infty)$$

(18) $$(1 + x^2)^{-1} = o(1) \qquad (x \to \infty)$$

(19) $$(1 + x^2)^{-1} = o(x^{-1}) \qquad (x \to \infty)$$

(20) $$(1 + x^2)^{-1} = O(x^{-2}) \qquad (x \to \infty)$$

(21) $$(1 + x^2)^{-1} \sim x^{-2} \qquad (x \to \infty)$$

(22) $$(1 + x^2)^{-1} = x^{-2} + o(x^{-2}) \qquad (x \to \infty)$$

(23) $$(1 + x^2)^{-1} = x^{-2} + o(x^{-3}) \qquad (x \to \infty)$$

(24) $$(1 + x^2)^{-1} = x^{-2} + O(x^{-4}) \qquad (x \to \infty)$$

(25) $$(1 + x^2)^{-1} = x^{-2} - x^{-4} + O(x^{-6}) \qquad (x \to \infty)$$

(26) $$n/(n + 1) \sim 1 \qquad (n \to \infty)$$

(27) $$\sin x \sim x \qquad (x \to 0)$$

(28) $$\cos x = 1 + O(x^2) \qquad (x \to 0)$$

(29) $$\sqrt{n^2 + 1} \sim n \qquad (n \to \infty)$$

(30) $$\sqrt{n^2 + 1} = n + o(1) \qquad (n \to \infty)$$

(31) $$\sqrt{n^2 + 1} = n + O(n^{-1}) \qquad (n \to \infty)$$

(32) $$(n/e)^n = O(n!) \qquad (n \to \infty)$$

(33) $$\sum_{n=1}^{\infty} x^n = O((1 - x)^{-1}) \qquad (x \to 1^-)$$

(34) $$\sum_{n=1}^{\infty} n^p x^n = O((1 - x)^{-p}) \qquad (x \to 1^-)$$

(35) $$\int_2^x \frac{dy}{y} = O(\log x) \qquad (x \to \infty)$$

(36) $$\int_0^{\pi} e^{-x^4} \sin nx \, dx = o(1) \qquad (n \to \infty)$$

From these examples it may be noticed that $f(x) = O(1)$ $(x \to x_0)$ means simply that $f(x)$ is bounded and that $f(x) = o(1)$ $(x \to x_0)$ means that $f(x)$ approaches zero as $x \to x_0$. Furthermore, we see that there is no point in putting two terms on the right of a "$\sim$" sign if one dominates the other, for example

$$f(x) \sim x + \sqrt{x} \qquad (x \to \infty)$$

conveys no more information than

$$f(x) \sim x$$

since the function $f(x) = x + x^{7/8} \log x$ satisfies both of them, as does $f(x) = x$ itself. As a final remark we note that it is possible to have

$$f(x) = O(x^{a+\varepsilon}) \qquad (x \to \infty)$$

for every $\varepsilon > 0$ and yet not have $f(x) = O(x^a)$. Indeed, for every $\varepsilon > 0$,

$$\sqrt{x} \log x = O(x^{1/2+\varepsilon}) \neq O(x^{1/2}) \qquad (x \to \infty).$$

Now let $\phi_0(x)$, $\phi_1(x)$, $\phi_2(x)$, ... be an infinite sequence of continuous functions, and let x_0 be a fixed point. We say that $\{\phi_n(x)\}_{n=0}^{\infty}$ is an *asymptotic sequence for* x_0 if for each fixed n we have

$$\phi_{n+1}(x) = o(\phi_n(x)) \qquad (x \to x_0).$$

For example, the sequence

$$1, \frac{1}{x}, \frac{1}{x^2}, \ldots$$

is an asymptotic sequence for ∞, and the sequence

$$1. x, x^2, \ldots$$

is an asymptotic sequence for 0.

Suppose $f(x)$ is a given function, and let $\{\phi_n(x)\}_{n=0}^{\infty}$ be an asymptotic sequence for x_0. A *formal* series

$$a_0\phi_0(x) + a_1\phi_1(x) + \cdots$$

is called an *asymptotic series for* $f(x)$ *at* x_0 if for each fixed integer n it is true that

$$f(x) = a_0\phi_0(x) + \cdots + a_n\phi_n(x) + o(\phi_n(x)) \qquad (x \to x_0). \tag{37}$$

The abbreviation

$$f(x) \approx \sum_{\nu=0}^{\infty} a_\nu\phi_\nu(x) \qquad (x \to x_0) \tag{38}$$

means that the formal series on the right side is an asymptotic series for $f(x)$ at x_0, in the sense of (37). It does not imply that the series converges, and in most of the interesting applications it will not converge. This means that for any *fixed* value of x, the series in (38) cannot be used for the *exact* calculation of $f(x)$, for the terms will decrease in size for a while but ultimately will increase to infinity. Nonetheless, such series are extremely useful for the *approximate* calculation of $f(x)$ because at the beginning the terms will usually decrease quite rapidly, and more rapidly the closer x is to x_0. In many cases just a few terms will give quite extraordinary accuracy.

In (2) we have seen an asymptotic expansion which is convergent. To illustrate the other kind, consider

$$f(x) = \int_0^{\infty} \frac{e^{-t}\, dt}{x + t} \tag{39}$$

when x is large and positive. Integrating once by parts, we get

$$f(x) = \frac{1}{x} - \int_0^\infty \frac{e^{-t}\,dt}{(x+t)^2}.$$

Generally, after integrating n times by parts, we find

$$f(x) = \frac{1}{x} - \frac{1}{x^2} + \frac{2!}{x^3} - \frac{3!}{x^4} + \cdots + (-1)^n \frac{n!}{x^{n+1}} + (-1)^{n+1}(n+1)! \int_0^\infty \frac{e^{-t}\,dt}{(x+t)^{n+2}}. \tag{40}$$

Denoting the remainder term by $R_n(x)$, we have

$$|R_n(x)| = (n+1)! \int_0^\infty \frac{e^{-t}\,dt}{(x+t)^{n+2}} = \frac{(n+1)!}{x^{n+1}} \int_0^\infty \frac{e^{-xy}}{(1+y)^{n+2}}\,dy \leqq \frac{(n+1)!}{x^{n+1}} \int_0^\infty e^{-xy}\,dy = \frac{(n+1)!}{x^{n+2}}. \tag{41}$$

Therefore, if we terminate the expansion (40) after the nth term—ignoring the remainder—the error we make is o of the last term kept, as required by the definition (37). Hence we may write

$$f(x) = \int_0^\infty \frac{e^{-t}\,dt}{x+t} \approx \sum_{\nu=0}^\infty (-1)^\nu \frac{\nu!}{x^{\nu+1}} \qquad (x \to \infty). \tag{42}$$

Actually the analysis in (41) shows that even more is true, namely that the magnitude of the error committed in stopping after n terms is less than the first term neglected. The series in (42) converges for no finite value of x. Indeed, if x is fixed, the νth term is obtained from the $(\nu - 1)$st by multiplying by ν/x (aside from sign). Therefore the terms decrease in size as long as $\nu/x < 1$, i.e., as long as $\nu < x$. For $\nu > x$ they rapidly increase, without bound, in magnitude. If (to simplify the argument) x is a fixed integer, the size of the smallest term in (42) is

$$\frac{x!}{x^{x+1}}$$

which, as we shall see presently, is

$$\sim \sqrt{2\pi/x}\, e^{-x} \qquad (x \to \infty).$$

This is the theoretical limit of accuracy in the use of (42). To put it more plainly, since $f(x) \sim x^{-1}$, the minimum relative error that can possibly be attained by using (42) for a fixed value of x is

$$\sim \sqrt{2\pi x}\, e^{-x} \qquad (x \to \infty)$$

which can be gotten by using x terms. The use of more terms will result in a larger error. On the other hand, notice that using only one term gives a relative error of $1/x$, which may be eminently acceptable if x is large, and is certainly preferable to the numerical evaluation of the integral in (39), which, in contrast to (42), gets more difficult as x gets larger.

With the above remarks we conclude our general discussion of asymptotic expansions and pass now to the question of obtaining such expansions in particular cases. These cases may be grouped, roughly, as (i) sums, (ii) integrals, and (iii) other. This will cover only a minute portion of the possible areas of application of asymptotic methods, but a respectable fraction of the areas for which there exist general rules of procedure.

4.2 SUMS

Let $f(x)$ be a given continuous function. Our objective is to study the rate of growth of

$$S(n) = \sum_{\nu=0}^{n} f(\nu) \tag{43}$$

as $n \to \infty$. It will probably come as no surprise, for example, to learn that

$$\sum_{\nu=0}^{n} \nu^3 \sim \frac{n^4}{4} \qquad (n \to \infty)$$

for the reader is, perhaps, used to comparing (43) with

$$\tilde{S}(n) = \int_1^n f(t)\, dt. \tag{44}$$

We wish to explore here the connection between the rates of growth of (43) and (44), with a view to writing down a complete asymptotic expansion, when possible, for (43), in which (44) will be the first term. We need, first, a certain amount of preliminary apparatus.

Let us start with the numbers

$$\zeta(2n) = \sum_{\nu=1}^{\infty} \frac{1}{\nu^{2n}} \qquad (n = 1, 2, \ldots) \tag{45}$$

and the function

$$\begin{aligned} g(z) &= \sum_{n=1}^{\infty} \zeta(2n) z^{2n} \qquad |z| < 1 \\ &= \sum_{n=1}^{\infty} z^{2n} \sum_{\nu=1}^{\infty} \frac{1}{\nu^{2n}} \\ &= \sum_{\nu=1}^{\infty} \sum_{n=1}^{\infty} \left(\frac{z^2}{\nu^2}\right)^n \\ &= \sum_{\nu=1}^{\infty} \frac{z^2/\nu^2}{1 - z^2/\nu^2} \\ &= z^2 \sum_{\nu=1}^{\infty} \frac{1}{\nu^2 - z^2} = \frac{z}{2} \sum_{\nu=1}^{\infty} \left\{\frac{1}{\nu - z} - \frac{1}{\nu + z}\right\}. \end{aligned} \tag{46}$$

Now this last series obviously converges for every fixed value of z, excepting only the nonzero integers. It therefore represents a function analytic in the whole plane except for simple poles at $z = \pm 1, \pm 2, \ldots$, with residue -1 at each pole. Another such function is

$$\frac{1}{z} - \pi \cot \pi z.$$

It can be shown that actually

$$\sum_{\nu=1}^{\infty} \left\{ \frac{1}{\nu - z} - \frac{1}{\nu + z} \right\} = \frac{1}{z} - \pi \cot \pi z.$$

The proof, while straightforward, is omitted here because it is rather lengthy.

Assuming this, it follows that

$$\sum_{n=1}^{\infty} \zeta(2n) z^{2n} = \frac{1}{2} - \frac{1}{2} \pi z \cot \pi z \tag{47}$$

or replacing z by iz/π,

$$\begin{aligned}
\sum_{n=1}^{\infty} \frac{\zeta(2n)}{\pi^{2n}} (-1)^n z^{2n} &= \frac{1}{2} \{1 - z \coth z\} \\
&= \frac{1}{2} - \frac{z}{2} \frac{e^z + e^{-z}}{e^z - e^{-z}} \\
&= \frac{1}{2} - \frac{z}{2} \frac{e^{2z} + 1}{e^{2z} - 1} \\
&= \frac{1}{2} - \frac{z}{2} \left\{ \frac{2}{e^{2z} - 1} + 1 \right\} \\
&= \frac{1}{2} - \frac{z}{e^{2z} - 1} - \frac{z}{2}.
\end{aligned}$$

Finally, replacing z by $z/2$, and transposing,

$$\frac{z}{e^z - 1} = 1 - \frac{z}{2} + 2 \sum_{n=1}^{\infty} (-1)^{n+1} \frac{\zeta(2n)}{(2\pi)^{2n}} z^{2n} \qquad (|z| < 2\pi). \tag{48}$$

So, if we define *Bernoulli's numbers* B_n $(n = 1, 2, \ldots)$ by

$$\frac{z}{e^z - 1} = \sum_{n=0}^{\infty} \frac{z^n}{n!} B_n \tag{49}$$

we see that

$$(50)\qquad \begin{aligned} B_0 &= 1 \\ B_1 &= -\tfrac{1}{2} \\ B_{2n+1} &= 0 \qquad (n = 1, 2, 3, \ldots) \\ B_{2n} &= (-1)^{n+1}(2n)!\,\frac{\zeta(2n)}{(2\pi)^{2n}} \cdot 2 \qquad (n = 1, 2, 3, \ldots). \end{aligned}$$

On the other hand, from (49),

$$(51)\qquad \begin{aligned} z &= e^z \sum_{n=0}^{\infty} \frac{z^n}{n!} B_n - \sum_{n=0}^{\infty} \frac{B_n}{n!} z^n \\ &= \sum_{n=0}^{\infty} \frac{z^n}{n!} \sum_{n=0}^{\infty} \frac{z^n}{n!} B_n - \sum_{n=0}^{\infty} \frac{B_n}{n!} z^n \\ &= \sum_{n=0}^{\infty} \frac{z^n}{n!} \left\{ \binom{n}{0} B_0 + \binom{n}{1} B_1 + \cdots + \binom{n}{n} B_n - B_n \right\} \\ &= \sum_{n=0}^{\infty} \frac{z^n}{n!} \left\{ \binom{n}{0} B_0 + \binom{n}{1} B_1 + \cdots + \binom{n}{n-1} B_{n-1} \right\}. \end{aligned}$$

Comparing the coefficients of like powers of z on both sides of (51), we see that Bernoulli's numbers can be calculated from the recurrence relation

$$(52)\qquad \binom{n}{0} B_0 + \binom{n}{1} B_1 + \binom{n}{2} B_2 + \cdots + \binom{n}{n-1} B_{n-1} = 0 \qquad (n = 2, 3, \ldots).$$

The first few are found, in this way, to be

$$(53)\qquad B_0 = 1,\ B_1 = -\frac{1}{2},\ B_2 = \frac{1}{6},\ B_3 = 0,\ B_4 = -\frac{1}{30},\ B_5 = 0,\ B_6 = \frac{1}{42}, \ldots$$

etc. Using (50), we have also,

$$\zeta(2) = \frac{\pi^2}{6}, \quad \zeta(4) = \frac{\pi^4}{90}, \quad \zeta(6) = \frac{\pi^6}{945}, \cdots$$

and so on, which are far from obvious, directly from (45).

Next we define the function $[x]$, the greatest integer contained in x, e.g., $[2] = 2$, $[3.165] = 3$, $[.71] = 0$, etc. Then $x - [x]$ is the fractional part of x and always lies between 0 and 1. Finally, the function $x - [x] - \frac{1}{2}$ lies, for each x, between $-\frac{1}{2}$ and $\frac{1}{2}$ and is easily seen from its graph (which the reader should sketch for himself) to be a periodic function of x, of period 1.

Now, let $f(x)$ be a given function which is continuously differentiable for positive x. Then

$$\int_1^n \left(x - [x] - \frac{1}{2}\right) f'(x)\,dx = \sum_{\nu=1}^{n-1} \int_\nu^{\nu+1} \left(x - [x] - \frac{1}{2}\right) f'(x)\,dx$$

$$= \sum_{\nu=1}^{n-1} \int_\nu^{\nu+1} \left(x - \nu - \frac{1}{2}\right) f'(x)\,dx$$

$$= \sum_{\nu=1}^{n-1} \left\{\int_\nu^{\nu+1} x f'(x)\,dx - \left(\nu + \frac{1}{2}\right)(f(\nu+1) - f(\nu))\right\}$$

$$= \sum_{\nu=1}^{n-1} \left\{[xf(x)]_\nu^{\nu+1} - \int_\nu^{\nu+1} f(x)\,dx - \left(\nu + \frac{1}{2}\right)(f(\nu+1) - f(\nu))\right\}$$

$$= \sum_{\nu=1}^{n-1} \left\{(\nu+1)f(\nu+1) - \nu f(\nu) - \int_\nu^{\nu+1} f(x)\,dx - \left(\nu + \frac{1}{2}\right) f(\nu+1) + \left(\nu + \frac{1}{2}\right) f(\nu)\right\}$$

$$= \sum_{\nu=1}^{n-1} \left\{\tfrac{1}{2} f(\nu+1) + \tfrac{1}{2} f(\nu) - \int_\nu^{\nu+1} f(x)\,dx\right\}$$

$$= \frac{1}{2}\{f(2) + f(1) + f(3) + f(2) + \cdots + f(n) + f(n-1)\} - \int_1^n f(x)\,dx$$

$$= \frac{1}{2}\{f(1) + 2f(2) + 2f(3) + \cdots + 2f(n-1) + f(n)\} - \int_1^n f(x)\,dx$$

$$= \sum_{\nu=1}^{n} f(\nu) - \tfrac{1}{2} f(1) - \tfrac{1}{2} f(n) - \int_1^n f(x)\,dx.$$

Transposing, we have shown

Theorem 1.† *Let $f(x)$ be continuously differentiable on the interval* $[1, n]$. *Then*

$$\sum_{\nu=1}^{n} f(\nu) = \int_1^n f(x)\,dx + \frac{1}{2}(f(1) + f(n)) + \int_1^n \left(x - [x] - \frac{1}{2}\right) f'(x)\,dx. \tag{54}$$

This is the first form of the Euler-Maclaurin sum formula. As an example of its application, take $f(x) = 1/x$ in (54), then

$$\sum_{\nu=1}^{n} \frac{1}{\nu} = \int_1^n \frac{dx}{x} + \frac{1}{2}\left(1 + \frac{1}{n}\right) - \int_1^n \left(x - [x] - \frac{1}{2}\right) \frac{dx}{x^2} \tag{55}$$
$$= \log n + \frac{1}{2} + \frac{1}{2n} - \int_1^n \left(x - [x] - \frac{1}{2}\right) \frac{dx}{x^2}.$$

† Euler [1], MacLaurin [1].

The integral on the right is $o(1)$ as $n \to \infty$, since $(x - [x] - \frac{1}{2})$ is less than $\frac{1}{2}$ in absolute value, and therefore

$$\sum_{\nu=1}^{n} \frac{1}{\nu} = \log n + O(1) \qquad (n \to \infty).$$

Actually we can say more about this sum because we can write

$$\sum_{\nu=1}^{n} \frac{1}{\nu} = \log n + \frac{1}{2} + \frac{1}{2n} - \int_1^\infty \left(x - [x] - \frac{1}{2}\right) \frac{dx}{x^2} + \int_n^\infty \left(x - [x] - \frac{1}{2}\right) \frac{dx}{x^2}$$

and define the constant (Euler's constant)

$$(56) \qquad \gamma = \frac{1}{2} - \int_1^\infty \left(x - [x] - \frac{1}{2}\right) \frac{dx}{x^2} = \lim_{n \to \infty} \left\{1 + \frac{1}{2} + \cdots + \frac{1}{n} - \log n\right\} = 0.5772 \cdots$$

Then

$$(57) \qquad 1 + \frac{1}{2} + \frac{1}{3} + \cdots + \frac{1}{n} = \log n + \gamma + o(1) \qquad (n \to \infty)$$

which is as far as we can go at this stage.

To refine Theorem 1 further, we would like to integrate by parts, differentiating $f(x)$ repeatedly, and integrating $(x - [x] - \frac{1}{2})$. To do this, we need first a reasonable choice for the indefinite integrals of $(x - [x] - \frac{1}{2})$.

Let us recall that the Fourier series

$$(58) \qquad P_1(x) = -\sum_{n=1}^{\infty} \frac{\sin 2n\pi x}{n\pi}$$

converges to $x - \frac{1}{2}$ in the interval (0, 1). Since obviously $P_1(x)$ is periodic with period one, it must represent $x - [x] - \frac{1}{2}$ for every nonintegral value of x. Therefore one reasonable choice of a function whose derivative is $x - [x] - \frac{1}{2}$ is

$$(59) \qquad P_2(x) = \sum_{n=1}^{\infty} \frac{2 \cos 2n\pi x}{(2n\pi)^2}$$

which is also periodic, of period one, and actually continuous everywhere, since its Fourier series converges absolutely and uniformly on any finite interval, in contrast to (58). On (0, 1), $P_2(x) = x^2/2 - x/2 + \frac{1}{12}$. In general, if we define

$$P_{2k}(x) = (-1)^{k+1} \sum_{n=1}^{\infty} \frac{2 \cos 2n\pi x}{(2n\pi)^{2k}} \tag{60}$$

$$P_{2k+1}(x) = (-1)^{k+1} \sum_{n=1}^{\infty} \frac{2 \sin 2n\pi x}{(2n\pi)^{2k+1}} \tag{61}$$

for $k = 1, 2, \ldots$, then these are all continuous periodic functions of period one satisfying

$$P'_{r+1}(x) = P_r(x) \qquad (r = 1, 2, \ldots) \tag{62}$$

$$\begin{aligned} P_{2r}(0) &= (-1)^{r+1} \sum_{n=1}^{\infty} \frac{2}{(2n\pi)^{2r}} \\ &= (-1)^{r+1} \frac{2}{(2\pi)^{2r}} \sum_{n=1}^{\infty} \frac{1}{n^{2r}} \\ &= (-1)^{r+1} \frac{2}{(2\pi)^{2r}} \zeta(2r) \\ &= \frac{B_{2r}}{(2r)!} = P_{2r}(1) = P_{2r}(2) = \cdots \qquad (r = 1, 2, \ldots). \end{aligned} \tag{63}$$

Thus armed, we take the integral on the right side of (54) and integrate by parts twice:

$$\begin{aligned} \int_1^n P_1(x) f'(x)\, dx &= [f'(x) P_2(x)]_1^n - \int_1^n P_2(x) f''(x)\, dx \\ &= f'(n) \frac{B_2}{2!} - f'(1) \frac{B_2}{2!} - [P_3(x) f''(x)]_1^n + \int_1^n P_3(x) f'''(x)\, dx \\ &= \frac{B_2}{2!} (f'(n) - f'(1)) + \int_1^n P_3(x) f'''(x)\, dx. \end{aligned}$$

This process may be repeated as many times as $f(x)$ is differentiable, the result being the extended form of the Euler-MacLaurin sum formula, which we state as

Theorem 2. *Let $f(x)$ be $2k + 1$ times continuously differentiable in $[1, n]$. Then*

$$\begin{aligned} \sum_{\nu=1}^{n} f(\nu) = \int_1^n f(x)\, dx &+ \tfrac{1}{2}(f(1) + f(n)) \\ &+ \frac{B_2}{2!} (f'(n) - f'(1)) + \frac{B_4}{4!} (f'''(n) - f'''(1)) \\ &+ \cdots + \frac{B_{2k}}{(2k)!} (f^{(2k-1)}(n) - f^{(2k-1)}(1)) \\ &+ \int_1^n P_{2k+1}(x) f^{(2k+1)}(x)\, dx. \end{aligned} \tag{64}$$

It should be remarked that (64) is not in itself an asymptotic expansion, an infinite series, or indeed anything but an identity which is useful for producing asymptotic expansions if and only if the remainder term

$$P_k(n) = \int_1^n P_{2k+1}(x) f^{(2k+1)}(x)\, dx \tag{65}$$

can be conveniently estimated. Some applications follow.

4.3 STIRLING'S FORMULA

We are interested here in an asymptotic expansion for $n!$ when n is large. If we take

$$f(x) = \log x$$

the left side of (64) is

$$\begin{aligned} \sum_{\nu=1}^{n} \log \nu &= \log \prod_{\nu=1}^{n} \nu \\ &= \log n! \end{aligned} \tag{66}$$

On the right side we have

$$\begin{aligned} &\int_1^n \log x\, dx + \frac{1}{2}\log n + \frac{B_2}{2!}\left(\frac{1}{n} - 1\right) + \frac{B_4}{4!}\left(\frac{2}{n^3} - 2\right) \\ &\quad + \frac{B_6}{6!}\left(\frac{4!}{n^5} - 4!\right) + \cdots + \frac{B_{2k}}{(2k)!}\left(\frac{(2k-2)!}{n^{2k-1}} - (2k-2)!\right) \\ &\quad + (2k)!\int_1^n P_{2k+1}(x)\frac{dx}{x^{2k+1}} \\ &= \left(n + \frac{1}{2}\right)\log n - (n-1) + \frac{B_2}{2!}\frac{1}{n} + \frac{B_4}{3\cdot 4n^3} + \frac{B_6}{5\cdot 6n^5} \\ &\quad + \cdots + \frac{B_{2k}}{(2k)(2k-1)}\frac{1}{n^{2k-1}} - \frac{B_2}{1\cdot 2} - \frac{B_4}{3\cdot 4} - \cdots - \frac{B_{2k}}{(2k)(2k-1)} \\ &\quad + (2k)!\int_1^\infty P_{2k+1}(x)\frac{dx}{x^{2k+1}} - (2k)!\int_n^\infty P_{2k+1}(x)\frac{dx}{x^{2k+1}}. \end{aligned}$$

If we define the constant

$$\lambda_k = 1 - \frac{B_2}{1\cdot 2} - \frac{B_4}{3\cdot 4} - \cdots - \frac{B_{2k}}{(2k)(2k-1)} + (2k)!\int_1^\infty P_{2k+1}(x)\frac{dx}{x^{2k+1}} \tag{67}$$

then we have

$$\begin{aligned} \log n! &= \left(n + \frac{1}{2}\right)\log n - n + \lambda_k + \frac{B_2}{1\cdot 2}\frac{1}{n} + \frac{B_4}{3\cdot 4}\frac{1}{n^3} + \cdots \\ &\quad + \frac{B_{2k}}{(2k)(2k-1)}\frac{1}{n^{2k-1}} - (2k)!\int_n^\infty P_{2k+1}(x)\frac{dx}{x^{2k+1}}. \end{aligned} \tag{68}$$

Next, we claim that λ_k is independent of k, for (68) shows clearly that

$$\lambda_k = \lim_{n\to\infty}\left\{\log n! - \left(n+\frac{1}{2}\right)\log n + n\right\}$$

and the right-hand side has nothing to do with k. We will show in a later section of this chapter that actually

$$\lambda_k = \log\sqrt{2\pi}$$

and will accept this, for the moment, without proof. Finally, in the remaining integral in (68), substitute $x = ny$. Then

$$\left|\int_n^\infty P_{2k+1}(x)\frac{dx}{x^{2k+1}}\right| = \left|\frac{1}{n^{2k}}\int_1^\infty P_{2k+1}(ny)\frac{dy}{y^{2k}}\right|$$
$$\le \frac{\text{const.}}{n^{2k}}\int_1^\infty \frac{dy}{y^2} = \frac{\text{const.}}{n^{2k}}$$

where we have used only the fact that since $P_{2k+1}(x)$ is continuous and periodic, it is uniformly bounded, in absolute value, by some fixed constant (depending on k). Therefore, if k is fixed, the remainder in (68) is $O(n^{-2k})(n \to \infty)$, and thus for every fixed k it is true that

(69)

$$\log n! = \left(n+\frac{1}{2}\right)\log n - n + \log\sqrt{2\pi} + \frac{B_2}{1\cdot 2}\frac{1}{n}$$
$$+ \frac{B_4}{3\cdot 4}\frac{1}{n^3} + \cdots + \frac{B_{2k}}{(2k)(2k-1)}\frac{1}{n^{2k-1}} + O(n^{-2k}) \quad (n\to\infty)$$

which is the complete asymptotic expansion for $\log n!$. We may say that

(70)
$$\log n! \approx \left(n+\frac{1}{2}\right)\log n - n + \log\sqrt{2\pi}$$
$$+\sum_{k=1}^{\infty}\frac{B_{2k}}{(2k-1)(2k)}\frac{1}{n^{2k-1}} \qquad (n\to\infty)$$

though the series converges for no fixed value of n.

Bearing in mind the result of exercise 2, since by (69) we surely have

(71)
$$\log n! = \left(n+\frac{1}{2}\right)\log n - n + \log\sqrt{2\pi} + o(1) \qquad (n\to\infty)$$

we may take exponentials and deduce that

(72)
$$n! \sim \sqrt{2\pi n}\left(\frac{n}{e}\right)^n \qquad (n\to\infty)$$

which is Stirling's formula. Notice how much less informative (72) is than (69).

4.4 SUMS OF POWERS

Let p be a fixed positive integer. We wish to investigate the rate of growth of

$$\sum_{\nu=1}^{n} \nu^p = 1^p + 2^p + \cdots + n^p. \tag{73}$$

as $n \to \infty$.

Suppose we write $f(x) = x^p$ in the Euler-MacLaurin sum formula. In this case we get an unexpected bonus because all derivatives of $f(x)$ of order higher than the pth vanish identically, and therefore the expansion terminates. What we have then is an evaluation of the sum (73) in closed form, in descending powers of n, the coefficients involving Bernoulli numbers.

Because of the derivatives evaluated at $x = 1$ in the sum formula, it is easier to work with the function $f(x) = (x - 1)^p$. Furthermore, for reasons of symmetry, we will insert the Bernoulli numbers with odd subscripts, even though they are zero. We find then,

$$\begin{aligned}
\sum_{\nu=1}^{n} (\nu - 1)^p &= \int_1^n (x - 1)^p\,dx + \frac{1}{2}(n - 1)^p + \frac{B_2}{2!}p(n - 1)^{p-1} \\
&\quad + \frac{B_3}{3!}p(p - 1)(n - 1)^{p-2} + \cdots \\
&= \frac{(n - 1)^{p+1}}{p + 1} + \frac{1}{2}(n - 1)^p + \frac{B_2}{2!}p(n - 1)^{p-1} \\
&\quad + \frac{B_3}{3!}p(p - 1)(n - 1)^{p-2} + \cdots \\
&= \frac{1}{p + 1}\Big\{(n - 1)^{p+1} + \frac{p + 1}{2}(n - 1)^p \\
&\qquad + \frac{B_2}{2!}(p + 1)p(n - 1)^{p-1} + \cdots\Big\} \\
&= \frac{1}{p + 1}\Big\{(n - 1)^{p+1} + \binom{p + 1}{1}\frac{1}{2}(n - 1)^p \\
&\qquad + \binom{p + 1}{2}B_2(n - 1)^{p-1} + \cdots\Big\} \\
&= \sum_{\nu=1}^{n-1} \nu^p
\end{aligned}$$

Now replace $n - 1$ by n, and transpose the term n^p from the left side of the equation to the right. Then

$$\begin{aligned}
&1^p + 2^p + \cdots + (n - 1)^p \\
&\quad = \frac{1}{p + 1}\Big\{n^{p+1} + \binom{p + 1}{1}B_1 n^p + \binom{p + 1}{2}B_2 n^{p-1} + \cdots\Big\}
\end{aligned} \tag{74}$$

One more remark is necessary concerning (74). We stated previously that the expansion (64) terminates because all derivatives higher than the pth are zero. Even more is true, however, for the integral in (64) even vanishes if $f^{(2k+1)}(x)$ is a nonzero constant. To see this, observe that the integral is over n full periods of the periodic function $P_{2k+1}(x)$, whereas from the Fourier series (61), it is evident that the integral over any full period is zero. From this observation it follows that the expansion (74) is to be terminated not at the constant term but at the last term containing a positive power of n. In other words,

$$(75)\qquad 1^p + 2^p + \cdots + (n-1)^p = \frac{1}{p+1}\sum_{\nu=0}^{p}\binom{p+1}{\nu}B_\nu n^{p+1-\nu}.$$

The first few of these formulas give

$$(76)\qquad 1 + 2 + 3 + \cdots + n = \frac{n^2}{2} + \frac{n}{2}$$

$$(77)\qquad 1^2 + 2^2 + 3^2 + \cdots + n^2 = \frac{n^3}{3} + \frac{n^2}{2} + \frac{n}{6}$$

$$(78)\qquad 1^3 + 2^3 + 3^3 + \cdots + n^3 = \frac{n^4}{4} + \frac{n^3}{2} + \frac{n^2}{4}$$

$$(79)\qquad 1^4 + 2^4 + 3^4 + \cdots + n^4 = \frac{n^5}{5} + \frac{n^4}{2} + \frac{n^3}{3} - \frac{n}{30}.$$

Next we consider the sums of nonintegral powers of the integers, where the series no longer terminates. Taking $f(x) = x^{-s}$ in (64), we find

$$\begin{aligned}(80)\qquad \sum_{\nu=1}^{n}\nu^{-s} &= \int_1^n x^{-s}\,dx + \frac{1}{2}(1+n^{-s}) + \frac{B_2}{2!}\left(s - \frac{s}{n^{s+1}}\right)\\ &\quad + \cdots + \frac{B_{2k}}{2k}\binom{s+2k-2}{2k-1}\left(1 - \frac{1}{n^{s+2k-1}}\right)\\ &\quad - (2k+1)!\binom{s+2k}{2k+1}\int_1^n \frac{P_{2k+1}(x)\,dx}{x^{s+2k+1}}\\ &= \left\{\frac{1}{s-1} + \frac{1}{2} + \frac{B_2}{2!}\binom{s}{1} + \cdots + \frac{B_{2k}}{2k}\binom{s+2k-2}{2k-1}\right.\\ &\quad \left. - (2k+1)!\binom{s+2k}{2k+1}\int_1^\infty \frac{P_{2k+1}(x)}{x^{s+2k+1}}\,dx\right\}\\ &\quad + \frac{1}{(1-s)n^{s-1}} + \frac{1}{2n^s} - \frac{B_2}{2}\binom{s}{1}\frac{1}{n^{s+1}} - \cdots\\ &\quad - \frac{B_{2k}}{2k}\binom{s+2k-2}{2k-1}\frac{1}{n^{s+2k-1}}\\ &\quad + (2k+1)!\binom{s+2k}{2k+1}\int_n^\infty \frac{P_{2k+1}(x)\,dx}{x^{s+2k+1}}.\end{aligned}$$

where k is taken large enough so that the integrals converge.

In this last equation the expression in braces is independent of n and apparently depends on k. However, suppose s is any complex number satisfying $\mathbf{Re}\, s > 1$. Keeping s fixed, we can make $n \to \infty$ and all terms involving n on the right side approach zero. Hence the expression in braces must be equal to

$$\zeta(s) = \sum_{\nu=1}^{\infty} \nu^{-s} \qquad (\mathbf{Re}\, s > 1) \tag{81}$$

where $\zeta(s)$ is the ζ-function of Riemann. We have previously encountered this function evaluated at $s = 2, 4, 6, \ldots$ in equation (45). Hence if we regard $\zeta(s)$ as a known function, equation (80) takes the form

$$\begin{aligned}\sum_{\nu=1}^{n} \nu^{-s} = \zeta(s) &+ \frac{1}{(1-s)n^{s-1}} + \frac{1}{2n^s} - \frac{B_2}{2}\binom{s}{1}\frac{1}{n^{s+1}} \\ &- \cdots - \frac{B_{2k}}{2k}\binom{s+2k-2}{2k-1}\frac{1}{n^{s+2k-1}} \\ &+ (2k+1)!\binom{s+2k}{2k+1}\int_n^{\infty}\frac{P_{2k+1}(x)\,dx}{x^{s+2k+1}}.\end{aligned} \tag{82}$$

On the other hand, we notice that although the series in (81) fails to converge when $\mathbf{Re}\, s \geqq 1$, the original quantity in braces in (80) represents an analytic function of s for every value of s except $s = 1$. Therefore the relation

$$\begin{aligned}\zeta(s) = \frac{1}{s-1} + \frac{1}{2} + \frac{B_2}{2!}\binom{s}{1} &+ \cdots + \frac{B_{2k}}{2k}\binom{s+2k-2}{2k-1} \\ -(2k+1)!\binom{s+2k}{2k+1}&\int_1^{\infty}\frac{P_{2k+1}(x)\,dx}{x^{s+2k+1}} \qquad (k > -\tfrac{1}{2}\,\mathbf{Re}\, s).\end{aligned} \tag{83}$$

furnishes an analytic continuation of $\zeta(s)$ throughout the plane, aside from the simple pole at $s = 1$. With this understanding, equation (82) remains valid for all $s \neq 1$. For example, if $s = -\frac{1}{2}$,

$$\sum_{\nu=1}^{n} \nu^{1/2} = \tfrac{2}{3}n^{3/2} + \tfrac{1}{2}n^{1/2} + \zeta(-\tfrac{1}{2}) + o(1) \qquad (n \to \infty) \tag{84}$$

while for $s = \frac{1}{2}$,

$$\sum_{\nu=1}^{n} \nu^{-1/2} = 2n^{1/2} + \zeta(\tfrac{1}{2}) + o(1) \qquad (n \to \infty). \tag{85}$$

Generally one sees that the term involving the ζ-function appears between the last ascending power of n and the first descending power.

4.5 THE FUNCTIONAL EQUATION OF $\zeta(s)$

The ζ-function of Riemann satisfies a remarkable functional equation which, aside from its considerable intrinsic mathematical importance, can be of value in identifying some of the constants which occur in asymptotic expansions.

First, taking $k = 0$ in (83) gives

$$\zeta(s) = \frac{1}{s-1} + \frac{1}{2} - s\int_1^\infty \frac{P_1(x)\,dx}{x^{s+1}}. \tag{86}$$

This relation is obviously valid if $\mathbf{Re}\, s > 0$, since the integral converges absolutely for such values of s. We claim that the integral actually converges if only $\mathbf{Re}\, s > -1$, for

$$\int_1^A \frac{P_1(x)\,dx}{x^{s+1}} = \frac{P_2(x)}{x^{s+1}}\Bigg]_1^A + (s+1)\int_1^A \frac{P_2(x)\,dx}{x^{s+2}},$$

and making $A \to \infty$ we see that the claim is correct. Now if $\mathbf{Re}\, s < 0$, we have

$$s\int_0^1 \frac{x - [x] - \frac{1}{2}}{x^{s+1}}\,dx = -\frac{1}{2} - \frac{1}{s-1} \tag{87}$$

by a trivial calculation. Hence if $-1 < \mathbf{Re}\, s < 0$, (86) and (87) hold simultaneously, and if we subtract (87) from (86),

$$\zeta(s) = -s\int_0^\infty \frac{P_1(x)}{x^{s+1}}\,dx \qquad -1 < \mathbf{Re}\, s < 0. \tag{88}$$

Substituting (58) in (88) and integrating term-by-term,

$$\begin{aligned}
\zeta(s) &= s\int_0^\infty \frac{dx}{x^{s+1}} \sum_{n=1}^\infty \frac{\sin 2n\pi x}{n\pi} \\
&= s\sum_{n=1}^\infty \frac{1}{n\pi}\int_0^\infty \frac{\sin 2n\pi x}{x^{s+1}}\,dx \\
&= \frac{s}{\pi}(2\pi)^s \left\{\sum_{n=1}^\infty n^{s-1}\right\} \int_0^\infty \frac{\sin y\,dy}{y^{s+1}} \\
&= \frac{s}{\pi}(2\pi)^s \zeta(1-s) \int_0^\infty \frac{\sin y\,dy}{y^{s+1}} \\
&= \frac{s}{\pi}(2\pi)^s \zeta(1-s)\left\{-\sin\frac{s\pi}{2}\,\Gamma(-s)\right\} \\
&= \frac{1}{\pi}(2\pi)^s \zeta(1-s)\Gamma(1-s)\sin\frac{s\pi}{2}.
\end{aligned}$$

The justification of termwise integration is not to be found in uniform convergence, since the series involved does not converge uniformly on every finite interval. Instead, one uses the fact that the series converges "boundedly," i.e., the partial sums are uniformly bounded. For a proof of the sufficiency of bounded convergence for the interchange of integration and summation, the reader is referred to any text in the theory of functions of a real variable.

The evaluation

$$\int_0^\infty \frac{\sin y\, dy}{y^{s+1}} = -\sin\frac{s\pi}{2}\Gamma(-s) \qquad (\mathbf{Re}\, s > -1)$$

results from a suitable deformation of the contour in equation (122) of Chapter 5.

Theorem 3. *The ζ-function of Riemann satisfies the functional equation*

$$\zeta(s) = 2^s\pi^{s-1}\sin\frac{s\pi}{2}\Gamma(1-s)\zeta(1-s). \tag{89}$$

We remark that although (89) was proved only for $-1 < \mathbf{Re}\, s < 0$, it continues to hold throughout the plane except at $s = 1$, by the principle of the permanence of functional equations, for both sides are analytic except at $s = 1$.

The principal importance of (89) from our immediate point of view is that it permits us to express, for example, $\zeta(-\frac{5}{2})$ in terms of $\zeta(\frac{7}{2})$ and known functions. Thus, whereas $\zeta(-\frac{5}{2})$ itself has only rather complicated integral representations, $\zeta(\frac{7}{2})$ is readily accessible to calculation from (81), the series converging at quite a satisfactory rate.

To illustrate these considerations we ask for the asymptotic rate of growth of the numbers

$$\gamma_n = 1^1\, 2^2\, 3^3 \cdots n^n \tag{90}$$

for large n. Since

$$\log \gamma_n = \sum_{\nu=1}^{n} \nu \log \nu \tag{91}$$

we take $f(x) = x \log x$ in (64), with the result that

$$\begin{aligned}
\log \gamma_n &= \int_1^n x\log x\, dx + \frac{n}{2}\log n + \frac{B_2}{2!}\log n - \int_1^n P_3(x)\frac{dx}{x^2}\\
&= \frac{n^2\log n}{2} - \frac{n^2}{4} + \frac{1}{4} + \tfrac{1}{2}n\log n + \frac{1}{12}\log n - \int_1^\infty P_3(x)\frac{dx}{x^2}\\
&\quad + \int_n^\infty P_3(x)\frac{dx}{x^2}\\
&= \frac{n^2\log n}{2} - \frac{n^2}{4} + \frac{n\log n}{2} + \frac{1}{12}\log n + \frac{1}{4} - \int_1^\infty \frac{P_3(x)\, dx}{x^2}\\
&\quad + o(1) \qquad (n\to\infty).
\end{aligned}$$

Taking exponentials,

$$(92)\quad \gamma_n \sim n^{n^2/2+n/2+1/12}\, e^{-(n^2-1)/4} \exp\left\{-\int_1^\infty P_3(x)\,\frac{dx}{x^2}\right\} \qquad (n \to \infty)$$

and we would like to evaluate the argument of the exponential in some more reasonable form.

Now with $k = 1$, (83) reads

$$(93)\qquad \zeta(s) = \frac{1}{s-1} + \frac{1}{2} + \frac{s}{12} - s(s+1)(s+2)\int_1^\infty \frac{P_3(x)\,dx}{x^{s+3}}.$$

Differentiating, and putting $s = -1$, there results

$$(94)\qquad \int_1^\infty P_3(x)\,\frac{dx}{x^2} = \zeta'(-1) + \frac{1}{6},$$

and (92) is

$$(95)\quad \gamma_n \sim n^{n^2/2+n/2+1/12} \exp\left(-\frac{n^2}{4} + \frac{1}{12} - \zeta'(-1)\right) \qquad (n \to \infty)$$

which is somewhat more presentable than (92). However, the functional equation, after differentiation, relates $\zeta'(-1)$ to $\zeta'(2)$. Indeed logarithmic differentiation of (89) gives

$$\frac{\zeta'(-1)}{\zeta(-1)} = \log 2\pi - \frac{\Gamma'(2)}{\Gamma(2)} - \zeta'(2)\,\frac{6}{\pi^2}.$$

From (93) with $s = -1$, we find $\zeta'(-1) = -\frac{1}{12}$, whereas equation (127) of Chapter 5 gives the value

$$\frac{\Gamma'(2)}{\Gamma(2)} = 1 - \gamma,$$

where γ is Euler's constant. Hence

$$(96)\qquad \zeta'(-1) = -\frac{1}{12}\{\log 2\pi - 1 + \gamma\} - \frac{1}{2\pi^2}\sum_{n=1}^\infty \frac{\log n}{n^2}.$$

Putting this value in (95) we get, finally, the interesting expansion

$$(97)\quad 1^1 2^2 3^3 \cdots n^n \sim (2\pi)^{1/12}\{1^1 2^{1/4} 3^{1/9} 4^{1/16} \cdots\}^{1/2\pi^2}\, e^{-n^2/4+\gamma/12}\, n^{n^2/2+n/2+1/12} \qquad (n \to \infty).$$

4.6 THE METHOD OF LAPLACE FOR INTEGRALS

From the asymptotic expansion of sums we turn next to certain kinds of integrals which lend themselves readily to analysis. We consider first the behavior of

$$(98)\qquad G(x) = \int_{-\infty}^\infty e^{xh(y)}\,dy.$$

when x is large. The content of the following theorem is, in essence, that the major contribution to the integral arises from points near the maximum of the function $h(y)$.

Theorem 4.† *Suppose*

(a) *$h(y)$ is real valued and continuous.*

(b) *$h(0) = 0$ and $h(y) < 0$ for $y \neq 0$.*

(c) *there are numbers $\alpha > \beta$ such that $h(y) \leqq -\alpha$ when $|y| \geqq \beta$.*

(d) *there is a neighborhood of $y = 0$ in which $h(y)$ is twice differentiable and $h''(0) < 0$.*

(e) *the integral in (98) is finite for each fixed $x > 0$.*

Then

$$G(x) \sim \left\{\frac{2\pi}{-xh''(0)}\right\}^{1/2} \qquad (x \to \infty). \tag{99}$$

Lemma. If $\epsilon > 0$ is given, and if $h(y)$ satisfies the hypotheses of the theorem, then there is a $\delta > 0$ such that

$$\left| h(y) - \frac{y^2}{2} h''(0) \right| \leqq \varepsilon y^2, \tag{100}$$

for all $|y| \leqq \delta$.

Proof of Lemma. Let

$$\psi(y) = h(y) - \frac{y^2}{2} h''(0).$$

Then $\psi(0) = \psi'(0) = \psi''(0) = 0$, since $h'(0) = 0$ by (b). Hence

$$\psi''(0) = \lim_{y \to 0} \frac{\psi'(y) - \psi'(0)}{y} = 0,$$

that is, $\psi'(y) = o(y)$ $(y \to 0)$. By the mean value theorem,

$$\begin{aligned} \psi(y) - \psi(0) &= y\psi'(\theta y) \qquad (0 < \theta < 1) \\ &= yo(\theta y) \\ &= o(y^2) \qquad (y \to 0). \end{aligned}$$

Therefore

$$\lim_{y \to 0} \frac{h(y) - (y^2/2)h''(0)}{y^2} = 0,$$

which is exactly what the lemma asserts.

Proof of Theorem 4.

Let $\delta > 0$ be given. Then there is a number ρ such that $h(y) \leqq -\rho$ when $-\infty < y \leqq -\delta$ or $\delta \leqq y < \infty$. To see this, suppose first that $\delta > \beta$, β

† Laplace [1].

being defined by hypothesis (c). Then take $\rho = \alpha$. If $\delta < \beta$, $h(y)$ is contінous on the interval $\delta \leqq y \leqq \beta$ and therefore attains its maximum value at a point ξ of that interval. Then the number

$$\rho = \max(|h(\xi)|, \alpha)$$

will obviously do (see figure 4.1 below).

For the given value of δ, we have

$$(101)\quad \int_{-\infty}^{-\delta} e^{xh(y)}\,dy + \int_{\delta}^{\infty} e^{xh(y)}\,dy = \int_{-\infty}^{-\delta} e^{h(y)}e^{(x-1)h(y)}\,dy + \int_{\delta}^{\infty} e^{h(y)}e^{(x-1)h(y)}\,dy$$
$$\leqq e^{-\rho(x-1)}\int_{-\infty}^{-\delta} e^{h(y)}\,dy + e^{-\rho(x-1)}\int_{\delta}^{\infty} e^{h(y)}\,dy$$
$$\leqq e^{-\rho(x-1)}\int_{-\infty}^{\infty} e^{h(y)}\,dy.$$

Next, let $\epsilon > 0$ be given. Choose δ to be the number whose existence was proved in the lemma, so that (100) holds in the interval $-\delta \leqq y \leqq \delta$. Then the contribution of the integral over $-\delta, \delta$ can be estimated as follows.

First, by (100)

$$\frac{y^2}{2}(h''(0) - 2\varepsilon) \leqq h(y) \leqq \frac{y^2}{2}(h''(0) + 2\varepsilon)$$

on $(-\delta, \delta)$. Therefore

$$(102)\quad \int_{-\delta}^{\delta} \exp\left[\tfrac{1}{2}xy^2(h''(0) - 2\varepsilon)\right] dy \leqq \int_{-\delta}^{\delta} e^{xh(y)}\,dy$$
$$\leqq \int_{-\delta}^{\delta} \exp\left[\tfrac{1}{2}xy^2(h''(0) + 2\varepsilon\right] dy.$$

Next we claim that in each of these integrals the limits $-\delta, \delta$ may be replaced by $-\infty, \infty$ with an error which is $O(e^{-kx})$ as $x \to \infty$, where k is a

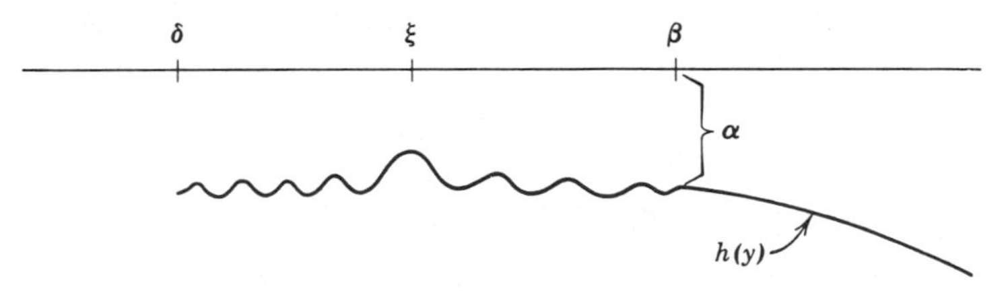

Figure 4.1

certain constant >0. Indeed, for the middle integral in (102),

$$\int_{-\delta}^{\delta} = \int_{-\infty}^{\infty} - \left\{\int_{\delta}^{\infty} + \int_{-\infty}^{-\delta}\right\}$$
$$= \int_{-\infty}^{\infty} + O(e^{-\rho x}) \qquad (x \to \infty)$$

according to (101). The other two integrals in (102) can be similarly handled. Hence from (102) we deduce that

$$\int_{-\infty}^{\infty} \exp\left[\tfrac{1}{2}xy^2(h''(0) - 2\varepsilon)\right] dy + O(e^{-k_1 x}) < \int_{-\infty}^{\infty} e^{xh(y)}\, dy.$$
$$< \int_{-\infty}^{\infty} \exp\left[\tfrac{1}{2}xy^2(h''(0) + 2\varepsilon)\right] dy + O(e^{-k_2 x}) \qquad (x \to \infty)$$

The first and third integrals are easily evaluated in closed form, the result being that

$$\left\{\frac{2\pi}{[-h''(0) + 2\varepsilon]x}\right\}^{1/2} + O(e^{-k_1 x}) < \int_{-\infty}^{\infty} e^{xh(y)}\, dy \cdot$$
$$< \left\{\frac{2\pi}{[-h''(0) - 2\varepsilon]x}\right\}^{1/2} + O(e^{-k_2 x}) \qquad (x \to \infty).$$

Hence for all sufficiently large x,

$$\left\{\frac{2\pi}{[-h''(0) + 3\varepsilon]x}\right\}^{1/2} < G(x) < \left\{\frac{2\pi}{[-h''(0) - 3\varepsilon]x}\right\}^{1/2},$$

or

$$\frac{1}{\sqrt{1 - 3\varepsilon/h''(0)}} < \left\{\frac{-xh''(0)}{2\pi}\right\}^{1/2} G(x) < \frac{1}{\sqrt{1 + 3\varepsilon/h''(0)}}. \tag{103}$$

Since ε was arbitrary, (103) is just the definition of the assertion

$$\lim_{x \to \infty} \left\{\frac{-xh''(0)}{2\pi}\right\}^{1/2} G(x) = 1$$

which was to be shown.

To illustrate the use of Theorem 4, we return to Stirling's formula for $n!$, this time writing

$$n! = \int_0^\infty e^{-t} t^n\, dt.$$

Replacing n by the continuous variable x,

$$x! = \Gamma(x + 1) = \int_0^\infty e^{-t} t^x\, dt.$$

The maximum value of the integrand occurs at $t = x$. Hence to reduce this to the standard case of Theorem 4, we make the substitution $t = x(y + 1)$, getting

$$x! = x^{x+1}e^{-x}\int_{-1}^{\infty} \exp\{x[\log(1 + y) - y]\}\, dy, \tag{104}$$

which is of the form (98), where

$$h(y) = \log(1 + y) - y \tag{105}$$

satisfies all the conditions (a)–(e) of the theorem. The fact that the integration extends only over $(-1, \infty)$, instead of $(-\infty, \infty)$, is quite irrelevant from the point of view of asymptotic behavior, for only the neighborhood of $y = 0$ is important. (This assertion can easily be made rigorous, along the lines of the proof of the theorem.) Therefore, from (99), since $h''(0) = -1$,

$$x! = \Gamma(x + 1) \sim \sqrt{2\pi x}\left(\frac{x}{e}\right)^x \qquad (x \to \infty), \tag{106}$$

which is again Stirling's formula. We have here the rigorous proof, which was promised earlier, for the evaluation of the constant λ_k occurring in (68), since any other value would produce an asymptotic expansion not agreeing with (106). It is interesting to observe that the Euler-Maclaurin formula is often capable of producing complete asymptotic expansions except that certain constants occur in obscure forms. Any independent method, such as the above, for producing only one term in that expansion will often lead to an evaluation of the constant in question by comparison of the expansions and thereby to a complete expansion in usable form.

Entirely trivial modifications of the proof of Theorem 4 show that under the hypotheses (a)–(e),

$$\int_{-\infty}^{\infty} g(y)e^{xh(y)}\, dy \sim g(0)\left\{\frac{2\pi}{-xh''(0)}\right\}^{1/2} \qquad (x \to \infty), \tag{107}$$

if $g(y)$ is continuous in a neighborhood of the origin and $g(0) \neq 0$, the idea being that since only the point $y = 0$ matters, $g(y)$ might as well be $g(0)$ everywhere.

4.7 THE METHOD OF STATIONARY PHASE

We turn next to oscillatory integrals of the type

$$G = \int_a^b e^{ih(y)}\, dy \tag{108}$$

where $h(y)$ is real. In this case the main theorem, below, states that the principal contribution to the integral comes from the points at which the "phase" $h(y)$ is stationary, i.e., where $h'(y) = 0$, or, if no such points exist, from the endpoints (a, b) of the interval.

Lemma 1 (*Abel's lemma*). Let $f(x) \geqq 0$ be bounded, and nonincreasing on (a, b). Then

$$mf(a) \leqq \int_a^b f(x)g(x)\,dx \leqq Mf(a)$$

where m, M are the lower and upper bounds for

$$G(x) = \int_a^x g(y)\,dy$$

on (a, b).
Proof.

$$\begin{aligned}\int_a^b f(x)g(x)\,dx &= \int_a^b f(x)\,dG(x)\\ &= f(b)G(b) - \int_a^b G(x)\,df(x).\end{aligned}$$

Thus,

$$m\left\{f(b) - \int_a^b df(x)\right\} \leqq \int_a^b f(x)g(x)\,dx \leqq M\left\{f(b) - \int_a^b df(x)\right\}$$

or

$$mf(a) \leqq \int_a^b f(x)g(x)\,dx \leqq Mf(a)$$

as required.

Lemma 2 (*Second Mean Value Theorem*). Let $f(x) \geqq 0$ be bounded and nonincreasing on (a, b). Then there is a number ξ, $a < \xi < b$ such that

$$f(a)\int_a^\xi g(x)\,dx = \int_a^b f(x)g(x)\,dx. \tag{109}$$

Proof. If $f(x) = 0$, this is obvious. Otherwise, since $f(x)$ is nonincreasing, $f(a) > 0$. By the previous lemma,

$$\frac{1}{f(a)}\int_a^b f(x)g(x)\,dx \tag{110}$$

lies between the greatest and least values of $\int_a^x g(y)\,dy = G(x)$. But $G(x)$ is continuous and therefore takes all values between m and M. In particular it takes the value (110) at some point ξ.

Lemma 3. Let $h(x)$ be differentiable on (a, b), and suppose $h'(x)$ is monotonic, with either $h'(x) \geqq m > 0$ or $h'(x) \leqq -m < 0$ in (a, b). Then

$$\left| \int_a^b e^{ih(y)} \, dy \right| \leqq \frac{4}{m}. \tag{111}$$

Proof. Suppose, for concreteness, that $h'(x)$ is positive and nondecreasing. By Lemma 2,

$$\begin{aligned} \int_a^b \cos h(y) \, dy &= \int_a^b \frac{h'(y) \cos h(y) \, dy}{h'(y)} \\ &= \frac{1}{h'(a)} \int_a^\xi h'(y) \cos h(y) \, dy = \frac{\sin h(\xi) - \sin h(a)}{h'(a)} \end{aligned}$$

and

$$\left| \int_a^b \cos h(y) \, dy \right| \leqq \frac{2}{m}.$$

Applying the same argument to the imaginary part proves the lemma. The other cases are handled similarly.

A trivial generalization of Lemma 3 is

Lemma 4. Let $g(x)$, $h(x)$ be differentiable on (a, b), with $g(x)/h'(x)$ monotonic, and suppose that $|g(x)/h'(x)| \geqq m > 0$.
Then

$$\left| \int_a^b g(y) e^{ih(y)} \, dy \right| \leqq \frac{4}{m}.$$

The proof is the same as that of Lemma 3.

Theorem 5. (*The Principle of Stationary Phase.*) *Let $h(y)$ be real and three times continuously differentiable on (a, b). Let ξ be a point of (a, b) at which $h'(\xi) = 0$. Further suppose that either*

$$0 < \gamma \leqq h''(y) \leqq M\gamma \tag{112}$$

or

$$0 < \gamma \leqq -h''(y) \leqq M\gamma \tag{13}$$

throughout (a, b). Finally, let

$$|h'''(y)| \leqq M\gamma_1 \qquad a \leqq y \leqq b.$$

Then

$$\begin{aligned} \int_a^b e^{ih(y)} \, dy &= \sqrt{2\pi} \, \frac{e^{\pm i(\pi/4) + ih(\xi)}}{\sqrt{|h''(\xi)|}} + O(\gamma^{-4/5} \gamma_1^{1/5}) \\ &\quad + O\left(\frac{1}{|h'(a)|}\right) + O\left(\frac{1}{|h'(b)|}\right) \end{aligned} \tag{114}$$

the choice of the plus or minus sign depending on whether (112) *or* (113) *respectively, holds.*

Before proceeding to the proof, notice that the O symbols are used without any comment about a variable tending to infinity. The meaning of (114) is that the difference between the left-hand side and the first term on the right can be split into three parts, the first of which is smaller than some absolute constant times $\gamma^{-4/5}\gamma_1^{1/5}$, etc. These terms are of importance when $h(y)$ depends on certain parameters and one wishes to know when the first term on the right correctly gives the behavior of the integral when the parameters are large.

Proof. Write

$$\int_a^b e^{ih(y)}\,dy = \int_a^{\xi-\delta} e^{ih(y)}\,dy + \int_{\xi-\delta}^{\xi+\delta} e^{ih(y)}\,dy + \int_{\xi+\delta}^{b} e^{ih(y)}\,dy$$

where $a + \delta \leqq \xi \leqq b - \delta$. By Lemma 3,

$$(115)\qquad \int_{\xi+\delta}^{b} e^{ih(y)}\,dy = O\left\{\frac{1}{|h'(\xi+\delta)|}\right\} = O\left\{\frac{1}{\left|\int_{\xi}^{\xi+\delta} h''(y)\,dy\right|}\right\}$$

$$= O\left\{\frac{1}{\delta\gamma}\right\}$$

and in the same way,

$$\int_a^{\xi-\delta} e^{ih(y)}\,dy = O\left(\frac{1}{\delta\gamma}\right)$$

Next, as in the proof of Laplace's theorem,

$$\int_{\xi-\delta}^{\xi+\delta} e^{ih(y)}\,dy = \int_{\xi-\delta}^{\xi+\delta} \exp\left\{i\left[h(\xi) + (y-\xi)h'(\xi) + \frac{1}{2}(y-\xi)^2h''(\xi)\right.\right.$$
$$\left.\left. + \frac{1}{6}(y-\xi)^3h'''(\xi+\theta(y-\xi))\right]\right\}dy$$
$$= e^{ih(\xi)}\int_{\xi-\delta}^{\xi+\delta} \exp\left\{i\left[\frac{1}{2}(y-\xi)^2h''(\xi)\right]\right\}$$
$$\times \exp\left\{\frac{i}{6}(y-\xi)^3h'''[\xi + C(y-\xi)]\right\}dy$$
$$= e^{ih(\xi)}\int_{\xi-\delta}^{\xi+\delta} \exp\left[\frac{i}{2}(y-\xi)^2h''(\xi)\right]\{1 + O[(y-\xi)^3\gamma_1]\}\,dy$$
$$= e^{ih(\xi)}\int_{\xi-\delta}^{\xi+\delta} \exp\left[\frac{i}{2}(y-\xi)^2h''(\xi)\right]dy + O(\delta^4\gamma_1).$$

Now let

$$u = \frac{1}{2}(y-\xi)^2h''(\xi)$$

in the integral, and suppose $h''(\xi) > 0$. Then for the integral we have

$$\int_{\xi-\delta}^{\xi+\delta} \exp\left[\frac{i}{2}(y-\xi)^2 h''(\xi)\right] dy$$

$$= \sqrt{2/h''(\xi)} \int_0^{(\delta^2/2)h''(\xi)} e^{iu}\,(du/\sqrt{u})$$

$$= \sqrt{2/h''(\xi)}\left[\int_0^{\infty} e^{iu}\,(du/\sqrt{u}) + O\left(\frac{1}{\delta\sqrt{\gamma}}\right)\right] \qquad \text{(Lemma 4)}$$

$$= \sqrt{2\pi/h''(\xi)}\, e^{i\pi/4} + O\left(\frac{1}{\delta\gamma}\right).$$

Hence, combining these results,

$$\int_a^b e^{ih(y)}\,dy = \sqrt{2\pi/h''(\xi)} \exp\left\{i\left[h(\xi) + \frac{\pi}{4}\right]\right\} + O\left(\frac{1}{\delta\gamma}\right) + O(\delta^4\gamma_1). \tag{116}$$

We choose δ now, if possible, so that these last two terms are of the same order, i.e., so that

$$\frac{1}{\delta\gamma} = \delta^4\gamma_1.$$

This means that we take $\delta = (\gamma\gamma_1)^{-1/5}$, if possible, in which case (116) becomes

$$\int_a^b e^{ih(y)}\,dy = \sqrt{2\pi/h''(\xi)} \exp\left\{i\left[h(\xi) + \frac{\pi}{4}\right]\right\} + O(\gamma_1^{1/5}\gamma^{-4/5}). \tag{117}$$

However, we assumed at the outset that $a + \delta \leqq \xi \leqq b - \delta$, which might not be true for the particular δ just chosen. In other words, ξ might be too close to an endpoint of the interval. If, for instance, $b - \delta < \xi \leqq b$, then we have actually integrated beyond the right end of the interval in (115) and have thereby committed an error of

$$e^{ih(\xi)} \int_b^{\xi+\delta} \exp\left[\frac{1}{2} i(y-\xi)^2 h''(\xi)\right] dy = O\left\{\frac{1}{(b-\xi)\gamma}\right\}$$

$$= O\left\{\frac{1}{|h'(b)|}\right\}$$

the first equality arising from Lemma 3 and the second by the same argument as the one appearing on the right side of (115). If the difficulty appears at the other endpoint, a, the argument is identical, and the theorem is proved.

As an example, suppose $h(y) = t\varphi(y)$, where $\varphi(y)$ is independent of t, and t is a parameter. If $\varphi(y)$ satisfies the hypotheses required of $h(y)$ in the theorem, then, as $t \to \infty$, the first term in (114) is of order $t^{-1/2}$, the next being $O(t^{-3/5})$, and each of the last two being $O(t^{-1})$. Hence, in this case,

$$\int_a^b e^{it\varphi(y)}\,dy = \sqrt{2\pi}\,\frac{\exp\left(\pm(i\pi/4) + it\phi(\xi)\right)}{\sqrt{t\,|\varphi''(\xi)|}} + O(t^{-3/5}) \qquad (t \to \infty) \tag{118}$$

the formula being valid if $\phi'(\xi) = 0$ at an interior point, ξ, of (a, b). As before, if $g(y)$ is continuous at ξ, we have also

$$\int_a^b g(y)e^{it\phi(y)}\,dy = \sqrt{2\pi}\,g(\xi)\,\frac{\exp\left(\pm(i\pi/4) + it\phi(\xi)\right)}{\sqrt{t\,|\varphi''(\xi)|}} + O(t^{-3/5}) \qquad (t \to \infty) \tag{119}$$

under the same conditions on $\varphi(y)$, if $g(\xi) \neq 0$.

4.8 RECURRENCE RELATIONS

It often happens that a sequence of numbers $a_0, a_1, a_2, \ldots$, is given, not explicitly, but instead by a recurrence relation with given initial values. The problem of determining the rate of growth of such a sequence can be quite troublesome. We give here a method which works on a reasonably wide class of such problems.

To illustrate the ideas involved, consider Fibonacci's sequence

$$\begin{aligned} a_{n+2} &= a_{n+1} + a_n \qquad (n = 0, 1, 2, \ldots) \\ a_0 &= a_1 = 1. \end{aligned} \tag{120}$$

This sequence has the appearance

$$1, 1, 2, 3, 5, 8, 13, 21, 34, 55, 89, \ldots$$

and we wish to estimate the size of the nth member a_n. For the a_n defined in (120), consider the function

$$f(z) = \sum_{n=0}^{\infty} \frac{a_n}{n!} z^n. \tag{121}$$

Then

$$\begin{aligned} f'(z) + f(z) &= \sum_{n=1}^{\infty} \frac{a_n z^{n-1}}{(n-1)!} + \sum_{n=0}^{\infty} \frac{a_n z^n}{n!} \\ &= \sum_{n=0}^{\infty} \frac{a_{n+1} + a_n}{n!} z^n \\ &= \sum_{n=0}^{\infty} \frac{a_{n+2} z^n}{n!} = f''(z). \end{aligned}$$

Solving the differential equation with constant coefficients,

$$f''(z) = f'(z) + f(z) \tag{122}$$

with

$$f(0) = f'(0) = 1$$

one finds

$$f(z) = \frac{1}{2}\left(1 + \frac{1}{\sqrt{5}}\right) \exp\left(\frac{1+\sqrt{5}}{2}z\right) + \frac{1}{2}\left(1 - \frac{1}{\sqrt{5}}\right) \exp\left(\frac{1-\sqrt{5}}{2}z\right). \tag{123}$$

Matching coefficients between (121) and (123), we find an explicit closed expression for the Fibonacci numbers

$$a_n = \frac{1}{\sqrt{5}}\left\{\left(\frac{1+\sqrt{5}}{2}\right)^{n+1} - \left(\frac{1-\sqrt{5}}{2}\right)^{n+1}\right\} \qquad (n = 0, 1, \ldots) \tag{124}$$

and, of course, the asymptotic relation

$$a_n \sim \frac{1}{\sqrt{5}}\left(\frac{1+\sqrt{5}}{2}\right)^{n+1} \qquad (n \to \infty). \tag{125}$$

To justify the formal process leading from (121) to (122) it is enough to observe that the final series obtained actually has a nonzero radius of convergence, and we shall henceforth ignore this question. The function $f(z)$ in (123) is called the *generating function* of the sequence (120).

More generally, consider the recurrence

$$a_{n+p} = c_1 a_{n+p-1} + c_2 a_{n+p-2} + \cdots + c_p a_n \tag{126}$$

with

$$a_0, a_1, \ldots, a_{p-1} \quad \text{given}. \tag{127}$$

With the same form of the generating function (121) as before, we find this time

$$f^{(p)}(z) = c_1 f^{(p-1)}(z) + \cdots + c_p f(z) \tag{128}$$

the values $f(0), f'(0), \ldots, f^{(p-1)}(0)$ being given.

Since (128) is again a linear differential equation with constant coefficients, the solution will give $f(z)$ as a linear combination of terms of the form $(\alpha_0 + \alpha_1 z + \cdots + \alpha_{m_j-1} z^{m_j-1})e^{r_j}$, where the r_j run through the roots of the equation

$$r^p = c_1 r^{p-1} + c_2 r^{p-2} + \cdots + c_p \tag{129}$$

of degree p, and m_j is the multiplicity of the root r_j. Assuming these roots to have been found, the solution of the recurrence (126) will be of the form

$$a_n = \sum_j \gamma_j(\alpha_0 + \alpha_1 n + \cdots + \alpha_{m_j-1} n^{m_j-1}) r_j^{\,n} \qquad (n = 0, 1, \ldots) \tag{130}$$

where the sum is extended over the distinct roots of (129) and the γ_j are constants determined from the initial data.

We now state a number of results which are all obvious by inspection of (130).

Theorem 6. *If all the roots of* (129) *lie in the circle* $|z| \leqq R$, *then for every* $\varepsilon > 0$ *we have*

$$a_n = O((R + \varepsilon)^n) \qquad (n \to \infty). \tag{131}$$

Theorem 7. *In order that* $a_n = o(1)$ $(n \to \infty)$ *for every set of initial values* (127), *it is necessary and sufficient that all the roots of* (129) *lie in the interior of the unit circle.*

Theorem 8. *In order that* $a_n = O(1)$ $(n \to \infty)$ *for every set of initial values* (127), *it is necessary and sufficient that all the roots of* (129) *lie in* $|z| \leqq 1$ *and that all roots of modulus unity be simple.*

The theorems of the preceding chapter are of obvious utility in deciding such questions.

If the coefficients $c_1, c_2, \ldots, c_p$ are not constant, i.e., if they depend on n, the differential equation satisfied by $f(z)$ can still be found, but its coefficients will depend on z. Hence a solution in closed form may not be possible.

Let us introduce the notation

$$f(z) \leftrightarrow a_n$$

to mean that $f(z)$ is related to the constants a_n by means of (121). Then

$$\begin{aligned} \frac{d}{dz} f(z) &\leftrightarrow a_{n+1} \\ &\vdots \\ \left(\frac{d}{dz}\right)^k f(z) &\leftrightarrow a_{n+k} \end{aligned} \tag{132}$$

and

$$\begin{aligned} na_n &\leftrightarrow \left(z \frac{d}{dz}\right) f(z) \\ &\vdots \\ n^k a_n &\leftrightarrow \left(z \frac{d}{dz}\right)^k f(z). \end{aligned}$$

Hence if $P(n)$ is a polynomial in n with constant coefficients,

$$P(n)a_n \leftrightarrow P\left(z\frac{d}{dz}\right)f(z).$$

Combining these, we see that if the a_n satisfy the recurrence

$$a_{n+k} = P_1(n)a_{n+k-1} + P_2(n)a_{n+k-2} + \cdots + P_k(n)a_n \qquad (n = 0, 1, \ldots) \tag{133}$$

the generating function of the a_n satisfies the differential equation

$$f^{(k)}(z) = P_1\left(z\frac{d}{dz}\right)f^{(k-1)}(z) + P_2\left(z\frac{d}{dz}\right)f^{(k-2)}(z) + \cdots + P_k\left(z\frac{d}{dz}\right)f(z) \tag{134}$$

if the $P_j(n)$ are polynomials in n with constant coefficients.

An example of a different kind results from noticing that

$$f(z) \leftrightarrow a_n \tag{135}$$

implies

$$e^{\alpha z}f(z) \leftrightarrow \sum_{\nu=0}^{n}\binom{n}{\nu}a_\nu\alpha^{n-\nu} \tag{136}$$

which is often of assistance in solving recurrence relations involving binomial coefficients.

For instance, we may work backward from (52) to (49), for if $f(z)$ is the generating function of the B_n, which satisfy (52), then

$$f(z) \leftrightarrow B_n$$

$$e^z f(z) \leftrightarrow \sum_{\nu=0}^{n}\binom{n}{\nu}B_\nu$$

$$e^z f(z) - f(z) \leftrightarrow \begin{cases} \sum_{\nu=0}^{n-1}\binom{n}{\nu}B_\nu = 0 & (n \geqq 2) \\ 1 & (n = 1) \\ 0 & (n = 0) \end{cases}$$

which is to say that

$$e^z f(z) - f(z) = z$$

and (49) follows.

Bibliography

There has recently appeared the first book devoted to the study of asymptotics for its own sake, namely,

1. N. G. DeBruijn, *Asymptotic Methods in Analysis*, Interscience Publishers, New York, 1958.

This book is highly recommended to the serious worker in all branches of the "mathematical sciences." In particular, the most powerful method for dealing with integrals of a general kind, the saddle-point method, was ignored in these pages largely because of the definitive discussion of the method in the above reference.

The asymptotic expansion of delicately oscillating sums and integrals has been studied most extensively by the analytic number theorists. When confronted with such problems, one should, perhaps, turn first to

2. E. C. Titchmarsh, *The Theory of the Riemann Zeta Function*, Oxford University Press, New York, 1951,

on which our discussion of the elusive principle of stationary phase was based.

For an abbreviated discussion, which because of its large bibliography is quite valuable, see

3. A. Erdélyi, *Asymptotic Expansions*, Dover Publications, New York, 1956.

Finally, a recent development of great generality and importance is contained in the paper of W. K. Hayman [1] listed in the Journal References at the end of this book. This is a method of finding the asymptotic behavior of numbers which are the coefficients of a rather wide class of analytic functions. Several kinds of problems in asymptotics which are superficially unrelated to such functions can nonetheless be transformed into problems of the type considered by Hayman, and his results and methods are readily accessible to the reader who has, for example, carefully studied the material in this chapter.

Exercises

1. Of the following assertions, some are true and some false. Which?

(*a*) $\sqrt{2x^2 + 1} \sim x \qquad (x \to \infty)$

(*b*) $\sin x = O(x^{3/2}) \qquad (x \to 0)$

(*c*) $\displaystyle\int_1^x \cos y^2 \frac{dy}{y} = O(\log x) \qquad (x \to \infty)$

(*d*) $\left(1 + \dfrac{1}{n}\right)^{\sqrt{n}} = O(1) \qquad (n \to \infty).$

2. Give an example which shows that

$$f(x) \sim g(x) \qquad (x \to \infty)$$

does not necessarily imply

$$e^{f(x)} \sim e^{g(x)}.$$

Prove that the conclusion does follow if the stronger relation

$$f(x) = g(x) + o(1) \qquad (x \to \infty)$$

holds.

3. Show that if

$$f(x) \sim g(x) \qquad (x \to \infty)$$

then not only does

$$\log f(x) \sim \log g(x) \qquad (x \to \infty)$$

but actually

$$\log f(x) = \log g(x) + o(1) \qquad (x \to \infty).$$

4. Show that

$$\sum_{n=1}^{\infty} \frac{x}{n^2} e^{-\frac{x}{n^2}} = \frac{1}{2} \sqrt{\pi x} + O(1) \qquad (x \to \infty)$$

5. The Bessel function $J_n(x)$ is defined by the series

$$J_n(x) = \sum_{k=0}^{\infty} \frac{(-1)^k (x/2)^{n+2k}}{k!\,(n+k)!}.$$

Show that

$$J_n(\sqrt{n}) \sim \sqrt{2/\pi}\, e^{-5/4} \left(\frac{e}{2\sqrt{n}}\right)^{n+1} \qquad (n \to \infty).$$

6. Prove that

$$B_{2n} \sim (-1)^{n+1} 4\sqrt{n\pi} \left(\frac{n}{\pi e}\right)^{2n} \qquad (n \to \infty).$$

7. Find the complete asymptotic series for

$$1 + \frac{1}{2} + \cdots + \frac{1}{n} \qquad (n \to \infty).$$

8. For Landau's numbers†

$$G_n = 1 + (\tfrac{1}{2})^2 + \left(\frac{1 \cdot 3}{2 \cdot 4}\right)^2 + \cdots + \left(\frac{1 \cdot 3 \cdot 5 \cdots (2n-1)}{2 \cdot 4 \cdots 2n}\right)^2$$

show that (use the result of exercise 14)

$$G_n \sim \frac{1}{\pi} \log n \qquad (n \to \infty).$$

9. Let $0 < A < \infty$ and suppose $f(x)$ is bounded on $[0, A]$ and continuous at A. If $f(A) \neq 0$, then

$$\int_0^A t^n f(t)\, dt \sim \frac{A^{n+1}}{n} f(A) \qquad (n \to \infty).$$

10. From Bessel's integral

$$J_0(x) = \frac{1}{2\pi} \int_0^{2\pi} \cos(x \sin\theta)\, d\theta$$

show that

$$J_0(x) = \sqrt{\frac{2}{\pi x}} \cos\left(x - \frac{\pi}{4}\right) + O(x^{-3/2}) \qquad (x \to \infty).$$

11. For the sequence

$$\begin{cases} \alpha_n = \sum_{\nu=0}^{n-1} \binom{n}{\nu} \alpha_\nu & (n = 1, 2, \ldots) \\ \alpha_0 = 1 \end{cases}$$

† Landau [1].

(*a*) Find the generating function $f(z)$.

(*b*) Show that for every $\epsilon > 0$, and no $\epsilon < 0$, we have

$$\alpha_n = O\left(\left(\left\{\frac{1}{\log 2} + \varepsilon\right\}\frac{n}{e}\right)^n\right) \qquad (n \to \infty).$$

12. Prove that (135) implies (136) and (132).

13. If

$$\lambda_n = \int_0^{\pi} x^n \sin x \, dx$$

then

$$\lambda_n = \frac{\pi^{n+2}}{(n+1)(n+2)}\left\{1 + \sum_{\nu=1}^{\infty}(-1)^{\nu}\frac{\pi^{2\nu}}{(n+3)(n+4)\cdots(n+2\nu+2)}\right\}$$

which is first convergent evaluation of λ_n in closed form, and second an asymptotic expansion for $n \to \infty$. (*Hint.* Find the generating function and remember that $\sin \pi = 1 + \cos \pi = 0$).

14. If

I $$a_\nu \sim b_\nu \qquad (\nu \to \infty)$$

II $$a_\nu \geqq 0, \quad b_\nu \geqq 0 \qquad (\nu = 1, 2, \ldots)$$

III $$\sum_{\nu=1}^{\infty} b_\nu = +\infty$$

then

$$\sum_{1}^{n} a_\nu \sim \sum_{\nu=1}^{n} b_\nu \qquad (n \to \infty).$$

chapter 5

Ordinary differential equations

5.1 INTRODUCTION

In this chapter we study some of the properties of the solutions of ordinary differential equations. The range of possible subject matter is so vast that we shall not attempt to do more than indicate some of the directions in which interesting results lie. We assume that the reader has already been exposed to a variety of special methods of obtaining solutions of equations of certain types. Our concern rests first of all with the existence and uniqueness of solutions of first order equations and therefore, by an easy extension, for equations of arbitrary order. Next we discuss some numerical methods for solving such equations, one of which, incidentally, will be a by-product of the existence theorem and its proof.

Finally, a specialization is made to the case of a second order equation with variable coefficients. With these equations we are primarily interested in their function-theoretic properties in the complex plane. Application of some of these results to the higher transcendental functions is then made.

5.2 EQUATIONS OF THE FIRST ORDER

The most general initial value problem of the first order requires the determination of a function $y(x)$ such that

(1) $$F(x, y, y') = 0$$

and

(2) $$y(x_0) = y_0$$

where F is a given function of its arguments and y_0 is a given number. Instead of discussing this completely general situation we impose the mild restriction that (1) can be conveniently solved for y' in terms of x, y, and therefore we take as our standard case

$$y'(x) = f(x, y(x)) \tag{3}$$

$$y(x_0) = y_0. \tag{4}$$

The combination of the differential equation (3) and initial condition (4) will be called the *initial value problem.* To say that a certain function $y(x)$ is a *solution* of the initial value problem is to say that (4) is true and that there is a number $h > 0$ such that $y(x)$ exists and (3) holds at every point of the interval $[x_0, x_0 + h]$.

It is possible for an initial value problem to have no solution, one solution, finitely many different solutions, or infinitely many different solutions. A problem with no solution is (see exercise 1)

$$y'(x) = \begin{cases} 1/x & x \neq 0 \\ 0 & x = 0 \end{cases} \tag{5}$$

$$y(0) = 0. \tag{6}$$

The problem

$$\begin{cases} y'(x) = +\sqrt{y(x)} \\ y(0) = 0 \end{cases} \tag{7}$$

has two solutions, namely $y(x) = 0$, and $y(x) = \frac{1}{4}x^2$.

Finally, the pathological specimen†

$$\begin{cases} y'(x) = 2\sqrt{|y(x)|} \\ y(0) = 0 \end{cases} \tag{8}$$

has infinitely many distinct solutions, for if α and β are arbitrary positive numbers, the function

$$y(x) = \begin{cases} 0 & -\beta \leqq x \leqq \alpha \\ -(x + \beta)^2 & x \leqq -\beta \\ (x - \alpha)^2 & x \geqq \alpha \end{cases} \tag{9}$$

is everywhere continuous, differentiable, and a solution of (8).

In view of these extraordinary possibilities it is obviously of importance to have criteria which are capable of detecting in advance when there will be a solution, when it will be unique, and how far to the right of x_0 it will continue to be a solution. It should be emphasized at this point that situations in which several solutions may exist actually occur, with distressing frequency, in physical situations, generally because of the incomplete

† Bourbaki [1].

formulation of a problem. The blind use of computing machinery in such cases can lead only to chaos.

5.3 PICARD'S THEOREM

Theorem 1. *Let $f(x, y)$ be a real valued function of the real variables x, y, defined on an open region R of the x-y plane. Suppose*

(a) $f(x, y)$ is a continuous function of x and y on R.

(b) there is a number L such that for any two points (x, y_1), (x, y_2) of R we have

$$|f(x, y_1) - f(x, y_2)| \leqq L\,|y_1 - y_2|. \tag{10}$$

Then, for any fixed point (x_0, y_0) of R there is a number $b > 0$ and a function $y(x)$ such that

1. *$y'(x)$ exists and is continuous in $|x - x_0| \leqq b$.*
2. *$y'(x) = f(x, y(x))$ for $|x - x_0| \leqq b$.*
3. *$y(x_0) = y_0$.*
4. *$y(x)$ is the only function satisfying c1 − 3 at once.*

Before proceeding to the proof, notice first the reappearance of the Lipschitz condition (10) which we have already encountered in the theory of Fourier series. The essential content of Picard's theorem is that a differential equation whose right-hand side satisfies a Lipschitz condition in R has a unique solution in some neighborhood of x_0, if (x_0, y_0) lies in R. A final remark is that the proof which follows actually gives an explicit method for finding the solution, the method being readily adaptable to automatic calculation.

Proof. First, since R is open, there is a rectangular neighborhood of (x_0, y_0)

$$S: \quad |x - x_0| \leqq a, \quad |y - y_0| \leqq c \tag{11}$$

lying entirely in R. Next, since $f(x, y)$ is continuous on the compact set S, it is bounded, say

$$|f(x, y)| \leqq M \qquad (x, y) \text{ in } S. \tag{12}$$

Now let

$$b = \min\left(a, \frac{c}{M}\right), \tag{13}$$

and let S^* be the rectangle

$$S^*: \quad |x - x_0| \leqq b \quad |y - y_0| \leqq c. \tag{14}$$

We will show that this number b has the property stated in the conclusion of the theorem.

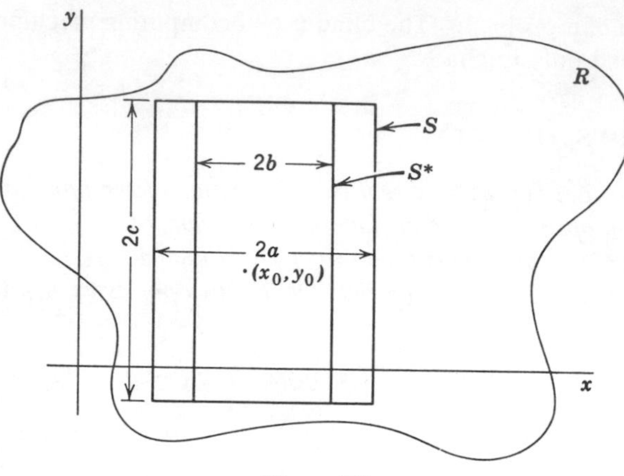

Figure 5.1

Let a sequence of functions $y_0(x), y_1(x), y_2(x), \ldots$ be defined for $|x - x_0| \leqq b$ by

$$y_0(x) = y_0 \tag{15}$$

$$y_{n+1}(x) = y_0 + \int_{x_0}^{x} f(t, y_n(t))\, dt \qquad (n = 0, 1, 2, \ldots). \tag{16}$$

Our first task is to show that these definitions actually make sense by proving that the point $(t, y_n(t))$ always lies inside the region R where $f(x, y)$ is defined. We show now that, in fact, $(t, y_n(t))$ lies always in S^*.

Indeed, for $|x - x_0| \leqq b$,

$$\begin{aligned} |y_1 - y_0| &= \left| \int_{x_0}^{x} f(t, y_0)\, dt \right| \\ &\leqq M\, |x - x_0| \\ &\leqq Mb \\ &\leqq c. \end{aligned}$$

Inductively, suppose it has been shown that

$$|y_k(x) - y_0| \leqq c$$

for $|x - x_0| \leqq b$. Then $(x, y_k(x))$ lies in S^*, and therefore $|f(x, y_k(x))| \leqq M$, from which

$$\begin{aligned} |y_{k+1}(x) - y_0| &= \left| \int_{x_0}^{x} f(t, y_k(t))\, dt \right| \\ &\leqq M\, |x - x_0| \leqq c \end{aligned}$$

and the proof is complete by induction.

Therefore the points $(x, y_n(x))$ lie in S^* if $|x - x_0| \leqq b$, for each $n = 0, 1, 2, \ldots$, the functions $f(x, y_n(x))$ are well defined, and (15), (16) do indeed define an infinite sequence of functions $\{y_n(x)\}_{n=0}^{\infty}$ in $|x - x_0| \leqq b$.

Next, we claim that

$$|y_n(x) - y_{n-1}(x)| \leqq ML^{n-1}\frac{(x - x_0)^n}{n!} \tag{17}$$

for $n = 1, 2, \ldots, x_0 \leqq x \leqq x_0 + b$. For,

$$\begin{aligned} |y_1(x) - y_0(x)| &= \left| \int_{x_0}^{x} f(t, y_0(t))\, dt \right| \\ &\leqq M(x - x_0) \end{aligned}$$

which proves (17) for $n = 1$. Supposing, as before, that (17) has been proved for $n = 1, 2, \ldots, k$, we have

$$\begin{aligned} |y_{k+1}(x) - y_k(x)| &= \left| \int_{x_0}^{x} [f(t, y_k(t)) - f(t, y_{k-1}(t))]\, dt \right| \\ &\leqq L\int_{x_0}^{x} |y_k(t) - y_{k-1}(t)|\, dt \\ &\leqq L\,\frac{ML^{k-1}}{k!}\int_{x_0}^{x} (t - x_0)^k\, dt \\ &= ML^k\,\frac{(x - x_0)^{k+1}}{(k + 1)!} \end{aligned}$$

and (17) is proved for every n. It follows that the terms of the series

$$y(x) = y_0 + \sum_{n=1}^{\infty} [y_n(x) - y_{n-1}(x)] \tag{18}$$

are dominated by the terms of a convergent series of constants, indeed

$$\begin{aligned} \left| y_0 + \sum_{n=1}^{\infty} [y_n(x) - y_{n-1}(x)] \right| &\leqq |y_0| + \sum_{n=1}^{\infty} |y_n(x) - y_{n-1}(x)| \\ &\leqq |y_0| + \sum_{n=1}^{\infty} ML^{n-1}\frac{(x - x_0)^n}{n!} \\ &\leqq |y_0| + \sum_{n=1}^{\infty} ML^{n-1}\frac{b^n}{n!} \\ &= |y_0| + \frac{M}{L}(e^{bL} - 1). \end{aligned}$$

Therefore the series (18) converges absolutely and uniformly for

$$x_0 \leqq x \leqq x_0 + b$$

to a sum $y(x)$, by the Weierstrass M-test. Since $y(x)$ is the uniform limit of a sequence of continuous functions, $y(x)$ is itself continuous in

$$x_0 \leqq x \leqq x_0 + b$$

Since the nth partial sum of the telescoping series in (18) is just $y_n(x)$, it follows that

$$\lim_{n\to\infty} y_n(x) = y(x) \tag{19}$$

uniformly on $[x_0, x_0 + b]$.

Next, if $\epsilon > 0$ is given, we can choose n_0 such that for $n \geqq n_0$ we have

$$|y_n(x) - y(x)| \leqq \frac{\epsilon}{L}$$

for all x in $x_0 \leqq x \leqq x_0 + b$.
Then, for $n \geqq n_0$,

$$\begin{aligned} |f(x, y(x)) - f(x, y_n(x))| &\leqq L|y(x) - y_n(x)| \\ &\leqq L \cdot \frac{\varepsilon}{L} \\ &= \varepsilon \end{aligned}$$

that is, $f(x, y_n(x)) \to f(x, y(x))$ uniformly in $[x_0, x_0 + b]$. That being so, we may write

$$y_n(x) = y_0 + \int_{x_0}^{x} f(t, y_n(t))\, dt$$

and take the limit under the integral sign on the right, getting

$$\begin{aligned} \lim_{n\to\infty} y_n(x) = y(x) &= y_0 + \lim_{n\to\infty} \int_{x_0}^{x} f(t, y_n(t))\, dt \\ &= y_0 + \int_{x_0}^{x} \lim_{n\to\infty} f(t, y_n(t))\, dt \\ &= y_0 + \int_{x_0}^{x} f(t, y(t))\, dt \end{aligned}$$

which is to say that $y(x)$ satisfies the equation

$$y(x) = y_0 + \int_{x_0}^{x} f(t, y(t))\, dt.$$

Hence, clearly, $y(x_0) = y_0$, and by differentiating,

$$y'(x) = f(x, y(x))$$

also, i.e., $y(x)$ is a solution of our initial value problem.

Next, we claim that $y(x)$ is unique, for suppose $\tilde{y}(x)$ is another solution of our problem,

$$\tilde{y}(x) = y_0 + \int_{x_0}^{x} f(t, \tilde{y}(t))\, dt.$$

Then

$$|\tilde{y}(x) - y_{n+1}(x)| \leq \int_{x_0}^{x} |f(t, \tilde{y}(t)) - f(t, y_{n+1}(t))| \, dt$$
$$\leq L \int_{x_0}^{x} |\tilde{y}(t) - y_n(t)| \, dt.$$

Taking $n = 0$,

$$|\tilde{y}(x) - y_1(x)| \leq Lc \, |x - x_0|$$

where c is an upper bound for $|\tilde{y}(t) - y_0|$ on $x_0 \leq t \leq x_0 + b$. Inductively, it is easy to show that for each n,

$$|\tilde{y}(x) - y_{n+1}(x)| \leq \frac{cL^{n+1}}{(n+1)!} (x - x_0)^{n+1}$$
$$\leq \frac{cL^{n+1}}{(n+1)!} b^{n+1}$$
$$= o(1)$$

as $n \to \infty$, which says that

$$\tilde{y}(x) = \lim_{n \to \infty} y_{n+1}(x) = y(x),$$

completing the proof.

5.4 REMARKS ON PICARD'S THEOREM, WINTNER'S METHOD

First, as an example of the explicit construction of a solution, consider

$$\begin{cases} y'(x) = y(x) \\ y(0) = 1. \end{cases} \tag{20}$$

Then the Picard iterates $y_n(x)$ are defined by

$$\begin{cases} y_{n+1}(x) = 1 + \int_0^x y_n(t) \, dt & (n = 0, 1, 2, \ldots) \\ y_0(x) = 1 \end{cases}$$

and we find easily

$$y_1(x) = 1 + x$$
$$y_2(x) = 1 + x + \frac{x^2}{2!}$$
$$\vdots \qquad \vdots$$
$$y_n(x) = 1 + x + \frac{x^2}{2!} + \cdots + \frac{x^n}{n!}$$

the nature of the convergence being obvious.

Next, we call attention to the local character of the theorem. It does not state that a function $y(x)$ exists which *everywhere* satisfies the differential equation and the initial data but only that $b > 0$ exists such that $y(x)$ is the solution in $x_0 \leqq x \leqq x_0 + b$. Indeed, consider

$$\begin{cases} y'(x) = y^2(x) \\ y(0) = 1 \end{cases} \tag{21}$$

which satisfies the conditions of the theorem. The solution $y(x)$ is clearly

$$y(x) = \frac{1}{1 - x} \tag{22}$$

which is a solution in $0 \leqq x \leqq 1 - \delta$ $(\delta > 0)$ but not in any interval containing $x = 1$.

The theorem actually provides an estimate of the size of the number b in explicit form, for, from (13),

$$b = \min\left(a, \frac{c}{M}\right)$$

where

$$M = \max_{(x,y) \text{ in } S} |f(x, y)| = M(a, c) \tag{23}$$

and S is any rectangle

$$S: \quad |x - x_0| \leqq a, \quad |y - y_0| \leqq c \tag{24}$$

in which $f(x, y)$ is defined and satisfies the hypotheses of the theorem. Now we may choose the numbers a and c in such a way as to get the best (i.e., largest) estimate for b, subject only to the restriction that the rectangle S lies in the region R of the theorem. Hence we have

Corollary 1. Let $f(x, y)$ satisfy the hypothesis of Theorem 1 in the region R. Let

$$b = \max_{S \text{ in } R} \min\left(a, \frac{c}{M(a, c)}\right) \tag{25}$$

where $M(a, c)$ is given by (23) and S is the rectangle (24). Then a unique solution of (3), (4) exists in $x_0 \leqq x \leqq x_0 + b$.

To illustrate the use of this result, let us try to estimate the number b for (21). Here R is any compact set in the plane, so the rectangle S is arbitrary. Further

$$\begin{aligned} M(a, c) &= \max_{S} |y^2| \\ &= (1 + c)^2 \end{aligned}$$

and (25) reads

$$b = \max_{a,c>0} \min \left(a, \frac{c}{(1 + c)^2} \right).$$

If $a \neq c(1 + c)^{-2}$, the min can be increased by increasing the smaller one. Hence the max occurs when $a = c(1 + c)^{-2}$ and is equal to the largest value of $c(1 + c)^{-2}$ for $c \geqq 0$, i.e.,

$$b = \frac{1}{4}$$

is the best estimate we can get directly from Picard's theorem. Since we know that a solution actually exists in (0, 1), the estimate is clearly quite conservative.

Because of the importance, in practice, of being able to estimate the largest number b directly from the given equation, we proceed now to develop a more refined estimator than that supplied by Corollary 1. We need first

Theorem 2. *Let $f(x, y)$, $g(x, y)$ be functions satisfying the hypotheses of Theorem 1. Suppose further*

(a) $|f(x, y)| \leqq g(x, y) \qquad (x, y)$ in R.

(b) $g(x, y)$ is, for each fixed x, a nondecreasing function of $|y|$, that is, if $|y_1| \leqq |y_2|$, then $g(x_1, y_1) \leqq g(x_1, y_2)$. Let $y(x)$, $Y(x)$ denote the solutions of

$$\begin{cases} y'(x) = f(x, y(x)) \\ y(0) = 0 \end{cases} \qquad \begin{cases} Y'(x) = g(x, Y(x)) \\ Y(0) = 0 \end{cases}$$

Then, if the point $(x, Y(x))$ lies in the rectangle

$$S\colon \ 0 \leqq x \leqq a, \qquad |y| \leqq c$$

so does the point $(x, y(x))$, and $|y(x)| \leqq Y(x)$.

Proof. The idea of the theorem is that if you dominate the right-hand side of an equation suitably, you also dominate the solution.

If we consider the two sets of Picard iterates

$$y_{n+1}(x) = \int_0^x f(t, y_n(t))\, dt$$

$$Y_{n+1}(x) = \int_0^x g(t, Y_n(t))\, dt$$

$$y_0(x) = Y_0(x) = 0$$

then we claim that for each value of $n = 0, 1, 2, \ldots$

$$|y_n(x)| \leqq Y_n(x)$$

when $0 \leqq x \leqq a$. This is obvious for $n = 0$. If it has been shown for $n = 0, 1, 2, \ldots, k$, then

$$\begin{aligned} |y_{k+1}(x)| &= \left| \int_0^x f(t, y_k(t))\, dt \right| \\ &\leqq \int_0^x |f(t, y_k(t))|\, dt \\ &\leqq \int_0^x g(t, y_k(t))\, dt \\ &\leqq \int_0^x g(t, Y_k(t))\, dt = Y_{k+1}(x) \end{aligned}$$

and the theorem is proved by an obvious limiting argument.

Now let $f(x, y)$ be given. A particularly simple choice of a dominating function $g(x, y)$ would be one which depends only on y, for then we could solve the dominating equation explicitly by a quadrature. But clearly the smallest function $\varphi(y)$ which satisfies (a) and (b) of Theorem 2, and depends on y alone, is

$$\varphi(y) = \max |f(x, s)|$$

the maximum being taken over the rectangle $0 \leqq x \leqq a$, $-|y| \leqq s \leqq |y|$. For this particular $\varphi(y)$, an equation which dominates the given equation $y' = f(x, y)$, $y(0) = 0$, is

$$Y'(x) = \varphi(Y(x))$$

$$Y(0) = 0$$

whose solution is given implicitly by

$$\int_0^{Y(x)} \frac{dt}{\varphi(t)} = x.$$

It follows that the point $(x, Y(x))$ will leave the rectangle $0 \leqq x \leqq a$, $|y| \leqq c$ when either $x > a$ or

$$x > \int_0^c \frac{dy}{\varphi(y)}.$$

Hence, if

$$x \leqq b = \min \left(a, \int_0^c \frac{dy}{\varphi(y)} \right)$$

we can be sure that $(x, Y(x))$, and therefore $(x, y(x))$ also, lies in this rectangle. Therefore we have

Theorem 3.† *If $f(x, y)$ satisfies the hypotheses of Theorem* 1 *in the rectangle*

$$|x - x_0| \leqq a, \quad |y - y_0| \leqq c$$

and we define

$$\varphi(y) = \max |f(x, s)| \text{ over } |x - x_0| \leqq a, \quad y_0 - |y| \leqq s \leqq y_0 + |y| \tag{26}$$

then there is a unique solution of the equation $y' = f(x, y), y(x_0) = y_0$, at least in the interval

$$x_0 \leqq x \leqq x_0 + \min\left(a, \int_0^c \frac{dy}{\varphi(y)}\right). \tag{27}$$

Returning to the example (21), we find

$$\begin{aligned} \varphi(y) &= \max |s^2| \qquad |x| \leqq a, \quad 1 - |y| \leqq s \leqq 1 + |y| \\ &= (1 + |y|)^2 \end{aligned}$$

and (27) gives the interval

$$0 \leqq x \leqq \min\left(a, \frac{c}{c+1}\right).$$

Again, to make this as large as possible we take $a = c/(c + 1)$, and therefore the solution exists in $0 \leqq x \leqq c/(c + 1)$ for every $c > 0$, i.e., in $0 \leqq x \leqq 1 - \delta$ for every $\delta > 0$, which is best possible. The unusual accuracy of the estimate in this particular case is, of course, traceable to the fact that the function $f(x, y)$ depends only on y.

5.5 NUMERICAL SOLUTION OF DIFFERENTIAL EQUATIONS

When all else fails, and the solution of a certain equation, or system of equations, must be found, recourse must be had to numerical methods, which we now discuss. Because of the enormity of the subject (books have been written on the numerical solution of ordinary differential equations), we shall attempt here to confine the discussion to some of the fundamental principles rather than to explore the variety of individual techniques.

We treat the (by now familiar) initial value problem

$$\begin{cases} y'(x) = f(x, y) \\ y(x_0) = y_0 \end{cases} \tag{28}$$

and suppose that a preliminary analysis has shown the existence and uniqueness of the solution in some interval. Our remarks will, virtually without exception, carry over to systems

$$\begin{cases} y_i'(x) = f_i(x, y_1(x), \ldots, y_n(x)) \\ y_i(x_0) = y_{i0} \qquad (i = 1, 2, \ldots, n) \end{cases}$$

† Wintner [1].

merely by formally regarding the letters f and y in (28) as n-dimensional vectors.

We suppose that a number $h > 0$, the mesh spacing, is chosen, and we choose points x_n ($n = 0, 1, 2, \ldots$) with spacing h, starting from x_0, i.e.,

$$x_n = x_0 + nh \qquad (n = 0, 1, 2, \ldots).$$

Our objective now is to calculate the values of the unknown function $y(x)$ at the points $x_0, x_1, \ldots$, i.e., the numbers

$$y_n = y(x_n) \qquad (n = 0, 1, 2, \ldots) \tag{29}$$

but we shall not attain this objective. In fact, we obtain only a sequence of numbers $u_0, u_1, u_2, \ldots$, and the relations

$$u_n = y_n \qquad (n = 0, 1, 2, \ldots)$$

will hold only if (i) the function $y(x)$ belongs to a certain class of functions for which our numerical integration formulas are exact and (ii) the calculations are carried out with infinite precision, that is, with infinitely many decimal places. In general, we know only that

$$y_n = u_n + T_n + R_n \qquad (n = 0, 1, 2, \ldots) \tag{30}$$

where T_n, the truncation error, results from the nonfulfillment of condition (i), and R_n, the round off error, from that of condition (ii). The central problem of the whole theory comes simply to this: given $\epsilon > 0$, and N; find a numerical integration method such that with the computing machinery available, we shall surely have

$$|y_n - u_n| = |T_n + R_n| \leqq \epsilon$$

for $n = 0, 1, 2, \ldots, N$.

A numerical integration formula is any relation which permits the recursive calculation of u_n from $u_{n-1}, u_{n-2}, \ldots, u_0$ and the differential equation, in such a way that if (i) $y(x)$ belongs to a certain class $\mathscr{F}$ of functions, and (ii) $u_0 = y_0, u_1 = y_1, \ldots, u_{n-1} = y_{n-1}$, and (iii) the calculations are performed with infinite precision, then $u_n = y_n$ also.

Perhaps the simplest nontrivial example of such a formula is Euler's method

$$u_{n+1} = u_n + hf(x_n, u_n) \tag{31}$$

which arises by expanding $y(x)$ in a Taylor's series about x_n,

$$y_{n+1} = y_n + hy_n' + \frac{h^2}{2} y_n'' + \cdots$$

and discarding all but the first two terms on the right. The class of functions for which (31) is exact is clearly the class of polynomials of degree ≤ 1, i.e., linear functions. Another example is the trapezoidal rule

$$u_{n+1} = u_n + \frac{h}{2}\{f(x_n, u_n) + f(x_{n+1}, u_{n+1})\} \tag{32}$$

which differs from (31) in the important respect that the unknown u_{n+1} appears on both sides of (32). This means, in practice, that we first guess u_{n+1} by some means, such as (31), substitute the guess on the right side of (32) and get a new guess, continuing to iterate until convergence occurs. Such a formula is called *implicit*, whereas a noniterative formula like (31) is *explicit*. A common feature of (31) and (32) is that in both cases u_{n+1} was calculated from u_n alone, the preceding values $u_{n-1}, u_{n-2}, \ldots$ being ignored. Hence each of these is called a *two-point formula*. In general, a *p-point formula* involves $u_{n+1}, u_n, u_{n-1}, \ldots, u_{n-p+2}$. An example of a three point formula is

$$u_{n+1} = u_{n-1} + 2hf(x_n, u_n). \tag{33}$$

All the formulas so far discussed have in common the fact that u_{n+1} is obtained as a linear combination of $u_n, u_{n-1}, \ldots$ and $f(x_{n+1}, u_{n+1})$, $f(x_n, u_n), \ldots$. Any such formula will be called *Lagrangian*. The most general Lagrangian formula is

$$\begin{aligned} u_{n+1} = {} & a_0 u_n + a_{-1} u_{n-1} + \cdots + a_{-p} u_{n-p} \\ & + h\{b_1 f(x_{n+1}, u_{n+1}) + b_0 f(x_n, u_n) \\ & + \cdots + b_{-p} f(x_{n-p}, u_{n-p})\} \end{aligned} \tag{34}$$

where h is the mesh spacing, and $a_0, \ldots, a_{-p}, b_1, \ldots, b_{-p}$ are given constants. We see that (34) is a $(p + 2)$-point formula which is implicit if $b_1 \neq 0$, explicit if $b_1 = 0$.

Because of their ready adaptability to automatic computing machinery, Lagrangian methods are among the most widely used of all integration formulas. Formerly, such methods were sometimes written in terms of differences because of the convenience of that notation for hand calculation, but for high speed electronic computation it is best to rearrange such formulas in the form (34), and we will not employ the difference notation here.

In the general formula (34) there appear $2p + 3$ preassigned constants $a_0, a_{-1}, \ldots, a_{-p}, b_1, b_0, \ldots, b_{-p}$ which may be chosen to satisfy $2p + 3$ conditions of exactness. For example, the constants could be chosen so as to make (34) exact for polynomials of degree $0, 1, 2, \ldots, 2p + 2$. This, however may be undesirable for a variety of reasons. Among the more obvious are that we may insist that $b_1 = 0$ to avoid iteration or we may want the constants to be simple rational numbers for ease in hand calculations (this is of no importance to a digital computer). Actually, as we shall see presently, the main reason is neither of the above but, roughly speaking, is that we lose control over the nature of the propagation of round off errors, that is, the rapidity with which they build up, if we impose too many conditions of exactness. We may say, in fact, that the requirements of low truncation error and of low round off amplification are opposing conditions, and compromises must be made.

5.6 TRUNCATION ERROR

With reference to the general Lagrangian formula (34), suppose that the backward values are exact, that is

$$u_n = y_n, \quad u_{n-1} = y_{n-1}, \ldots, \quad u_{n-p} = y_{n-p}.$$

Then the difference

$$T_n = y_{n+1} - u_{n+1} \tag{35}$$

is called the single-step truncation of the formula. Clearly this error depends on the unknown function $y(x)$ and on n. We wish to derive an explicit formula for the truncation error in terms of the derivatives of $y(x)$ if they exist. Let $y(x)$ have continuous derivatives to order $n + 1$ in the interval $[a, x]$. Then Taylor's formula is

$$y(x) = y(a) + (x - a)y'(a) + \cdots + \frac{(x - a)^n}{n!} y^{(n)}(a) + R_n \tag{36}$$

where the remainder R_n has the integral form

$$R_n = \frac{1}{n!} \int_a^x y^{(n+1)}(s)(x - s)^n \, ds. \tag{37}$$

Now in terms of the function

$$J_n(t) = \begin{cases} t^n & t \geqq 0 \\ 0 & t < 0 \end{cases} \tag{38}$$

we have

$$R_n = \frac{1}{n!} \int_a^\infty y^{(n+1)}(s) J_n(x - s) \, ds \tag{39}$$

and observe that

$$\frac{\partial}{\partial x} J_n(x - s) = n J_{n-1}(x - s). \tag{40}$$

Now, suppose that the coefficients $a_0, \ldots, a_{-p}, b_1, b_0, \ldots, b_{-p}$ in (34) have been determined so that (34) is exact if $y(x)$ is a polynomial of degree

0, 1, 2, . . . , k, that (34) is not exact for polynomials $y(x)$ of degree $\geqq k + 1$, and that $k \geqq 2$. Finally, let $\mathscr{L}_x\, \varphi(x)$ denote the result of performing the operation on the right side of (34) to the function $\varphi(x)$, that is,

$$\mathscr{L}_x\varphi(x) = a_0\varphi(x_n) + \cdots + a_{-p}\varphi(x_{n-p}) + h\{b_1\varphi'(x_{n+1}) + \cdots + b_{-p}\varphi'(x_{n-p})\}. \tag{41}$$

Then, if $a \leqq x_{n-p}$, from (36),

$$y(x) = P_k(x) + \frac{1}{k!}\int_a^\infty y^{(k+1)}(s)J_k(x - s)\, ds \tag{42}$$

where $P_k(x)$ is a polynomial of degree k. Applying the operator $\mathscr{L}_x$,

$$\begin{aligned} \mathscr{L}_x y(x) &\equiv \mathscr{L}_x P_k(x) + \frac{1}{k!}\int_a^\infty y^{(k+1)}(s)\mathscr{L}_x J_k(x - s)\, ds \\ &= P_k(x + h) + \frac{1}{k!}\int_a^\infty y^{(k+1)}(s)\mathscr{L}_x J_k(x - s)\, ds \end{aligned} \tag{43}$$

since $\mathscr{L}_x$ is exact on $P_k(x)$. Replacing x by $x + h$ in (42) and subtracting (43), we find for the truncation error

$$\begin{aligned} T(x + h) &= y(x + h) - \mathscr{L}_x y(x) \\ &= \frac{1}{k!}\int_a^\infty y^{(k+1)}(s)\{J_k(x + h - s) - \mathscr{L}_x J_k(x - s)\}\, ds \\ &\equiv \frac{1}{k!}\int_a^\infty y^{(k+1)}(s)G_k(s)\, ds \end{aligned} \tag{44}$$

where

$$G_k(s) = J_k(x + h - s) - \mathscr{L}_x J_k(x - s). \tag{45}$$

Now for $s \geqq x + h$, $G_k(s)$ vanishes, since both terms on the right of (45) vanish separately. If $s \leqq x - ph$, then at each of the values of x at which the operator $\mathscr{L}_x$ requires the evaluation of $J_k(x - s)$, its argument, $x - s$, will be positive, $J_k(x - s)$ will equal $(x - s)^k$, the operator $\mathscr{L}_x$ will produce exact results, and $G_k(s)$ will again vanish. Hence $G_k(s)$ vanishes outside the interval $x - ph \leqq s \leqq x + h$. It is convenient to choose $x = 0$, therefore, and define

$$G_k(s) = J_k(h - s) - \mathscr{L}_x J_k(x - s)]_{x=0} \qquad (-ph \leqq s \leqq h) \tag{46}$$

the characteristic function of the Lagrangian formula (34).

Now if $G_k(s)$ does not change sign, we may invoke the mean value theorem in (44), getting

$$\begin{aligned} T_k(h) &= \frac{1}{k!}\int_a^\infty y^{(k+1)}(s)G_k(s)\, ds \\ &= \frac{y^{(k+1)}(\xi)}{k!}\int_{-ph}^h G_k(s)\, ds \\ &= (\text{const.})y^{(k+1)}(\xi) \end{aligned} \tag{47}$$

where $-ph < \xi < h$, and the constant is independent of the function $y(x)$, depending only on the Lagrangian formula used.

A formula for which $G_k(s)$ does not change sign is called *definite*.

Theorem 4. *The single-step truncation error of a definite Lagrangian formula which is exact for* $1, x, \ldots, x^k$ *but not for* x^{k+1} *is*

$$T = (\text{const.})\, y^{(k+1)}(\xi) \tag{48}$$

where ξ lies in the range of points included in the formula, $y(x)$ is any $(k+1)$-times continuously differentiable function, and the constant does not depend on $y(x)$.

We illustrate this result with the trapezoidal rule

$$y_{n+1} = y_n + \frac{h}{2}[y_n' + y_{n+1}'] \tag{49}$$

where $k = 2$, $p = 0$. From (46), if $0 < s < h$, $J_2(h - s) = (h - s)^2$ and

$$\mathscr{L}_x J_2(x - s)]_{x=0} = J_2(-s) + \frac{h}{2}\left[\frac{\partial}{\partial x} J_2(x - s)\bigg|_{x=0} + \frac{\partial}{\partial x} J_2(x - s)\bigg|_{x=h}\right]$$
$$= 0 + \frac{h}{2}[0 + 2(h - s)] = h(h - s).$$

Hence

$$\begin{aligned} G_2(s) &= (h - s)^2 - h(h - s) \\ &= s(s - h) \qquad (0 \leqq s \leqq h) \end{aligned} \tag{50}$$

is the characteristic function of (49), and it does not change sign in $[0, h]$. Hence (49) is definite and (48) holds. From (47) the constant in (48) has the value

$$\frac{1}{2!}\int_0^h (s^2 - hs)\, ds = -\frac{h^3}{12}.$$

and we may complete (49) by writing

$$y_{n+1} = y_n + \frac{h}{2}[y_n' + y_{n+1}'] - \frac{h^3}{12} y'''(\xi) \qquad (x_n < \xi < x_{n+1}). \tag{51}$$

One sees, for instance, that if the mesh size is cut in half, the truncation error is reduced by a factor of eight.

5.7 PREDICTOR-CORRECTOR FORMULAS

If the Lagrangian formula

$$\begin{aligned} u_{n+1} = {} & a_0 u_n + a_{-1} u_{n-1} + \cdots + a_{-p} u_{n-p} \\ & + h\{b_1 f(x_{n+1}, x_{n+1}) + \cdots + b_{-p} f(x_{n-p}, u_{n-p})\} \end{aligned} \tag{52}$$

is implicit ($b_1 \neq 0$), we have already remarked that u_{n+1} must, in general, be found by iteration. In such cases it is desirable to use another formula, the predictor, to predict, or guess, the value u_{n+1}, to substitute this value in the right side of (52), getting a new value of u_{n+1}, and continue iterating to convergence. The predictor formula will of course be explicit and will normally be somewhat less accurate than the corrector formula, for it only provides a guess which is ultimately forgotten. The predictor should be reasonably accurate, however, or else a prohibitive number of iterations may be required for satisfactory convergence.

To study the iterative process, let $u_{n+1}^{(r)}$ denote the rth iterated value of u_{n+1}, $u_{n+1}^{(0)}$ being given from the predictor formula. Then

$$(53)\qquad \begin{aligned} u_{n+1}^{(r+1)} &= a_0 u_n + a_1 u_{n-1} + \cdots + a_{-p} u_{n-p} \\ &\quad + h\{b_1 f(x_{n+1}, u_{n+1}^{(r)}) + \cdots + b_{-p} f(x_{n-p}, u_{n-p})\} \end{aligned}$$

and

$$(54)\qquad \begin{aligned} u_{n+1} &= a_0 u_n + a_{-1} u_{n-1} + \cdots + a_{-p} u_{n-p} \\ &\quad + h\{b_1 f(x_{n+1}, u_{n+1}) + \cdots + b_{-p} f(x_{n-p}, u_{n-p})\} \end{aligned}$$

whence by subtraction,

$$(55)\qquad u_{n+1}^{(r+1)} - u_{n+1} = hb_1\{f(x_{n+1}, u_{n+1}^{(r)}) - f(x_{n+1}, u_{n+1})\}.$$

We can now prove

Theorem 5. *Let $f(x, y)$ satisfy a Lipschitz condition*

$$|f(x, \xi) - f(x, \eta)| \leqq L\,|\xi - \eta|$$

on a line segment $x = x_{n+1}$, $|y - u_{n+1}| \leqq c$. Suppose further that

$$(56)\qquad h < \frac{1}{|b_1|\,L}.$$

and that the point $(x_{n+1}, u_{n+1}^{(0)})$ is on the segment referred to. Then the iterative process (53) *converges to a solution, u_{n+1}, of* (54).

Proof. Put $\gamma_r = |u_{n+1}^{(r)} - u_{n+1}|$. By hypothesis, $\gamma_0 \leqq c$. From (55),

$$(57)\qquad \begin{aligned} \gamma_{r+1} &= |u_{n+1}^{(r+1)} - u_{n+1}| = h\,|b_1|\,|f(x_{n+1}, u_{n+1}^{(r)}) - f(x_{n+1}, u_{n+1})| \\ &\leqq h\,|b_1|\,L|u_{n+1}^{(r)} - u_{n+1}| \\ &= h\,|b_1|\,L\gamma_r \end{aligned}$$

provided $(x_{n+1}, u_{n+1}^{(r)})$ is on the segment. This is true for $r = 0$, and (57) with $r = 0$ shows that $\gamma_1 \leqq (h\,|b_1|\,L)\gamma_0 < c$; hence it is true for $r = 1$, etc., for $r \geqq 0$. But then, from (57),

$$\gamma_r \leqq (h\,|b_1|\,L)^r \gamma_0$$

hence $\gamma_r \to 0$ as $r \to \infty$, i.e., $u_{n+1}^{(r)} \to u_{n+1}$, which was to be shown.

This result brings to light an interesting computational aspect of the use of predictor-corrector formulas. Suppose we wish to integrate an initial-value problem over the range $0 \leqq x \leqq R$, where R is thought of as large. Let a certain predictor-corrector pair be specified to carry out the integration, and let $T(h)$ denote the total computation time required to get from $x = 0$ to $x = R$ if a mesh spacing h is used, R being fixed.

Clearly $T(h) \to +\infty$ as $h \to 0^+$ because the total number of points in the integration is $[R/h]$. On the other hand, there is an h_0 such that $T(h) \to +\infty$ as $h \to h_0^-$ for, if h is too large, the iterative process will fail to converge.

Since $T(h) > 0$ always, it follows that there is an optimal value of h for which the computation time is least. Naturally this optimal h may be unsatisfactory for other reasons, such as truncation or round off error buildup.

In the case of simultaneous equations considerations of convergence become more complicated. If we temporarily use subscripts to index the unknown functions, consider the *linear* system

$$
\begin{aligned}
y_1' &= a_{11}y_1 + a_{12}y_2 + \cdots + a_{1N}y_N \\
&\vdots \\
y_N' &= a_{N1}y_1 + a_{N2}y_2 + \cdots + a_{NN}y_N
\end{aligned}
\tag{58}
$$

where the a_{ij} are constants.

The analogue of (55) is clearly

$$
\begin{pmatrix} y_1^{(r+1)}(x_{n+1}) - y_1(x_{n+1}) \\ \vdots \\ y_N^{(r+1)}(x_{n+1}) - y_N(x_{n+1}) \end{pmatrix}
= hb_1 \begin{pmatrix} a_{11} & a_{12} & \cdots & a_{1N} \\ a_{21} & a_{22} & \cdots & a_{2N} \\ \vdots & & & \vdots \\ a_{N1} & a_{N2} & \cdots & a_{NN} \end{pmatrix}
\begin{pmatrix} y_1^{(r)}(x_{n+1}) - y_1(x_{n+1}) \\ \vdots \\ y_N^{(r)}(x_{n+1}) - y_N(x_{n+1}) \end{pmatrix}
$$

or, with obvious abbreviations,

$$E_{r+1} = hb_1AE_r.$$

For convergence, then, it is necessary and sufficient that the matrix

$$(hb_1A)^r \to 0 \qquad (r \to \infty).$$

Referring to Theorem 30 of Chapter 1, we need the eigenvalues of hb_1A to lie in the unit circle. But these eigenvalues are $hb_1\lambda_1, \ldots, hb_1\lambda_N$, where the eigenvalues of A are $\lambda_1, \ldots, \lambda_N$. Hence for convergence

$$h\,|b_1|\,|\lambda_i| < 1 \qquad (i = 1, 2, \ldots, N)$$

and we have

Theorem 6. *In order that the Lagrangian corrector formula* (52), *when applied to* (58), *should be convergent, it is necessary and sufficient that*

$$h < \frac{1}{|b_1|\,|\lambda_{\max}|}. \tag{59}$$

where $\lambda_{\max}$ is the dominant eigenvalue of A.

For the general *nonlinear* system

$$y_i(x) = f_i(x, y_1(x), \ldots, y_N(x)) \qquad (i = 1, 2, \ldots, N)$$

it is easy to see that if one takes for a_{ij} any numbers such that the partial Lipschitz conditions

$$|f_i(x, y_1, \ldots, y_{j-1}, \xi, y_{j+1}, \ldots, y_N) - f_i(x, y_1, \ldots, y_{j-1}, \eta, y_{j+1}, \ldots, y_N)| \\ \leqq a_{ij}\,|\xi - \eta| \qquad (i, j = 1, 2, \ldots, N)$$

are satisfied in a suitable domain, then (59) gives a sufficient condition for convergence, the proof being the obvious generalization of that of Theorem 5.

5.8 STABILITY

We begin our discussion of stability with a very specific example. Let us suppose that the equation

$$\begin{cases} y'(x) = -\dfrac{1}{L}\,y(x) \\ y(0) = 1 \end{cases} \qquad (L = \text{const.} > 0), \tag{60}$$

whose solution is

$$y(x) = e^{-x/L}, \tag{61}$$

is to be numerically integrated with the formula

$$u_{n+1} = u_{n-3} + \frac{4h}{3}\{2f(x_n, u_n) - f(x_{n-1}, u_{n-1}) + 2f(x_{n-2}, u_{n-2})\}. \tag{62}$$

This is an explicit Lagrangian formula whose truncation error

$$T_n = \frac{28}{90}\,h^5 y^{(5)}(\xi), \tag{63}$$

is quite low, i.e., (62) is quite a good formula from the point of view of accuracy. Now, putting $f(x, y) = -y/L$ in (62) we find

$$u_{n+1} = \frac{4\gamma}{3}\{-2u_n + u_{n-1} - 2u_{n-2}\} + u_{n-3}, \tag{64}$$

where we have set $\gamma = h/L$. We see that (64) is simply a recurrence relation for the numbers u_n. From Section 4.8 of the preceding chapter we know that the solution of (64) is of the form

$$u_n = c_1 r_1{}^n + c_2 r_2{}^n + c_3 r_3{}^n + c_4 r_4{}^n \tag{65}$$

where r_1, r_2, r_3, r_4 are the roots of the equation

$$r^4 + \frac{8\gamma}{3}r^3 - \frac{4\gamma}{3}r^2 + \frac{8\gamma}{3}r - 1 = 0. \tag{66}$$

If $\gamma = 0$, the roots of (66) are at $1, -1, i, -i$. Since these are distinct, it follows that the roots are analytic functions of γ for sufficiently small γ and therefore that we may develop the roots in convergent power series in γ.

Consider the root r_1 which is 1 when $\gamma = 0$. If we substitute the expansion

$$r_1 = 1 + \alpha_1\gamma + \alpha_2\gamma^2 + \cdots \tag{67}$$

into (66), we find easily that

$$r_1 = 1 - \gamma + \frac{\gamma^2}{2!} - \frac{\gamma^3}{3!} + \cdots \tag{68}$$

agrees with the series for $e^{-\gamma}$ through terms of order γ^4. Similarly, we find

$$r_2 = -1 - \frac{5\gamma}{3} + O(\gamma^2) \qquad (\gamma \to 0), \tag{69}$$

$$r_3 = i - \frac{\gamma}{3}i + O(\gamma^2) \qquad (\gamma \to 0), \tag{70}$$

$$r_4 = -i + \frac{\gamma}{3}i + O(\gamma^2) \qquad (\gamma \to 0). \tag{71}$$

Since r_1 is an approximation to $e^{-\gamma}$, $r_1{}^n$ is an approximation to

$$e^{-n\gamma} = e^{-nh/L} = e^{-x_n/L},$$

the exact solution (61). Hence the first term in (65) is a good approximation to (61). The remaining terms in (65) have no significance as far as the differential equation (61) is concerned. They are, so to speak, the price we pay for having such an excellent approximation to $e^{-\gamma}$, in r_1. Ideally, therefore, we would want the three extra terms in (65), the so-called "parasitic solutions," to damp out as $n \to \infty$ more rapidly than the first term.

This is far from being the case, however, since it is clear from (69) that for all sufficiently small γ (i.e., for all sufficiently small h) r_2 is actually outside the unit circle, and therefore the second term in (65) *grows* in magnitude as $n \to \infty$, while the desirable first term *and* the true solution (61) decrease.

Hence, even though the initial data will make c_1 in (65) large compared to c_2, c_3, c_4, as the calculation proceeds the parasitic solution $c_2 r_2{}^n$ will surely eventually dominate, leading to useless results. Furthermore, the introduction of round-off error will at any step tend to be magnified as n increases rather than to be damped out, and for large n, the numerical "solution" may be expected to consist mostly of round off error accumulation and parasitic solutions of exponential growth. The situation cannot be saved by taking a smaller mesh width h, for r_2 lies outside the unit circle and hence dominates r_1, for every fixed $h > 0$. The difficulty is fundamental and lies with the choice of the formula (62), which, we shall say, is unstable.

We now study the result of applying the general Lagrangian formula

$$\begin{aligned} u_{n+1} = a_0 u_n + a_{-1} u_{n-1} + \cdots + a_{-p} u_{n-p} \\ + h\{b_1 f(x_{n+1}, u_{n+1}) + \cdots + b_{-p} f(x_{n-p}, u_{n-p})\} \end{aligned} \tag{72}$$

to the same differential equation (60). Then

$$\begin{aligned} u_{n+1} = a_0 u_n + \cdots + a_{-p} u_{n-p} \\ + h\left\{-\frac{b_1}{L} u_{n+1} - \cdots - \frac{b_{-p}}{L} u_{n-p}\right\} \end{aligned}$$

or

$$(1 + \gamma b_1) u_{n+1} = (a_0 - \gamma b_0) u_n + \cdots + (a_{-p} - \gamma b_{-p}) u_{n-p} \tag{73}$$

where, again, $\gamma = h/L$.

The solution is again of the form

$$u_n = c_1 r_1{}^n + \cdots + c_{p+1} r_{p+1}^n, \tag{74}$$

the r_j being the roots of

$$\psi(r, \gamma) = (1 + \gamma b_1) r^{p+1} - (a_0 - \gamma b_0) r^p - \cdots - (a_{-p} - \gamma b_{-p}) = 0. \tag{75}$$

Now

$$\psi(1, 0) = 1 - a_0 - \cdots - a_{-p}, \tag{76}$$

which, as can be seen by putting $f(x, u) = 0$ in (72), vanishes if the Lagrangian formula (72) is exact for constant $y(x)$. Hence (75) has a root $r = 1$ when $\gamma = 0$. For $\gamma \neq 0$ but small, the *principal root* of (75) will be that root which moves into $r = 1$ when $\gamma = 0$. We call the principal root r_1; then

$$r_1 = 1 + o(1) \qquad (\gamma \to 0). \tag{77}$$

Next,

$$\left.\frac{\partial\psi(r,\gamma)}{\partial r}\right|_{r=1,\gamma=0} = (p+1) - pa_0 - (p-1)a_{-1} - \cdots - a_{-p+1}. \tag{78}$$

To evaluate the right side, suppose (72) is exact for linear functions $y(x)$ also. Putting $f(x, y) = 1$, $y(x) = x$, in (72), there follows

$$x_{n+1} = a_0 x_n + a_{-1} x_{n-1} + \cdots + a_{-p} x_{n-p} + h\{b_1 + \cdots + b_{-p}\}$$

or

$$x_0 + (n+1)h = a_0(x_0 + nh) + \cdots + a_{-p}(x_0 + (n-p)h) + h\{b_1 + \cdots + b_{-p}\}.$$

The terms independent of h vanish because (76) is zero. The coefficient of h must also vanish, which gives

$$\begin{aligned} n + 1 &= na_0 + \cdots + (n-p)a_{-p} + \{b_1 + \cdots + b_{-p}\} \\ &= (n-p)a_{-p} + (n-p+1)a_{-p+1} + \cdots + (n-p+p)a_0 \\ &\quad + \{b_1 + \cdots + b_{-p}\} \\ &= (n-p)(a_{-p} + a_{-p+1} + \cdots + a_0) + a_{-p+1} + \cdots + pa_0 \\ &\quad + \{b_1 + \cdots + b_{-p}\} \\ &= (n-p) + a_{-p+1} + \cdots + pa_0 + (b_1 + \cdots + b_{-p}) \end{aligned}$$

and therefore

$$p + 1 = a_{-p+1} + \cdots + pa_0 + b_1 + \cdots + b_{-p}.$$

Comparing with (78), we find

Theorem 7. *Let the formula* (72) *be exact for linear functions of* x. *Then* $r_1 = 1$ *is a simple root of* (75) *with* $\gamma = 0$, *if and only if*

$$b_1 + \cdots + b_{-p} \neq 0. \tag{77}$$

If (77) is satisfied, then the root $r_1 = r_1(\gamma)$ will be an analytic function of γ for all sufficiently small γ, and therefore the expansion

$$r_1 = 1 + \alpha_1\gamma + \alpha_2\gamma^2 + \cdots \tag{78}$$

will converge. If one substitutes (78) into (75), it is not hard to see that (78) agrees with $e^{-\gamma}$ through terms of order γ^k where k is the largest integer such that (72) correctly integrates the equation $y(x) = kx^{k-1}$, as we saw in (68). Hence $r_1(\gamma)$ is the "desirable" root, the others being parasitic. For stability, we want the parasitic roots to be smaller, in modulus, than $r_1(\gamma)$, for all small enough γ. Since $r_1(0) = 1$, this will surely be the case if the other roots of $\psi(r, 0) = 0$ lie strictly interior to the unit circle.

Hence a conservative definition of stability is

Definition 1. We say that the Lagrangian formula (72) is *stable* if the roots of the polynomial equation

$$\frac{\psi(r, 0)}{r - 1} = \frac{r^{p+1} - a_0 r^p - \cdots - a_{-p}}{r - 1} = 0 \tag{79}$$

lie in the circle $|z| \leq \frac{1}{2}$.

Definition 2. The *radius of stability*, Λ, of (72) is the largest value of $\gamma = h/L$ for which

$$|r_j(\gamma)| \leq r_1(\gamma) \qquad (j = 2, 3, \ldots, p + 1). \tag{80}$$

Let us recall from the test equation (60), that the number L is the so-called relaxation length of the solution, i.e., the distance over which the solution drops by a factor of e. Hence to say that γ is small is to say that the mesh spacing h is small compared to the distance over which the true solution is changing appreciably. Furthermore, since the notion of stability was defined only with reference to the special equation (60), one should not expect stable formulas to behave well in all situations. Nonetheless, because many of the equations arising in practice have solutions which are, at least locally, of exponential character, (60) is perhaps the natural choice of a criterion for stability.

Examples of stable formulas are provided by the Adams-Bashforth family, which are uniquely determined by the conditions that (i) $a_0 = 1$, (ii) $a_{-1} = \cdots = a_{-p} = 0$, and (iii) they are exact for $1, x, \ldots, x^{p+1}$. Indeed, with such a formula, (79) is simply

$$\frac{\psi(r, 0)}{r - 1} = \frac{r^{p+1} - r^p}{r - 1} = r^p = 0,$$

whose roots are all at the origin. The two-point Adams-Bashforth formula is the trapezoidal rule. The three-point formula is

$$u_{n+1} = u_n + h\{\tfrac{5}{12} f(x_{n+1}, u_{n+1}) + \tfrac{2}{3} f(x_n, u_n) - \tfrac{1}{12} f(x_{n-1}, u_{n-1})\} \tag{81}$$

etc. A four-point formula which is stable and which has a somewhat more favorable truncation error than the four-point Adams-Bashforth formula is Hamming's method

$$u_{n+1} = \tfrac{9}{8} u_n - \tfrac{1}{8} u_{n-2} + \frac{3h}{8} \{f(x_{n+1}, u_{n+1}) + 2f(x_n, u_n) - f(x_{n-1}, u_{n-1})\}. \tag{82}$$

In conclusion we remark that the choice of an integration formula involves a compromise between truncation (accuracy), stability, cost (computation time), and round off error. The accuracy and cost both increase with the number of points in the formula. For fixed p, stability decreases with increasing accuracy. As the mesh h is decreased, cost and accuracy increase while round off increases. At the present state of knowledge, the best choice is perhaps more of an art than a science. For additional detail the interested reader is referred to any of the excellent books listed at the end of the chapter.

5.9 LINEAR EQUATIONS OF THE SECOND ORDER

We consider now an equation of the form

$$y''(z) + P(z)y'(z) + Q(z)y(z) = 0 \tag{83}$$

$$y(z_0) = y_0; \quad y'(z_0) = y_0'$$

where it is assumed that z_0 lies in a region $\mathscr{R}$ of the complex plane in which $P(z)$, $Q(z)$ are analytic except for finitely many poles.

Definition 1. z_0 is an *ordinary point* of (83) if $P(z)$ and $Q(z)$ are regular in a neighborhood of z_0.

Definition 2. z_0 is a *regular point* of (83) if $P(z)$, $Q(z)$ have poles at z_0 of order at most one and two, respectively, i.e., if

$$P(z) = O((z - z_0)^{-1}) \qquad (z \to z_0) \tag{84a}$$

$$Q(z) = O((z - z_0)^{-2}) \qquad (z \to z_0). \tag{84b}$$

Definition 3. z_0 is an *irregular point* of (83) if it is neither an ordinary point nor a regular point.

Theorem 8. *Let z_0 be an ordinary point of* (83). *Then there is a unique function $y(z)$ which satisfies* (83) *and is analytic in the largest circle centered at z_0 in which $P(z)$ and $Q(z)$ are analytic.*

The proof will be recognized as a simple variant of that of Picard's theorem.

Proof. We may suppose, without loss of generality, that $z_0 = 0$. In (83), put

$$y(z) = v(z) \exp\left\{-\tfrac{1}{2}\int_0^z P(\zeta)\, d\zeta\right\}. \tag{85}$$

It then takes the form

$$v''(z) + H(z)v(z) = 0 \tag{86}$$

where

$$H(z) = Q(z) - \tfrac{1}{2}P'(z) - \tfrac{1}{4}P^2(z). \tag{87}$$

Define, recursively, the functions

$$v_0(z) = \alpha + \beta z \tag{88}$$

$$v_n(z) = \int_0^z (\zeta - z)H(\zeta)v_{n-1}(\zeta)\,d\zeta \qquad (n = 1, 2, \ldots), \tag{89}$$

where α and β are constants to be determined later.

Now, let G be a circle centered at $z = 0$ in which $P(z)$, $Q(z)$ are regular, and put

$$M = \max_{z \text{ in } G} |H(z)| \tag{90}$$
$$m = \max_{z \text{ in } G} |\alpha + \beta z|.$$

Then we claim that, throughout G,

$$|v_n(z)| \leqq mM^n \frac{|z|^{2n}}{n!} \qquad (n = 0, 1, \ldots). \tag{91}$$

Indeed, this is clear when $n = 0$. If true for $n = 0, 1, \ldots, k - 1$, then

$$\begin{aligned}
|v_k(z)| &= \left| \int_0^z (\zeta - z)H(\zeta)v_{k-1}(\zeta)\,d\zeta \right| \\
&\leqq \frac{mM^{k-1}}{(k-1)!} \int_0^{|z|} |\zeta - z|\; M\; |\zeta|^{2k-2} |d\zeta| \\
&\leqq \frac{mM^k}{(k-1)!} |z| \int_0^{|z|} x^{2k-2}\,dx \\
&= \frac{mM^k}{(k-1)!} |z|^{2k} \frac{1}{2k-1} \\
&< \frac{mM^k |z|^{2k}}{k!},
\end{aligned}$$

the path of integration being the straight line joining 0 and z, and (91) is established. Hence, if ρ is the radius of G,

$$|v_n(z)| \leqq \frac{m(M\rho^2)^n}{n!},$$

and the series of analytic functions

$$v(z) = \sum_{n=0}^{\infty} v_n(z) \tag{92}$$

is absolutely and uniformly convergent, by the Weierstrass M-test, in G. Hence $v(z)$ is analytic in G, and summing both sides of (89) from $n = 1$ to ∞,

$$v(z) = v_0(z) + \int_0^z (\zeta - z)H(\zeta)v(\zeta)\,d\zeta, \tag{93}$$

and differentiating twice, (86) follows. In addition, from (93), $v(0) = \alpha$, $v'(0) = \beta$, while from (85), $y(0) = v(0)$, $y'(0) = v'(0) - \frac{1}{2}P(0)v(0)$. Therefore, if we take $\alpha = y_0$, $\beta = y_0' + \frac{1}{2}P(0)y_0$, the function (85) is a solution of (83) with the required properties. If $\tilde{v}(z)$ is another analytic solution of (86), then $\varphi(z) = v(z) - \tilde{v}(z)$ satisfies

$$\varphi''(z) + H(z)\varphi(z) = 0 \tag{94}$$

$$\varphi(0) = \varphi'(0) = 0. \tag{95}$$

But these imply $\varphi''(0) = 0$, and by differentiating (94) repeatedly and putting $z = 0$, that $\varphi^{(k)}(0) = 0$ for each $k = 0, 1, \ldots$. Hence $\varphi(z) = 0$, and $v(z)$, and therefore $y(z)$, is unique.

5.10 SOLUTION NEAR A REGULAR POINT

If the point z_0 is a regular point of the equation, then, in general, there do not exist solutions analytic in a neighborhood of z_0. Rather, we usually will find a solution with a branch point at z_0. We suppose again that $z_0 = 0$. Then we may multiply (83) through by z^2 and get

$$z^2y''(z) + z(zP(z))y'(z) + z^2Q(z)y(z) = 0 \tag{96}$$

where

$$zP(z) = p_0 + p_1z + p_2z^2 + \cdots \tag{97}$$

$$z^2Q(z) = q_0 + q_1z + q_2z^2 + \cdots \tag{98}$$

the expansions converging in some circle $|z| \leqq \rho$, $\rho > 0$. Let us look for a solution in the form

$$y(z) = z^\lambda\{1 + a_1z + a_2z^2 + \cdots\} \tag{99}$$

where the index λ and coefficients $a_1, a_2, \ldots$ are to be found.

Formal substitution of (99) into (96) yields

$$\begin{aligned} z^\lambda\left\{\lambda(\lambda - 1) + \sum_{n=1}^{\infty} a_n(n + \lambda)(n + \lambda - 1)z^n\right\} \\ + z^\lambda\{zP(z)\}\left\{\lambda + \sum_{n=1}^{\infty} a_n(n + \lambda)z^n\right\} \\ + z^\lambda\{z^2Q(z)\}\left\{1 + \sum_{n=1}^{\infty} a_nz^n\right\} = 0. \end{aligned} \tag{100}$$

If we next substitute (97), (98) into (100), carry out the multiplications of series indicated, and set the coefficient of each power of z to equal to zero, we obtain the formulas

$$\lambda^2 + (p_0 - 1)\lambda + q_0 = 0 \tag{101}$$

$$a_1\{q_0 + (p_0 - 1)(\lambda + 1) + (\lambda + 1)^2\} + \lambda p_1 + q_1 = 0 \tag{102}$$

etc., the general equation being

$$\{(n + \lambda)^2 + (p_0 - 1)(n + \lambda) + q_0\}a_n + \sum_{k=1}^{n-1} a_{n-k}\{(n + \lambda - k)p_k + q_k\} + \lambda p_n + q_n = 0 \qquad (n = 1, 2, \ldots). \tag{103}$$

Equation (101) is a quadratic equation in λ, the so-called *indicial equation*, and determines two (possibly equal) values of λ, say λ_1, λ_2. If one of these values is chosen, the succeeding equations (103) determine $a_1, a_2, a_3, \ldots$ recursively, provided that the coefficient of a_n in (103) is never zero.

But this coefficient is just $\varphi(\lambda + n)$, where $\varphi(\lambda)$ is the left-hand side of (101). If $\lambda = \lambda_1$, the root of (101) with larger real part, then, clearly, $\varphi(\lambda_1 + n) \neq 0$, and all the a_n can be found. For $\lambda = \lambda_2$, to say that $\varphi(\lambda_2 + n) = 0$ is to say that $\lambda_2 + n = \lambda_1$, i.e., that $\lambda_1 - \lambda_2$ is an integer. We may summarize by saying that (96) is satisfied by two different formal power series of the type (99) if the roots of (101) do not differ by an integer ($= 0, 1, 2, \ldots$), with only one formal solution of that type otherwise, corresponding to the root λ_1 with larger real part.

5.11 CONVERGENCE OF THE FORMAL SOLUTION

To justify the formal procedure of the preceding section it is enough to show that the series in (99) converges in some circle of positive radius, centered at the origin, to an analytic function. Again, let $\lambda_1, \lambda_2, \mathbf{Re}\,\lambda_1 \geqq \mathbf{Re}\,\lambda_2$ be the solutions of the indicial equation $\varphi(\lambda) = 0$. If $\lambda_1 - \lambda_2 = \sigma$, then

$$\begin{aligned} \varphi(\lambda_1 + n) &= (\lambda_1 + n)^2 + (p_0 - 1)(\lambda_1 + n) + q \\ &= (\lambda_1^2 + (p_0 - 1)\lambda_1 + q_0) + (2\lambda_1 + p_0 - 1)n + n^2 \\ &= n\{n + 2\lambda_1 + p_0 - 1\} \\ &= n(n + \sigma). \end{aligned} \tag{104}$$

Next, by Cauchy's inequality, there is a number $M > 1$ such that

$$|p_n| \leqq \frac{M}{r^n}; \quad |q_n| \leqq \frac{M}{r^n}; \quad |\lambda_1 p_n + q_n| \leqq \frac{M}{r^n} \qquad (n = 0, 1, 2, \ldots) \tag{105}$$

for all small enough r.

Now, take $\lambda = \lambda_1$, and let the numbers $a_1, a_2, \ldots$ be recursively calculated from (103). We claim that

$$|a_n| \leqq \left(\frac{M}{r}\right)^n. \tag{106}$$

Indeed, for $n = 1$,

$$\begin{aligned} |a_1| &= \left| \frac{q_1 + \lambda_1 p_1}{\varphi(\lambda_1 + 1)} \right| \\ &\leqq \frac{M}{r|1 + \sigma|} \\ &\leqq \frac{M}{r} \end{aligned}$$

since $\mathbf{Re}\ \sigma \geqq 0$. Inductively, suppose (106) true for $n = 1, 2, \ldots, k - 1$. Then

$$\begin{aligned} |a_k| &= \left| \frac{\sum_{\nu=1}^{k-1} a_{k-\nu}[(\lambda_1 + k - \nu]p_\nu + q_\nu] + \lambda_1 p_k + q_k}{\varphi(\lambda_1 + k)} \right| \\ &\leqq \frac{1}{k|k + \sigma|} \left\{ \sum_{\nu=1}^{k-1} |\lambda_1 p_\nu + q_\nu|\, |a_{k-\nu}| \right. \\ &\qquad \left. + \sum_{\nu=1}^{k-1} (k - \nu)|a_{k-\nu}|\, |p_\nu| + |\lambda_1 p_k + q_k| \right\} \\ &\leqq \frac{1}{k|k + \sigma|} \left\{ \sum_{\nu=1}^{k-1} \frac{M}{r^\nu} \frac{M^{k-\nu}}{r^{k-\nu}} + \sum_{\nu=1}^{k-1} (k - \nu) \frac{M^{k-\nu}}{r^{k-\nu}} \frac{M}{r^\nu} + \frac{M}{r^k} \right\} \\ &= \frac{M^k}{r^k k|k + \sigma|} \left\{ \sum_{\nu=1}^{k-1} \frac{1}{M^{\nu-1}} + \sum_{\nu=1}^{k-1} (k - \nu) \frac{1}{M^{\nu-1}} + \frac{1}{M^{k-1}} \right\} \\ &\leqq \frac{M^k}{r^k k|k + \sigma|} \left\{ (k - 1) + \frac{k(k - 1)}{2} + 1 \right\} \\ &= \frac{(k + 1)}{2k|1 + \sigma/k|} \left(\frac{M}{r}\right)^k \\ &\leqq \frac{k + 1}{2k} \left(\frac{M}{r}\right)^k \\ &\leqq \left(\frac{M}{r}\right)^k \end{aligned}$$

where we have used the fact that $\mathbf{Re}\ \sigma > 0$ implies $|1 + \sigma/k| \geqq 1$, and (106) is proved for every n. By comparison with the geometric series, then, the series in (99) converges uniformly for $|z| < r/M$, and therefore the function

$y(z)$ in (99) is analytic in that circle aside from a branch point at $z = 0$ if λ_1 is not an integer.

The same analysis can be made for the smaller exponent with the same result if σ is not a positive integer.

Theorem 9. *If $z = 0$ is a regular point of* (96) *and λ_1, λ_2,* **Re** $(\lambda_1 - \lambda_2) \geqq 0$ *are the roots of* (101), *then there are either one or two solutions of* (96) *in the form* (99), *depending on whether $\lambda_1 - \lambda_2$ is or is not a non-negative integer, respectively. A solution so obtained is analytic in a circle about $z = 0$ except for a branch point at the origin if λ is not an integer.*

5.12 A SECOND SOLUTION IN THE EXCEPTIONAL CASE

Suppose now that $\lambda_1 - \lambda_2$ is a non-negative integer n. The method of the preceding section then gives us one solution, $y_1(z)$, corresponding to λ_1. Returning to (96),

$$z^2y''(z) + z(zP(z))y'(z) + (z^2Q(z))y(z) = 0$$

put

$$y(z) = y_1(z)u(z) \tag{107a}$$

where $y_1(z)$ is the solution already found. Then the equation satisfied by $u(z)$ is

$$u''(z) + \left\{P(z) + 2\,\frac{y_1'(z)}{y_1(z)}\right\} u'(z) = 0 \tag{107b}$$

which is of the first order in $u'(z)$, and hence easily solved. The solution is

$$\begin{aligned}
u(z) &= c_0 + c_1\int^z \frac{1}{y_1^2(\zeta)} \exp\left\{-\int^\zeta P(\zeta')\, d\zeta'\right\} d\zeta \qquad (107c)\\
&= c_0 + c_1\int^z \frac{1}{y_1^2(\zeta)} \exp\left\{-\int^\zeta \left[\frac{p_0}{\zeta'} + p_1 + \cdots\right] d\zeta'\right\} d\zeta\\
&= c_0 + c_1\int^z \frac{\zeta^{-p_0}}{y_1^2(\zeta)} \exp\left\{-\int^\zeta [p_1 + p_2\zeta' + \cdots]\, d\zeta'\right\} d\zeta\\
&= c_0 + c_1\int^z \frac{\zeta^{-p_0}}{[\zeta^{\lambda_1}(1 + a_1\zeta + \ldots)]^2} \exp\left\{-\int^\zeta [p_1 + p_2\zeta'\right.\\
&\qquad\qquad \left. + \cdots]\, d\zeta'\right\} d\zeta\\
&= c_0 + c_1\int^z \zeta^{-p_0-2\lambda_1}\, h(\zeta)\, d\zeta
\end{aligned}$$

where we have written

$$h(\zeta) = [1 + a_1\zeta + \cdots]^{-2} \exp\left\{-\int^{\zeta} [p_1 + p_2\zeta' + \cdots] d\zeta'\right\} \tag{108}$$

which is analytic in any circle centered at the origin in which $zP(z)$ is regular. Now, suppose

$$h(\zeta) = 1 + \sum_{\nu=1}^{\infty} h_\nu \zeta^\nu \tag{109}$$

and recall that $p_0 + 2\lambda_1 = 1 + \lambda_1 - \lambda_2 = n + 1$. Then (107) gives, if $n \neq 0$,

$$\begin{aligned} u(z) &= c_0 + c_1 \int^z \zeta^{-(n+1)} \left\{1 + \sum_{\nu=1}^{\infty} h_\nu \zeta^\nu\right\} d\zeta \\ &= c_0 + c_1 \left\{-\frac{1}{nz^n} - \sum_{\nu=1}^{n-1} \frac{h_\nu}{n - \nu} z^{\nu - n} + h_n \log z \right. \\ &\qquad \left. + \sum_{\nu=n+1}^{\infty} \frac{h_\nu}{\nu - n} z^{\nu - n}\right\}. \end{aligned}$$

Referring back to the substitution (107), our second solution is

$$\begin{aligned} y(z) &= y_1(z)u(z) \\ &= c_0 y_1(z) + c_1 z^{\lambda_1}[1 + a_1 z + \cdots] \\ &\quad \times \left\{-\frac{1}{nz^n} - \sum_{\nu=1}^{n-1} \frac{h_\nu}{n - \nu} z^{\nu - n} + h_n \log z \right. \\ &\quad \left. + \sum_{\nu=n+1}^{\infty} \frac{h_\nu}{\nu - n} z^{\nu - n}\right\} \\ &= c_0 y_1(z) + c_1 y_1(z) h_n \log z \\ &\quad + c_1[1 + a_1 z + a_2 z^2 + \cdots]\left\{-\frac{z^{\lambda_2}}{n} - \sum_{\nu=1}^{n-1} \frac{h_\nu}{n - \nu} z^{\nu + \lambda_2} \right. \\ &\quad \left. + \sum_{\nu=n+1}^{\infty} \frac{h_\nu}{\nu - n} z^{\nu + \lambda_2}\right\} \\ &= c_0 y_1(z) + c_1 y_1(z) h_n \log z + c_1 z^{\lambda_2}\left\{-\frac{1}{n} + \sum_{\nu=1}^{\infty} \gamma_\nu z^\nu\right\}. \end{aligned}$$

If $n = 0$, the same calculation shows

Theorem 10. *If the roots of the indicial equation differ by an integer n, the general solution of* (96) *is of the form*

$$y(z) = c_0 y_1(z) + c_1\{h_n y_1(z) \log z + z^{\lambda_2}\psi(z)\} \tag{110}$$

if $n \neq 0$, *where* $\psi(z)$ *is analytic and* $\psi(0) = -1/n$. *If* $n = 0$, *the solution is*

$$y(z) = c_0 y_1(z) + c_1\{y_1(z) \log z + z^{\lambda_2}\psi(z)\} \tag{111}$$

where $\psi(z)$ *is analytic and* $\psi(0) = 0$.

5.13 THE GAMMA FUNCTION

As before, let

$$\gamma = \lim_{n\to\infty} \left\{1 + \frac{1}{2} + \cdots + \frac{1}{n} - \log n\right\} \tag{112}$$

denote Euler's constant, and consider the product

$$P(z) = \prod_{n=1}^{\infty} \left\{\left(1 + \frac{z}{n}\right)e^{-z/n}\right\}. \tag{113}$$

We claim that this product converges, for every z, uniformly in any finite circle. Indeed, let N be a fixed integer, suppose $|z| \leqq N/2$, and that $n > N$. Then

$$\begin{aligned}\left|\log\left(1 + \frac{z}{n}\right) - \frac{z}{n}\right| &= \left|-\frac{z^2}{2n^2} + \frac{z^3}{3n^3} - \cdots\right| \\ &\leqq \frac{|z|^2}{n^2}\left\{1 + \left|\frac{z}{n}\right| + \left|\frac{z}{n}\right|^2 + \cdots\right\} \\ &\leqq \frac{N^2}{4n^2}\left\{1 + \frac{1}{2} + \frac{1}{4} + \cdots\right\} \\ &= \frac{N^2}{2n^2}.\end{aligned}$$

Now

$$\sum_{n=N+1}^{\infty} \frac{N^2}{2n^2} < \infty,$$

and therefore the series

$$\sum_{n=N+1}^{\infty} \left\{\log\left(1 + \frac{z}{n}\right) - \frac{z}{n}\right\}$$

converges absolutely and uniformly for $|z| \leqq N/2$. Hence so does the series

$$\sum_{n=1}^{\infty} \left\{\log\left(1 + \frac{z}{n}\right) - \frac{z}{n}\right\} \tag{114}$$

converge to a function which is analytic for $|z| \leqq N/2$. Therefore the exponential of (114), namely,

$$\prod_{n=1}^{\infty} \left\{\left(1 + \frac{z}{n}\right)e^{-z/n}\right\}, \tag{115}$$

is also analytic for $|z| \leqq N/2$. But N was arbitrary; hence the product in (113) converges everywhere and so represents an entire function. We define the Gamma function $\Gamma(z)$ by

$$\frac{1}{\Gamma(z)} = ze^{\gamma z} \prod_{n=1}^{\infty} \left\{\left(1 + \frac{z}{n}\right)e^{-z/n}\right\} \tag{116}$$

and we have proved

Theorem 10. *The function* $1/\Gamma(z)$, *defined by* (116), *is an entire function.*

It follows that $\Gamma(z)$ itself is analytic except at the points where $1/\Gamma(z)$ is zero. From (116), these points are clearly at $z = 0, -1, -2, -3, \ldots$, and we have

Theorem 11. *The function* $\Gamma(z)$ *itself is regular everywhere except at* $z = 0, -1, -2, \ldots$ *where it has simple poles.*

The product in (116) is the Weierstrass product for $1/\Gamma(z)$. Another product representation for this function is due to Euler and is obtained from (116) as follows:
From (112) and (116),

(117)

$$\begin{aligned}
\frac{1}{\Gamma(z)} &= ze^{\gamma z}\prod_{n=1}^{\infty}\left\{\left(1+\frac{z}{n}\right)e^{-z/n}\right\} \\
&= z\left\{\lim_{N\to\infty}\exp\left[1+\frac{1}{2}+\cdots+\frac{1}{N}-\log N\right]z\right\} \\
&\quad\times\left\{\lim_{N\to\infty}\prod_{n=1}^{N}\left[\left(1+\frac{z}{n}\right)e^{-z/n}\right]\right\} \\
&= z\lim_{N\to\infty}\left\{\exp\left[1+\frac{1}{2}+\cdots+\frac{1}{N}-\log N\right]z\right\}\prod_{n=1}^{N}\left\{\left(1+\frac{z}{n}\right)e^{-z/n}\right\} \\
&= z\lim_{N\to\infty}\left\{\exp z\left[1+\frac{1}{2}+\cdots+\frac{1}{N}-\log N\right]\left[\prod_{n=1}^{N}\left(1+\frac{z}{n}\right)\right]\right. \\
&\qquad\qquad \left.\exp -z\left[1+\frac{1}{2}+\cdots+\frac{1}{N}\right]\right\} \\
&= z\lim_{N\to\infty}\left\{e^{-z\log N}\prod_{n=1}^{N}\left(1+\frac{z}{n}\right)\right\} \\
&= z\lim_{N\to\infty}\left\{N^{-z}\prod_{n=1}^{N}\left(1+\frac{z}{n}\right)\right\}
\end{aligned}$$

However, from the identity

$$\begin{aligned}
\prod_{n=1}^{N-1}\left(1+\frac{1}{n}\right) &= (1+1)\left(1+\frac{1}{2}\right)\left(1+\frac{1}{3}\right)\cdots\left(1+\frac{1}{N-1}\right) \\
&= 2\cdot\frac{3}{2}\cdot\frac{4}{3}\cdots\frac{N}{N-1} \\
&= N
\end{aligned}$$

(117) gives

$$\frac{1}{\Gamma(z)} = z \lim_{N\to\infty} \left\{ \prod_{n=1}^{N-1} \left(1 + \frac{1}{n}\right)^{-z} \prod_{n=1}^{N} \left(1 + \frac{z}{n}\right) \right\}$$

$$= z \lim_{N\to\infty} \left\{ \prod_{n=1}^{N} \left[\left(1 + \frac{1}{n}\right)^{-z} \left(1 + \frac{z}{n}\right) \right] \left(1 + \frac{1}{N}\right)^{z} \right\}$$

$$= z \prod_{n=1}^{\infty} \left\{ \left(1 + \frac{1}{n}\right)^{-z} \left(1 + \frac{z}{n}\right) \right\}$$

or

$$\Gamma(z) = \frac{1}{z} \prod_{n=1}^{\infty} \left\{ \left(1 + \frac{1}{n}\right)^{z} \left(1 + \frac{z}{n}\right)^{-1} \right\}, \tag{118}$$

which is Euler's result.

We next derive the functional equation of $\Gamma(z)$.

Theorem 12. *If z is not a negative integer, then*

$$\Gamma(z + 1) = z\Gamma(z). \tag{119}$$

Proof. In (118), replace z by $z + 1$, and divide, getting

$$\frac{\Gamma(z+1)}{\Gamma(z)} = \frac{z}{z+1} \lim_{N\to\infty} \frac{\prod_{n=1}^{N} \left\{ \left(1 + \frac{1}{n}\right)^{z+1} \left(1 + \frac{z+1}{n}\right)^{-1} \right\}}{\prod_{n=1}^{N} \left\{ \left(1 + \frac{1}{n}\right)^{z} \left(1 + \frac{z}{n}\right)^{-1} \right\}}$$

$$= \frac{z}{z+1} \lim_{N\to\infty} \prod_{n=1}^{N} \left\{ \left(1 + \frac{1}{n}\right) \frac{n+z}{n+1+z} \right\}$$

$$= \frac{z}{z+1} \lim_{N\to\infty} \left\{ \left[2 \cdot \frac{z+1}{z+2} \right] \left[\frac{3}{2} \frac{z+2}{z+3} \right] \cdots \left[\frac{N+1}{N} \frac{z+N}{z+N+1} \right] \right\}$$

$$= \frac{z}{z+1} \lim_{N\to\infty} \left\{ \frac{(N+1)(z+1)}{z+N+1} \right\}$$

$$= z \lim_{N\to\infty} \frac{N+1}{z+N+1}$$

$$= z.$$

Theorem 13. *If n is a positive integer, $\Gamma(n) = (n-1)!$*

Proof. From (118) it is obvious that $\Gamma(1) = 1$. If the result has been proved for $n = 1, 2, \ldots, k$, then $\Gamma(k + 1) = k\Gamma(k) = k(k - 1)! = k!$

Next we have a formula for reflecting $\Gamma(z)$ in the line $\mathbf{Re}\, z = \frac{1}{2}$.

Theorem 14. *If z is not an integer,*

$$\Gamma(z)\Gamma(1 - z) = \frac{\pi}{\sin \pi z}. \tag{120}$$

Proof. From (116),

$$\begin{aligned}\Gamma(z)\Gamma(-z) &= -\frac{1}{z^2}\prod_{n=1}^{\infty}\left\{\left(1+\frac{z}{n}\right)e^{-z/n}\right\}^{-1}\prod_{n=1}^{\infty}\left\{\left(1-\frac{z}{n}\right)e^{z/n}\right\}^{-1}\\ &= -\frac{1}{z^2}\prod_{n=1}^{\infty}\left(1-\frac{z^2}{n^2}\right)^{-1}\\ &= -\frac{1}{z^2}\left\{\prod_{n=1}^{\infty}\left(1-\frac{z^2}{n^2}\right)\right\}^{-1}\\ &= -\frac{1}{z^2}\left\{\frac{\sin \pi z}{\pi z}\right\}^{-1} = -\frac{\pi}{z\sin \pi z}.\end{aligned}$$

Now (119) gives

$$\Gamma(1-z) = -z\Gamma(-z)$$

and the result follows.

Also due to Euler is

Theorem 15. *If* z *is not* $0, -1, -2, \ldots$, *then*

$$\Gamma(z) = \lim_{n\to\infty}\left\{\frac{1\cdot 2\cdots(n-1)}{z(z+1)\cdots(z+n-1)}n^z\right\}. \tag{121}$$

Proof. We have

$$\begin{aligned}&\lim_{n\to\infty}\left\{\frac{1\cdot 2\cdots(n-1)}{z(z+1)\cdots(z+n-1)}n^z\right\}\\ &\qquad= \lim_{n\to\infty}\left\{\frac{1\cdot 2\cdots(n-1)}{z(z+1)\cdots(z+n-1)}\frac{2^z}{1^z}\frac{3^z}{2^z}\cdots\frac{n^z}{(n-1)^z}\right\}\\ &\qquad= \lim_{n\to\infty}\frac{1}{z}\prod_{\nu=1}^{n-1}\left\{\frac{\nu}{z+\nu}\frac{(\nu+1)^z}{\nu^z}\right\}\\ &\qquad= \frac{1}{z}\prod_{\nu=1}^{\infty}\left\{\left(1+\frac{1}{\nu}\right)^z\left(1+\frac{z}{\nu}\right)^{-1}\right\}\\ &\qquad= \Gamma(z)\end{aligned}$$

by (118).

Next we have an integral representation in

Theorem 16. *If* **Re** $z > 0$, *then*

$$\Gamma(z) = \int_0^\infty e^{-t}t^{z-1}\,d\tau \tag{122}$$

where t^{z-1} *means* $\exp\{(z-1)\log t\}$.

To prove this we shall need two preliminary results.

Lemma 1. For $0 \leqq t \leqq n$, $n \geqq 2$, we have

$$\psi(t) = e^t\left(1-\frac{t}{n}\right)^{n-1} \leqq \frac{e}{2}.$$

Proof. For,

$$\frac{\psi'(t)}{\psi(t)} = \frac{1-t}{n-t},$$

hence $\psi'(t)$ vanishes only at $t = 1$. The maximum of $\psi(t)$ is the largest of the three numbers $\psi(0) = 1$, $\psi(1) = e(1 - 1/n)^{n-1}$, $\psi(n) = 0$. Hence $\psi(t) \leqq \psi(1) \leqq e/2$ for $n \geqq 2$.

Lemma 2. For $0 \leqq t \leqq n$, $n \geqq 2$, we have

$$0 \leqq e^{-t} - \left(1 - \frac{t}{n}\right)^n \leqq \frac{e}{4}\frac{t^2}{n} e^{-t}. \tag{123}$$

Proof. Let

$$\varphi(t) = e^{-t} - \left(1 - \frac{t}{n}\right)^n.$$

Then

$$\varphi'(t) + \varphi(t) = \frac{t}{n}\left(1 - \frac{t}{n}\right)^{n-1},$$

and

$$\begin{aligned}\varphi(t) &= \frac{e^{-t}}{n}\int_0^t \tau e^{\tau}\left(1 - \frac{\tau}{n}\right)^{n-1} d\tau \\ &= \frac{e^{-t}}{n}\int_0^t \tau\psi(\tau)\, d\tau \\ &\leqq \frac{e^{-t}}{n}\frac{e}{2}\int_0^t \tau\, d\tau \\ &= \frac{et^2}{4n} e^{-t}\end{aligned}$$

by Lemma 1.

Proof of Theorem 16.

Let

$$\begin{aligned}H_n(z) &= \int_0^n \left(1 - \frac{t}{n}\right)^n t^{z-1}\, dt \\ &= n^z \int_0^1 (1 - y)^n y^{z-1}\, dy.\end{aligned} \tag{124}$$

Integrating by parts repeatedly,

$$\begin{aligned}\int_0^1 (1-y)^n y^{z-1}\, dy &= \left[\frac{1}{z} y^z (1-y)^n\right]_{y=0}^1 + \frac{n}{z}\int_0^1 (1-y)^{n-1} y^z\, dy \\ &= \cdots \\ &= \frac{n!}{z(z+1)\cdots(z+n-1)}\int_0^1 y^{n+z-1}\, dy\end{aligned}$$

and therefore

$$H_n(z) = \frac{n!}{z(z+1)\cdots(z+n)}\, n^z. \tag{125}$$

By (121), then,

$$\begin{aligned} \lim_{n\to\infty} H_n(z) &= \Gamma(z) \\ &= \lim_{n\to\infty} \int_0^n \left(1 - \frac{t}{n}\right)^n t^{z-1}\, dt. \end{aligned} \tag{126}$$

Define, for $\mathbf{Re}\, z > 0$,

$$\tilde{\Gamma}(z) = \int_0^\infty e^{-t} t^{z-1}\, dt,$$

and we will show that $\tilde{\Gamma}(z) = \Gamma(z)$. Indeed, subtracting from (126),

$$\begin{aligned} \tilde{\Gamma}(z) - \Gamma(z) &= \lim_{n\to\infty} \left\{ \int_0^n t^{z-1}\left[e^{-t} - \left(1 - \frac{t}{n}\right)^n\right] dt + \int_n^\infty e^{-t} t^{z-1}\, dt \right\} \\ &= \lim_{n\to\infty} \left\{ O\left(n^{-1} \int_0^n t^{z+1} e^{-t}\, dt\right) + o(1) \right\} \\ &= \lim_{n\to\infty} \{O(n^{-1}) + o(1)\} = 0, \end{aligned}$$

completing the proof.

Next, let us take the logarithm of (116) and differentiate with respect to z, getting

$$\frac{\Gamma'(z)}{\Gamma(z)} = -\gamma - \frac{1}{z} + \lim_{n\to\infty} \sum_{m=1}^{n} \left(\frac{1}{m} - \frac{1}{z+m}\right). \tag{127}$$

The function on the left is frequently denoted by $\psi(z)$. If z is an integer, say $z = p$, the series on the right telescopes,

$$\begin{aligned} \psi(p) &= -\gamma - \frac{1}{p} + \lim_{n\to\infty} \left\{ 1 + \frac{1}{2} + \cdots + \frac{1}{n} - \frac{1}{p+1} - \cdots - \frac{1}{p+n} \right\} \\ &= -\gamma - \frac{1}{p} + \lim_{n\to\infty} \left\{ 1 + \frac{1}{2} + \cdots + \frac{1}{p} - \frac{1}{n+1} - \cdots - \frac{1}{n+p} \right\} \\ &= -\gamma + 1 + \frac{1}{2} + \cdots + \frac{1}{p-1} \end{aligned}$$

and

$$\psi(p+q) - \psi(p) = \frac{1}{p} + \cdots + \frac{1}{p+q-1} = \sum_{j=0}^{q-1} \frac{1}{p+j}. \tag{128}$$

This last relation is often valuable in dealing with finite sums of the kind occurring on the right side, when used in conjunction with tables of the ψ-function. We have also encountered the ψ-function in the previous chapter, (95), (96), in connection with asymptotic expansions.

Finally, we note a few special values of the Gamma function. First, putting $z = \frac{1}{2}$ in (120), we have

$$\Gamma\left(\frac{1}{2}\right) = \sqrt{\pi}. \tag{129}$$

Then the functional equation yields

$$\Gamma\left(n + \frac{1}{2}\right) = \frac{1 \cdot 3 \cdots (2n - 1)}{2^n} \sqrt{\pi} \qquad (n = 1, 2, \ldots). \tag{130}$$

Next, (120) gives

$$\Gamma\left(\frac{1}{2} - n\right)\Gamma\left(n + \frac{1}{2}\right) = (-1)^n \pi,$$

and from (130),

$$\Gamma\left(\frac{1}{2} - n\right) = \frac{(-2)^n \sqrt{\pi}}{1 \cdot 3 \cdot 5 \cdots (2n - 1)} \qquad (n = 1, 2, \ldots). \tag{131}$$

On the imaginary axis, let $z = it$, where t is real. Then from

$$\Gamma(z)\Gamma(1 - z) = \Gamma(z)\{-z\Gamma(-z)\} = \frac{\pi}{\sin \pi z}$$

we find by putting $z = it$,

$$\Gamma(it)\Gamma(-it) = \frac{\pi}{t \sinh \pi t} = \Gamma(it)\overline{\Gamma(it)}$$
$$= |\Gamma(it)|^2,$$

and therefore

$$|\Gamma(it)| = \sqrt{\frac{\pi}{t \sinh \pi t}} \qquad (t \text{ real}). \tag{132}$$

5.14 BESSEL FUNCTIONS

We consider now the differential equation

$$x^2 \frac{d^2y}{dx^2} + x \frac{dy}{dx} + (x^2 - p^2)y = 0 \tag{133}$$

known as Bessel's equation. This is of the form (83), with

$$P(z) = \frac{1}{z}; \quad Q(z) = 1 - \frac{p^2}{z^2}, \tag{134}$$

and we see that every point is an ordinary point of (133) except $z = 0$, which is a regular point. To solve (133) by the method of Frobenius, we first have equations (97), (98) in the form

$$zP(z) = 1 \tag{135}$$

$$z^2Q(z) = -p^2 + z^2 \tag{136}$$

whence $p_0 = 1$, $p_j = 0(j \geqq 1)$, $q_0 = -p^2$, $q_1 = 0$, $q_2 = 1$, $q_j = 0(j \geqq 3)$. The indicial equation (101) is

$$\lambda^2 - p^2 = 0 \tag{137}$$

so the indices are $\lambda_1 = p$, $\lambda_2 = -p$, and we are sure of one solution corresponding to $\lambda_1 = p$, whereas the exceptional case when $\lambda_1 - \lambda_2 = 2p$ is an integer will have to be studied separately.

To get the solution corresponding to $\lambda_1 = p$, the recurrence (103) is in this case

$$[(s+p)^2 - p^2]a_s + \sum_{k=1}^{s-1} a_{s-k}[(s+p-k)p_k + q_k] + pp_s + q_s = 0 \qquad (s = 1, 2, \ldots)$$

and putting in the values of the p_k, q_k,

$$s(2p+s)a_s + a_{s-2} + pp_s + q_s = 0 \qquad (s \geqq 2). \tag{138}$$

Now these equations can be satisfied by taking a_0 arbitrary, $a_1 = 0$, and determining the succeeding coefficients recursively. However, it is clear that we shall find $a_1 = a_3 = a_5 = \cdots = 0$, and only the even-indexed coefficients remain. With $s = 2$,

$$a_2 = -\frac{1 + a_0}{4(p+1)},$$

while for $s > 2$, (138) is simply

$$a_s = -\frac{a_{s-2}}{s(2p+s)}.$$

From these it is easy to see that

$$a_{2k} = (-1)^k \frac{(a_0 + 1)}{k!\, 4^k(p+1)\cdots(p+k)} \qquad (k = 1, 2, \ldots). \tag{139}$$

Since

$$(p+1)(p+2)\cdots(p+k) = \frac{\Gamma(p+k+1)}{\Gamma(p+1)},$$

we have found a solution of (133) in the form

$$\begin{aligned} y(x) &= (1 + a_0)\Gamma(p+1) \sum_{k=0}^{\infty} \frac{(-1)^k x^{2k+p}}{4^k k!\, \Gamma(k+p+1)} \\ &= 2^p(1 + a_0)\Gamma(p+1) \sum_{k=0}^{\infty} \frac{(-1)^k (x/2)^{2k+p}}{k!\, \Gamma(k+p+1)} \\ &= (\text{const.})\, J_p(x) \end{aligned} \tag{140}$$

where

$$J_p(x) = \sum_{k=0}^{\infty} \frac{(-1)^k (x/2)^{2k+p}}{k!\, \Gamma(k+p+1)} \tag{141}$$

is the Bessel function of the first kind, of order p. It is easy to see that, for p fixed, the series in (141) has radius of convergence $+\infty$ and therefore represents a function of x which is everywhere regular except, if p is not an integer, for a branch point at the origin. On the other hand, if x is fixed, we know that $[\Gamma(z)]^{-1}$ is an entire function and therefore, considered as a function of p, (141) is a uniformly convergent series of analytic functions and is therefore analytic. We state this as

Theorem 17. *The function $z^{-p}J_p(z)$ is, for fixed p, an entire function of z, and for fixed z, an entire function of p.*

If p is not an integer, the procedure yields another solution corresponding to $\lambda_2 = -p$, which is

$$J_{-p}(x) = \sum_{k=0}^{\infty} \frac{(-1)^k(x/2)^{2k-p}}{k!\,\Gamma(k-p+1)} \tag{142}$$

and the general solution of (133), when p is not an integer, is therefore

$$y(x) = c_1J_p(x) + c_2J_{-p}(x).$$

If $p = 0$, (141) and (142) are obviously identical, whereas, if p is a positive integer, recalling that $[\Gamma(z)]^{-1}$ vanishes at the negative integers, we see that the terms $k = 0, 1, \ldots, p - 1$ do not contribute in (142), and we have, for integer $p > 0$,

$$\begin{aligned} J_{-p}(x) &= \sum_{k=p}^{\infty} \frac{(-1)^k(x/2)^{2k-p}}{k!\,\Gamma(k+1-p)} \\ &= \sum_{k=0}^{\infty} \frac{(-1)^{p+k}(x/2)^{2k+p}}{k!\,\Gamma(k+p+1)} \\ &= (-1)^pJ_p(x). \end{aligned} \tag{143}$$

Hence, for such values of p ($p = 0, 1, 2, \ldots$), the two solutions so far found are manifestly not independent. We omit the computational details,these being tedious but straightforward, and present, instead, the second independent solution of Bessel's equation in the form

$$Y_n(x) = \lim_{p\to n} \left\{\frac{\cos p\pi J_p(x) - J_{-p}(x)}{\sin p\pi}\right\} \tag{144}$$

when n is $0, 1, 2, \ldots$. We do not record the full expansion of $Y_n(x)$, except to note that, as we expect from (110), $Y_n(x)$ is not regular at $x = 0$, but is of the form

$$Y_n(x) = \frac{2}{\pi} \log x\, J_n(x) + x^{-n}A(x) \tag{145}$$

where $A(x)$ is entire.

In the small space allotted we cannot hope to do justice to the many facets of the theory of Bessel functions. In making a choice of subjects to discuss we were therefore guided by the topics which are of interest from the *function-theoretic* point of view. These are (a) analytic character, (b) rate of growth or asymptotic behavior and, (c) location of zeros. The first of these has already been settled in Theorem 17.

Next, consider the function $e^{\frac{1}{2}z(t-1/t)}$. This is, for fixed z, a function of t which is regular in any annulus $0 < \rho_1 \leqq |t| \leqq \rho_2$. It therefore admits a Laurent series expansion

$$\exp\left[\tfrac{1}{2}z\left(t - \frac{1}{t}\right)\right] = \sum_{n=-\infty}^{\infty} \varphi_n(z)t^n. \tag{146}$$

The coefficients $\varphi_n(z)$ are, as usual, given by the integral

$$\varphi_n(z) = \frac{1}{2\pi i}\int_C t^{-n-1}\exp\left[\tfrac{1}{2}z\left(t - \frac{1}{t}\right)\right] dt \tag{147}$$

where the contour encloses the origin. Putting $t = 2u/z$,

$$\varphi_n(z) = \frac{1}{2\pi i}\left(\frac{z}{2}\right)^n \int_C u^{-n-1}\exp\left[u - \frac{z^2}{4u}\right] du,$$

where C may be taken as the unit circle. Expanding the exponential,

$$\varphi_n(z) = \frac{1}{2\pi i}\sum_{r=0}^{\infty}\frac{(-1)^r}{r!}\left(\frac{z}{2}\right)^{n+2r}\int_C u^{-n-r-1}e^u\, du.$$

The residue of the integrand at $u = 0$, i.e., the coefficient of u^{-1} in the expansion of $u^{-n-r-1}e^u$ in a Laurent series, is obviously $[(n + r)!]^{-1}$, if $n + r$ is zero or a positive integer, and zero otherwise.

Hence by the residue theorem, we find

Theorem 18. *For $t \neq 0$ we have*

$$\exp\left[\tfrac{1}{2}z\left(t - \frac{1}{t}\right)\right] = \sum_{n=-\infty}^{\infty} J_n(z)t^n. \tag{148}$$

Further, if we put $t = e^{i\theta}$ in (147), we find

$$\begin{aligned} J_n(z) &= \frac{1}{2\pi}\int_{-\pi}^{\pi}\exp(-in\theta + iz\sin\theta)\, d\theta \\ &= \frac{1}{2\pi}\left\{\int_{-\pi}^{0} + \int_0^{\pi}\right\}\exp(-in\theta + iz\sin\theta)\, d\theta \\ &= \frac{1}{2\pi}\left\{\int_0^{\pi}\exp(in\theta - iz\sin\theta)\, d\theta + \int_0^{\pi}\exp(-in\theta + iz\sin\theta)\, d\theta\right\} \end{aligned} \tag{149}$$

or finally

$$J_n(z) = \frac{1}{\pi}\int_0^{\pi} \cos(n\theta - z\sin\theta)\, d\theta \tag{150}$$

which is Bessel's integral formula.

Now, exactly as in exercise 10 of the preceding chapter, the method of stationary phase yields

Theorem 19. *We have as $x \to \infty$, n fixed,*

$$J_n(x) = \sqrt{2/\pi x}\cos\left(x - \frac{n\pi}{2} - \frac{\pi}{4}\right) + O(x)^{-3/5}. \tag{151}$$

Proof. Equation (150) is

$$J_n(x) = \frac{1}{\pi}\,\mathbf{Re}\int_0^{\pi} \exp\,[i(n\theta - x\sin\theta)]\, d\theta \tag{152}$$

which is of the form considered in the method of stationary phase with

$$h(\theta) = n\theta - x\sin\theta,$$

Now, $h'(\xi) = 0$ at the point

$$\xi = \cos^{-1}\frac{n}{x}$$

lying between 0 and π, and not elsewhere in that interval. Handling the endpoints just as in the exercise referred to, we find from (114) of the previous chapter,

$$J_n(x) = \sqrt{2/\pi}\,\frac{\cos\left(\frac{\pi}{4} + n\cos^{-1}\frac{n}{x} - \sqrt{x^2 - n^2}\right)}{(x^2 - n^2)^{1/4}} + O(x^{-3/5}).$$

However, as $x \to \infty$, n being fixed,

$$n\cos^{-1}\frac{n}{x} = n\frac{\pi}{2} + o(1)$$

$$\sqrt{x^2 - n^2} = x + o(1)$$

$$(x^2 - n^2)^{1/4} = \sqrt{x} + o(1)$$

and the result follows.

Since $J_n(-x) = (-1)^n J_n(x)$ when x is an integer, the expansion (151) holds also for $x \to -\infty$ if x is replaced by $|x|$, and $(-1)^n$ inserted in front.

Now, the positive real axis is decidedly not the direction in which $J_n(z)$ grows most rapidly. Indeed, from inspection of the power series (141) it is evident that when z is purely imaginary, all the terms have like sign, and maximum rate of growth is therefore achieved as $z \to \pm i\infty$.

To find this rate of growth, put $z = it$, t real, in (150), getting

$$J_n(it) = \frac{1}{2\pi}\int_0^\pi \left\{e^{in\theta}e^{t\sin\theta} + e^{-in\theta}e^{-t\sin\theta}\right\} d\theta. \tag{153}$$

If we regard t as large and positive, then, since $\sin\theta \geqq 0$ in the range considered, the second term under the integral sign in (153) surely remains bounded as $t \to \infty$. For the first term we use Laplace's method (equation (109) of the preceding chapter) and find

$$\begin{aligned}\frac{1}{2\pi}\int_0^\pi e^{in\theta}e^{t\sin\theta}\,d\theta &= \frac{1}{2\pi}e^t\int_0^\pi e^{in\theta}e^{t(\sin\theta-1)}\,d\theta \\ &\sim \frac{1}{2\pi}e^t e^{(in\pi/2)}\sqrt{2\pi/t} \qquad (t\to\infty) \\ &= \frac{i^n}{\sqrt{2\pi}}\frac{e^t}{\sqrt{t}}.\end{aligned}$$

Hence

$$J_n(it) \sim \frac{i^n}{\sqrt{2\pi t}}e^t \qquad (t\to\infty) \tag{154}$$

so that $J_n(z)$ grows exponentially along the positive imaginary axis, and this is the line of most rapid growth.

If z is fixed, the power series expansion of $J_\nu(z)$ is an asymptotic expansion for large ν, and we find

$$J_\nu(x) \sim \frac{1}{\Gamma(\nu+1)}\left(\frac{x}{2}\right)^\nu \qquad (\nu\to\infty). \tag{155}$$

Next we propose to show that the zeros of $J_\nu(z)$ are all real provided the order ν is suitably restricted. First, we need an identity involving an integral of the product of $J_\nu(\alpha z)$ and $J_\nu(\beta z)$. Let

$$u(z) = J_\nu(\alpha z)$$
$$v(z) = J_\nu(\beta z)$$

where α, β are fixed complex numbers. Then Bessel's differential equation reads

$$zu'' + u' + \left(\alpha^2 - \frac{\nu^2}{z^2}\right)zu = 0 \tag{156}$$

$$zv'' + v' + \left(\beta^2 - \frac{\nu^2}{z^2}\right)zv = 0. \tag{157}$$

Multiplying (156) by $v(z)$, (157) by $u(z)$ and subtracting, we get

$$z(u''v - v''u) + (u'v - v'u) = (\beta^2 - \alpha^2)zuv$$

or equivalently,

$$\frac{d}{dz}\{z(u'v - uv')\} = (\beta^2 - \alpha^2)zuv. \tag{158}$$

Now, near the origin,

$$u(z) = O(z^{\nu}) \qquad (z \to 0)$$
$$v(z) = O(z^{\nu}) \qquad (z \to 0)$$

and therefore

$$zu(z)v(z) = O(z^{2\nu+1}) \quad (z \to 0).$$

It follows that we may integrate (158) from $z = 0$ to $z = 1$ provided $2\nu + 1 > -1$, i.e., if $\nu > -1$. Hence, if $\nu > -1$, we obtain

$$\text{(159)} \quad \int_0^1 xJ_{\nu}(\alpha x)J_{\nu}(\beta x)\,dx = \frac{1}{\beta^2 - \alpha^2}\{\alpha J_{\nu}'(\alpha)J_{\nu}(\beta) - \beta J_{\nu}(\alpha)J_{\nu}'(\beta)\}.$$

Now suppose α is a complex zero of $J_{\nu}(z)$, and that $\nu > -1$. From the power series (141) it is evident that α cannot be purely imaginary. Thus if we put $\beta = \bar{\alpha}$ in (159), then $\beta^2 - \alpha^2 \neq 0$, and also $J_{\nu}(\bar{\alpha}) = 0$, since $J_{\nu}(z)$ has real coefficients. Then (159) gives

$$\begin{aligned}\int_0^1 xJ_{\nu}(\alpha x)J_{\nu}(\bar{\alpha}x)\,dx &= 0 \\ = \int_0^1 xJ_{\nu}(\alpha x)\overline{J_{\nu}(\alpha x)}\,dx & \\ = \int_0^1 x\,|J_{\nu}(\alpha x)|^2\,dx &\end{aligned}$$

which is patently impossible. Thus we have

Theorem 19. *Let $\nu > -1$. Then the zeros of $J_{\nu}(z)$ are all real. Further, if $x_{\nu n}$ denotes the* nth *positive zero of $J_{\nu}(z)$, arranged in increasing order of size, then, for ν fixed, we have*

$$\text{(160)} \qquad x_{\nu n} \sim n\pi \qquad (n \to \infty)$$

and

$$\text{(161)} \qquad x_{\nu,n+1} - x_{\nu,n} \sim \pi \qquad (n \to \infty).$$

The last two assertions follow at once from the asymptotic expansion (151), since we must have

$$x_{\nu n} - \frac{\nu\pi}{2} - \frac{\pi}{4} \sim (n - \tfrac{1}{2})\pi \qquad (n \to \infty)$$

which is (160).

Additional information about the zeros of $J_n(z)$ can be obtained from the recurrence formulas

$$\text{(162)} \qquad \frac{2n}{x}J_n(x) = J_{n-1}(x) + J_{n+1}(x)$$

$$\text{(163)} \qquad 2J_n'(x) = J_{n-1}(x) - J_{n+1}(x)$$

whose derivation is left as an exercise.

If we add (162), (163) and divide by 2, we get

$$J_{n-1}(x) = \frac{n}{x} J_n(x) + J_n'(x) \tag{164}$$

while subtracting and dividing by 2 yields

$$J_{n+1}(x) = \frac{n}{x} J_n(x) - J_n'(x). \tag{165}$$

The last two relations can be written as

$$(x^n J_n(x))' = x^n J_{n-1}(x) \tag{166}$$

$$(x^{-n} J_n(x))' = -x^{-n} J_{n+1}(x) \tag{167}$$

and we have

Lemma 4. Between consecutive positive zeros of $J_\nu(x)$ there is at least one zero of $J_{\nu+1}(x)$.
Proof. By Rolle's theorem, the derivative of $x^{-\nu}J_\nu(x)$ must vanish between its zeros, and the result is clear from (167).

Lemma 5. Between consecutive positive zeros of $J_\nu(x)$ there is at least one zero of $J_{\nu-1}(x)$.
Proof. Same reasoning applied to (166).

Theorem 20. *The zeros of $J_\nu(x)$ and of $J_{\nu+1}(x)$ are interlaced, that is, between consecutive positive zeros of one there is one and only one zero of the other.*
Proof. From Lemma 4, $J_{\nu+1}(x)$ has at least one zero between consecutive zeros of $J_\nu(x)$ and from Lemma 5, $J_\nu(x)$ has at least one zero between consecutive zeros of $J_{\nu+1}(x)$. It follows that between consecutive zeros of either, there is one and only one of the other, which was to be shown.

Bibliography

Existence and uniqueness theory for ordinary differential equations may be found in

1. F. J. Murray and K. S. Miller, *Existence Theorems*, New York University Press, 1954.
2. E. J. B. Goursat, *A Course in Mathematical Analysis*, Ginn and Co., New York, 1945.
3. E. A. Coddington and N. Levinson, *Theory of Ordinary Differential Equations*, McGraw-Hill Book Co., New York, 1955.
4. E. L. Ince, *Ordinary Differential Equations*, Dover Press, London, 1927.

The discussion of the size of a neighborhood in which a solution surely exists is based on

5. A. Wintner, *On the Process of Successive Approximation in Initial Value Problems*, *Annali Di Matematica*, vol. **41** (1956), 343–357.

The theory of linear equations of the second order is also treated in

6. E. J. Whittaker and G. N. Watson, *Modern Analysis*, Cambridge University Press, New York, 1927.

For numerical integration techniques, see

7. W. E. Milne, *The Numerical Solution of Differential Equations*, John Wiley and Sons, New York, 1953.
8. L. Collatz, *The Numerical Treatment of Differential Equations*, Springer-Verlag, Berlin, 1960.
9. L. Fox, *The Numerical Solution of Two-Point Boundary Problems*, Clarendon Press, Oxford, 1957.
10. F. B. Hildebrand, *An Introduction to Numerical Analysis*, McGraw-Hill Book Co., New York, 1956.
11. A. Ralston and H. S. Wilf, *Mathematical Methods for Digital Computers*, John Wiley and Sons, New York, 1960.

The Gamma function and Bessel functions are both discussed in reference 6 as well as in

12. E. Rainville, *Special Functions*, The Macmillan Co., New York, 1960.
13. A. Erdélyi et al., *Bateman Manuscript Project*, McGraw-Hill Book Co., New York, 1954.

Bessel functions are exhaustively treated in

14. G. N. Watson, *A Treatise on the Theory of Bessel Functions*, The Macmillan Co., New York, 1944.

Exercises

1. Which of the hypotheses of Theorem 1 is not satisfied in (5)–(6)? in (7)? in (8)? Prove your answers.
2. The solution of the equation

$$y'(x) = e^{-xy(x)}$$
$$y(1) = 1$$

satisfies

$$y(x) \leqq x$$

in some interval $1 \leqq x \leqq 1 + \beta$.

3. For the system

$$y' = -2xy^2$$
$$y(0) = 1$$

find, by Theorem 1 and then by Theorem 3, a neighborhood of the origin in which a solution exists. What is the actual neighborhood? Can you explain the large discrepancy?

4. Write down the conditions on the coefficients of a Lagrangian formula in order that it should be exact for $1, x, x^2, \ldots, x^k$.
5. We have already seen, in a previous chapter, an expression in closed form for the error involved in using the trapezoidal rule over n steps. Where?
6. (*a*) Parameterize all possible stable three-point formulas which are exact for 1 and x.

   (*b*) What is the largest integer k such that there exists a stable three-point formula which is exact for $1, x, \ldots, x^k$?

7. Determine the rate of growth of $\Gamma(z)$ on the line $z = \frac{1}{2} + it$. (*Hint*. Follow the argument leading to (132).)
8. A polynomial with only negative real zeros which is positive at the origin has positive coefficients, obviously. The function $[z\Gamma(z)]^{-1}$ has only negative real zeros and is positive at the origin. Does it have positive Taylor coefficients?
9. Show that

$$\frac{\sqrt{\pi}}{2} = \prod_{\nu=1}^{\infty} \left\{ \frac{\sqrt{\nu(\nu+1)}}{\nu + \frac{1}{2}} \right\}.$$

10. Show that

$$J_{\frac{1}{2}}(x) = \sqrt{\frac{2}{\pi x}} \sin x .$$

11. (*a*) By multiplying

$$e^{\frac{x}{2}(t-1/t)} \quad \text{and} \quad e^{\frac{y}{2}(t-1/t)}$$

show that

$$J_0(x+y) = J_0(x)J_0(y) - 2J_1(x)J_1(y) + 2J_2(x)J_2(y) \cdots$$

(*b*) From the result of part (*a*), show that

$$1 = J_0^2(x) + 2J_1^2(x) + 2J_2^2(x) + \cdots$$

and therefore that for x real, $|J_0(x)| \leq 1$, $|J_n(x)| \leq 1/\sqrt{2}$, for $n = 1, 2, \ldots$

chapter 6

Conformal mapping

6.1 INTRODUCTION

The study of conformal mapping is that of the mapping properties of analytic functions. For a physical scientist the subject derives its usefulness from the possibility of transforming a problem which naturally occurs in a rather difficult geometric setting into another in which the geometry is simpler. For a mathematician much of the interest of this subject arises from the study of the relationships between the analytical and the geometrical properties of analytic functions.

The aims of these two kinds of workers are apparently divergent, the first being primarily interested in constructing a map having certain desired properties, the second, in the function-theoretic restrictions which are imposed on classes of functions with certain broadly defined mapping properties. It must be remarked, however, that any detailed consideration of one of these kinds of questions can scarcely avoid involvement with the other.

In this chapter we are concerned, for the most part, with the study of broad classes of mapping functions rather than with details of technique. First we discuss some basic ideas and definitions and then proceed to the question of the kind of mapping that can be carried out by means of analytic functions. The fundamental theorem dealing with this question, the Riemann mapping theorem, will then be proved.

Following this we will consider a very general method for carrying out a mapping of one set onto another and then certain more special methods for accomplishing this object under more restrictive conditions. Some applications to physical problems follow. We conclude with a study of a few of the more interesting particular classes of mapping functions.

6.2 CONFORMAL MAPPING

We say that a set D in the complex plane is *connected* if any two of its points can be joined by a simple (i.e., not self-intersecting) continuous arc (Jordan arc). D is *simply connected* if for every Jordan curve C (closed Jordan arc) lying in D, the interior of C also lies in D. A connected set which is not simply connected is called *multiply connected.* A *domain* is a connected open set.

Now, let us visualize two complex planes, which we shall call, respectively, the z-plane and the w-plane.† Let D be a domain in the z-plane, and let $f(z)$ be a function which is regular in D. With each point z in D we associate the point $w = f(z)$ in the w-plane. The set of all points so obtained is called the image set, D', of D under the mapping $f(z)$, and we sometimes write, symbolically, $D' = f(D)$. It should be noticed that some values w in D' may be the images of several points of D, i.e., there may be distinct points of D, $z_1, z_2, \ldots, z_n$ say, such that

$$f(z_1) = f(z_2) = \cdots = f(z_n).$$

We will say that such a point w is n-times covered by the map $w = f(z)$ and that D' is n-times covered if this is true of every point w in D'. For instance, the map

$$w = z^n$$

carries the unit circle $|z| < 1$ onto the n-times covered unit circle $|w| < 1$.

Now, let D be a domain in the z-plane, and let C_1, C_2 be two differentiable arcs lying in D and intersecting at a point P of D. If $f(z)$ is a function regular in D, then, clearly, $f(C_1)$, $f(C_2)$ are differentiable arcs lying in $D' = f(D)$ and intersecting at a point $P' = f(P)$. We say that the mapping $w = f(z)$ is *conformal* at P if, for every such pair of arcs, the angle between C_1 and C_2 at P is equal to the angle between $f(C_1)$ and $f(C_2)$ at $f(P)$. The mapping is *conformal in D* if it is conformal at each point in D.

Theorem 1. *For the map $w = f(z)$ to be conformal at P, it is necessary and sufficient that $f'(P) \neq 0$.*

Proof. First we show that the condition is sufficient. Let the arcs C_1, C_2 be given parametrically by

$$z_1 = \varphi(t)$$
$$z_2 = \psi(t) \qquad 0 \leqq t \leqq 1$$

† This designation seems to have originated with the German words *zahl* (number) and *wert* (value).

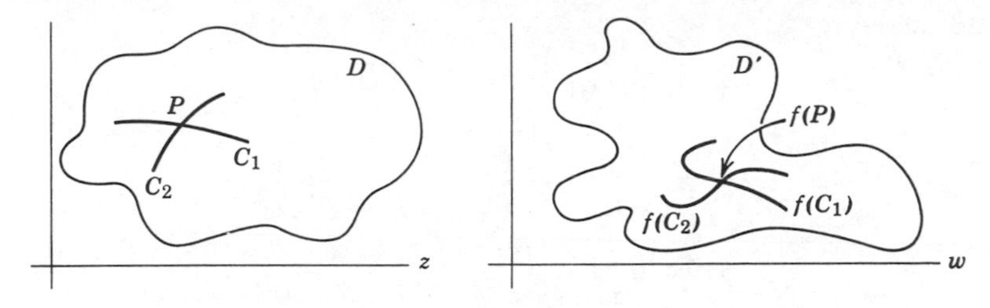

Figure 6.1

and suppose z_1, z_2 are points on C_1, C_2 at a distance d from P. Then

$$z_1 - P = de^{i\xi}, \quad z_2 - P = de^{i\eta}$$

$$\frac{z_2 - P}{z_1 - P} = e^{i(\eta - \xi)}.$$

Hence the angle between the chords $\overline{Pz_1}$ and $\overline{Pz_2}$ is $\eta - \xi$. As $d \to 0$, then, this must approach the angle γ between the curves. That is

$$\gamma = \lim_{d\to 0} \arg \left\{\frac{z_2 - P}{z_1 - P}\right\}.$$

For the angle γ' between $f(C_1)$ and $f(C_2)$ at $f(P)$, we have, similarly,

$$\begin{aligned}
\gamma' &= \lim_{d\to 0} \arg \left\{\frac{f(z_2) - f(P)}{f(z_1) - f(P)}\right\} \\
&= \lim_{d\to 0} \arg \left\{\frac{\dfrac{f(z_2) - f(P)}{z_2 - P}}{\dfrac{f(z_1) - f(P)}{z_1 - P}} \frac{z_2 - P}{z_1 - P}\right\} \\
&= \lim_{d\to 0} \arg \left\{\frac{f'(P)}{f'(P)}\right\} + \lim_{d\to 0} \arg \frac{z_2 - P}{z_1 - P} \\
&= 0 + \gamma = \gamma
\end{aligned}$$

since we supposed $f'(P) \neq 0$.

Conversely, we claim the condition $f'(P) \neq 0$ is necessary, for suppose it fails. Then, say

$$f'(P) = f''(P) = \cdots = f^{(q-1)}(P) = 0$$

for some $q \geqq 2$, whereas $f^{(q)}(P) \neq 0$. Then, near P, we have

$$f(z) = f(P) + O(z - z_0)^q \qquad (z \to z_0)$$

and therefore

$$\begin{aligned}\gamma' &= \lim_{d\to 0} \arg \left\{\frac{f(z_2) - f(P)}{f(z_1) - f(P)}\right\} \\ &= \lim_{d\to 0} \arg \left\{\frac{a_q(z_2 - P)^q + \dots}{a_q(z_1 - P)^q + \dots}\right\} \\ &= \lim_{d\to 0} \arg \left\{\left(\frac{z_2 - P}{z_1 - P}\right)^q\right\} \\ &= q \lim_{d\to 0} \arg \frac{z_2 - P}{z_1 - P} = q\gamma\end{aligned}$$

which shows that the angle is magnified by q, and therefore the map is not conformal.

6.3 UNIVALENT FUNCTIONS

For a given domain D in the z-plane we are now in possession of two properties of an analytic function $f(z)$ which are "nice" to have, namely, (i) $f(z)$ should yield a conformal map of D and (ii) $f(z)$ should take no value twice in D, so that $f(D)$ is only once covered.

A function $f(z)$, regular in a domain D, which takes no value twice in D is said to be *univalent in D*. An alternate definition is that

$$\frac{f(z_2) - f(z_1)}{z_2 - z_1} \neq 0, \quad z_1, z_2 \text{ in } D. \tag{1}$$

If $f(z)$ is univalent in D, then $f(D)$ is once covered and is called a *schlicht domain*. Sometimes one speaks of a schlicht function as equivalent to a univalent function, but we reserve the word "univalent" for functions and "schlicht" for domains.

Now, suppose $f(z)$ is regular and univalent in D; then we claim that $f'(z) \neq 0$ in D. For, suppose the contrary; then $f'(z_0) = 0$ for some z_0 in D. Then near z_0,

$$f(z) = f(z_0) + \frac{f''(z_0)}{2}(z - z_0)^2 + \cdots \tag{2}$$

and $f(z) - f(z_0)$ has a double zero at $z = z_0$, that is, $f(z)$ assumes the value $f(z_0)$ twice at z_0, contradicting the assumption of univalence. Actually, (2) shows that near z_0, $f(z) - f(z_0)$ behaves like $(z - z_0)^2$ and therefore assumes all values in a neighborhood of $f(z_0)$ at least twice.

Consequently, we have shown that a univalent map is conformal. The converse of this proposition is false, as can be seen, for instance, from the function

$$f(z) = (1 + z)^n - 1 \tag{3}$$

whose derivative clearly vanishes nowhere in the unit circle $|z| < 1$, but which takes the value $f(z_0)$ at each of the points $\omega z_0 - (1 - \omega)$, where ω is any one of the nth roots of unity. It is easy to see that several of these points can lie in the unit circle, so that this function is not univalent in $|z| < 1$. The property of univalence is considerably more sophisticated than that of conformality.

Theorem 2. *If $f(z)$ is regular and univalent in D, then $f(z)$ maps D conformally onto the schlicht domain $f(D)$, the converse being false, in general.*

6.4 FAMILIES OF FUNCTIONS REGULAR ON A DOMAIN

The purpose of this section is to generalize the well-known classical theorem which asserts that a continuous function defined on a compact set attains its maximum value at a point of the set. Our objective is a theorem about families of analytic functions to the effect that if we have a functional defined on a family of such functions, that is, a rule which assigns a number to each function in the family, then that functional attains its maximum value on a certain function of the family. Clearly we need first to generalize the notion of compactness to such situations.

Definition 1. Let F be a family of functions defined and regular on a domain D. The family F is *locally uniformly bounded* if for each point ζ of D, we can find a number M and a neighborhood U of ζ, such that

$$|f(z)| \leq M$$

for all z in U and all $f(z)$ in F.

Definition 2. The family F is said to be *equicontinuous* if for every compact subdomain G of D and $\varepsilon > 0$ we can find $\delta > 0$, such that

$$|f(z_1) - f(z_2)| < \varepsilon$$

if $|z_1 - z_2| < \delta$, z_1, z_2 in G, for every function $f(z)$ in F.

Definition 3. The family F is called a *Montel family* of functions in D if

(i) in any sequence $f_1(z), f_2(z), \ldots$ of functions in F one can find a subsequence $f_{k_1}(z), f_{k_2}(z), \ldots$ which is uniformly convergent in every compact subdomain of D.

(ii) every convergent sequence of functions of F has a limit in F.

Definitions 1 and 2 are simply definitions of bounded functions and continuous functions with the added proviso that the choices of the parameters can be made uniformly for all members of the family. Definition 3 is obviously a generalization of the idea of a compact set, where (i) is essentially a boundedness condition and (ii) a closure condition.

Now let F be a given family. A *functional* $H[f]$ on F is simply a rule which attaches a real number to each member of F. For instance, if F is the family of functions $f(z)$ defined and regular in $|z| < 1$ and satisfying there

$$|f(z)| \leq 1$$

then we may assign to each f in F the number

$$H[f] = \frac{1}{2\pi}\int_0^{2\pi} |f(\tfrac{1}{2}e^{i\theta})| \, d\theta.$$

A functional $H[f]$ is *continuous* if

$$\lim_{n\to\infty} f_n(z) = f(z)$$

where $f_1, f_2, \ldots, f$ are all in F, implies

$$\lim_{n\to\infty} H[f_n(z)] = H[f].$$

We can now prove

Theorem 3.† *A continuous functional defined on a Montel family F attains its maximum modulus on a function $f(z)$ in F.*

Proof. Let L denote the least upper bound of $|H[f]|$ as f ranges over F. Then there is a sequence $\{f_n(z)\}_1^\infty$ of functions of F such that

$$\lim_{n\to\infty} |H[f_n(z)]| = L.$$

Since F is a Montel family, there is a subsequence $\{f_{n_k}(z)\}_{k=1}^\infty$ which converges to a function $f(z)$ in F. Then

$$\begin{aligned} |H[f(z)]| &= \lim_{k\to\infty} |H[f_{n_k}(z)]| \\ &= L \end{aligned}$$

which shows, first, that $L < \infty$ and, second, that $H[f]$ attains its maximum on $f(z)$, proving the theorem.

Now, in the definition of a Montel family the condition (i) is rather difficult to verify, in particular cases, as it stands. Thus we require certain other, more accessible properties of families of functions which will imply that a convergent subsequence can be extracted from every infinite sequence. This more accessible property will turn out to be that of local uniform boundedness. We need first a few introductory results.

Lemma 1. Let the family F be locally uniformly bounded in D. Then the same is true of the family F' of derivatives of functions of F.

Proof. Let ζ be a fixed point of D. Then there is a number M and a neighborhood U of ζ such that $|f(z)| \leq M$ for all z in U, f in F. In U, centered at ζ, draw two concentric circles C_1, C_2 of radii δ, 2δ, respectively.

† Montel [1].

Then for $|z - \zeta| \leqq \delta$ we have

$$|f'(z)| = \left| \frac{1}{2\pi i} \int_{C_2} \frac{f(\eta)\, d\eta}{(\eta - z)^2} \right|$$

$$\leqq \frac{1}{2\pi} \frac{M}{\delta^2} \cdot 4\pi\delta$$

$$= \frac{2M}{\delta}$$

Hence with the point ζ we associate the neighborhood U': $|z - \zeta| \leqq \delta$ and the bound $M' = 2M/\delta$, and the lemma is proved.

Lemma 2. Let the family F be locally uniformly bounded in D. Then F is equicontinuous in D.

Proof. Let G be a closed subdomain of D. By Lemma 1, $|f'(z)| \leqq M$, where M does not depend on $f(z)$, throughout G; for otherwise there would be a sequence of points $\{z_\nu\}_1^\infty$ converging to a point ζ of G, and a sequence $\{f_\nu(z)\}_1^\infty$ of functions of F such that $|f_\nu'(z_\nu)|_{\nu=1}^\infty$ is unbounded. But then F' would not be uniformly bounded at ζ.

Thus, for fixed z_1 in G and all z_2 near z_1,

$$|f(z_1) - f(z_2)| = \left| \int_{z_2}^{z_1} f'(z)\, dz \right|$$

$$\leqq M\, |z_1 - z_2|$$

which can be made less than ε for all f in F by keeping z_2 close enough to z_1, the phrase "close enough" being independent of the particular f chosen.

Lemma 3. Let $\{f_\nu(z)\}_1^\infty$ be a locally uniformly bounded sequence of analytic functions in D, and suppose the sequence converges at a set of points which is dense in D. Then the sequence is uniformly convergent on every compact subdomain G of D.

Proof. Let $\varepsilon > 0$ be given. Around each point ζ of G draw a circle of small enough radius so that

$$|f_n(z) - f_n(\zeta)| < \varepsilon$$

for all z in the circle and for all n. This can be done because, by Lemma 2, the family is equicontinuous. Since these circles cover G and G is compact, we can, by the Heine-Borel theorem, extract a finite number of these circles, $C_1, \ldots, C_m$, say, which cover G. Since the sequence converges at a dense set of points, there is in each circle C_j a point ζ_j at which it converges. Since there are only m circles, it follows that for p and n large enough,

$$|f_n(\zeta_j) - f_p(\zeta_j)| < \varepsilon \qquad (j = 1, 2, \ldots, m).$$

Now, let z be any point of G. We claim the sequence converges at z. Indeed, z is in some circle, say C_j, centered at η_j. Then

$$\begin{aligned}|f_n(z) - f_p(z)| &= |f_n(z) - f_n(\eta_j) + f_n(\eta_j) - f_n(\zeta_j) \\ &\quad + f_n(\zeta_j) - f_p(\zeta_j) + f_p(\zeta_j) - f_p(\eta_j) \\ &\quad + f_p(\eta_j) - f_p(z)| \\ &\leqq |f_n(z) - f_n(\eta_j)| + |f_n(\eta_j) - f_n(\zeta_j)| \\ &\quad + |f_n(\zeta_j) - f_p(\zeta_j)| + |f_p(\zeta_j) - f_p(\eta_j)| \\ &\quad + |f_p(\eta_j) - f_p(z)|.\end{aligned}$$

The first two and the last two terms are $<\varepsilon$ because of the construction of the original circles. The third term is $<\varepsilon$ because the sequence converges at ζ_j. This shows that the sequence converges at all points of G. To show uniform convergence it is only necessary to observe that the size of p and n large enough to make $|f_n(z) - f_p(z)|$ small does not depend on which point z in the circle C_j was chosen. Since there are only finitely many circles, the largest p and n occurring in any of them will work uniformly for all points of G.

Theorem 4. (*Montel's Theorem*). *Let F be a family of functions regular in a domain D and satisfying* (i) *F is locally uniformly bounded in D and* (ii) *every convergent sequence of functions in F has a limit in F. Then F is a Montel family, and therefore every continuous functional defined on F assumes its maximum modulus on a function of F.*

Proof. Referring to the definition of a Montel family, what we have to prove is that every infinite sequence of functions of F has a convergent subsequence.

Let $z_1, z_2, \ldots$ be a sequence of points which is dense in D, for instance, the set of all points whose real and imaginary parts are rational numbers and which lie in D, and let $f_1(z), f_2(z), f_3(z), \ldots$ be an infinite sequence of functions of F.

For the point z_1 we can, by hypothesis, find a number M_1 such that $|f_n(z_1)| \leqq M_1$ for all n. But this means that the sequence of *numbers* $f_1(z_1), f_2(z_1), \ldots$ all lie in a finite circle, so there must be a subsequence of these numbers, say

$$f_{k_1}(z_1), \quad f_{k_2}(z_1), \ldots$$

which converges, that is, the sequence of *functions*

$$f_{k_1}(z), \quad f_{k_2}(z), \ldots \tag{4}$$

converges at the point z_1. Passing to the point z_2, we consider the functions (4) as before, and they must have a subsequence

$$f_{l_1}(z), \quad f_{l_2}(z), \ldots \tag{5}$$

which converges at z_2 and z_1 also. We continue this process, passing successively to $z_3, z_4, \ldots$, each time refining the sequence further.

Now we exhibit a single sequence which converges at all the points $z_1, z_2, \ldots$ at once. The first function of this sequence is the first function in (4), the second function is the second function in (5), . . . , the nth function is the nth function in the list of those surviving after point z_n, etc. We claim this sequence converges at each of $z_1, z_2, \ldots$, for let z_q be any fixed one of these points. Then every term in our sequence beyond the qth was also contained in the sequence of functions which remained after treating z_q. But that sequence converged at z_q; hence so does the refined sequence. Since the points $z_1, z_2, \ldots$ are dense, our theorem is proved by Lemma 3.

6.5 THE RIEMANN MAPPING THEOREM

We are now in possession of sufficient apparatus to deal with the central problem of the theory of conformal mapping: Given domains D, D', is there an analytic function which maps D onto D'? We restrict attention to the cases where D, D' are schlicht and simply connected.

Suppose D consists of the entire closed plane with the exception of a single point z_0 and that D' is the interior of the unit circle. We claim that there is no analytic function which maps D onto D'. For, suppose $f(z)$ were such a function; then $f(z)$ is regular everywhere in the plane except at z_0 and satisfies $|f(z)| \leqq 1$. But then

$$g(z) = f\left(\frac{zz_0}{z-1}\right)$$

is regular throughout the finite plane and satisfies

$$|g(z)| \leqq 1$$

everywhere; hence, by Liouville's theorem, $g(z)$ is a constant, which is impossible. Consequently, no such mapping exists. We may say then that a domain with just one boundary point is not conformally equivalent to the unit circle.

Now let D, D' be given schlicht simply connected domains. Suppose that each of the domains D, D' is conformally equivalent to the unit circle. Then we claim that they are conformally equivalent to each other. Indeed, if

$$w_1 = f_1(z)$$

maps D 1-1 onto $|w_1| < 1$ and

$$w_2 = f_2(z)$$

maps D' 1-1 onto $|w_2| < 1$, then, since $f_2(z)$ is univalent in D', the inverse function $f_2^{-1}(w_2)$ is well defined and regular in the unit circle $|w_2| < 1$ and maps it onto D'. Hence the function

$$w = f(z) = f_2^{-1}(f_1(z)) \qquad (z \text{ in } D)$$

maps D 1-1 onto D', as required. We are now ready to state

Theorem 5. (*The Riemann Mapping Theorem*). *Let D, D′ be schlicht, simply connected domains, each with more than one boundary point. Then there is a function f(z), regular and univalent in D, which maps D 1-1 conformally onto D′.*

The effect of the remarks preceding the theorem is that, first, the hypothesis concerning boundary points cannot be relaxed and, second, that in order to prove the theorem, it is enough to show that a domain D satisfying the conditions in the theorem can be mapped 1-1 onto the unit circle $|w| < 1$.

For the proof of Theorem 5 we need four preliminary results—three of which will be proved here—the fourth being a standard result in the theory of functions whose proof may be found in any text on the subject.

Lemma 4. (Hurwitz). Every zero of the limit function $f(z)$ of a convergent sequence of analytic functions is a limit point of zeros of the members of the sequence. Precisely, if $\{f_n(z)\}_1^\infty$ are regular in a domain D and converge to a nonconstant function $f(z)$ regular in D, then for every ζ in D at which $f(\zeta) = 0$, and every circle $|z - \zeta| < \delta$ about ζ, there is a zero of each $f_n(z)$, for $n > N$, in the circle.

Proof. By reducing δ, if necessary, we may suppose that $f(z)$ vanishes, in $|z - \zeta| \leqq \delta$, only at ζ, since $f(z)$ is not constant. Then on the circumference $|z - \zeta| = \delta$ we have, first, $|f(z)| > \mu > 0$ and, second, $|f_n(a) - f(z)| < \mu$, for all large enough n, and some number μ.

Now, write

$$f_n(z) = f(z) + f_n(z) - f(z)$$

and observe that on $|z - \zeta| = \delta$ we have

$$|f_n(z) - f(z)| < f(z)$$

and therefore, by Rouché's theorem, $f_n(z)$ and $f(z)$ have the same number of zeros inside $|z - \zeta| \leqq \delta$, namely, at least one, which was to be shown.

Lemma 5. Let F be the family of all functions regular and univalent in a domain D. Suppose F is locally uniformly bounded in D and that there is a point α in D and a number $\delta > 0$ such that

$$|f'(\alpha)| \geqq \delta$$

for all $f(z)$ in F. Then F is a Montel family in D.

Proof. In view of Theorem 4, what we have to show is that a convergent sequence of functions of F has a limit in F. Suppose $f_1(z), f_2(z), \ldots$ is such a sequence, and let $f(z)$ denote the limit function. Suppose $f(z)$ is not univalent in D. Then there exist z_1 in D, z_2 in D, $z_1 \neq z_2$ such that

$$f(z_1) = f(z_2)$$

Then the sequence of functions $f_n(z) - f_n(z_1)$ vanishes nowhere except at z_1, while the limit function $f(z) - f(z_1)$ vanishes also at z_2, contradicting Lemma 4 unless $f(z)$ is a constant. But this possibility is ruled out by the hypothesis $|f_n'(\alpha)| \geqq \delta > 0$ $(n = 1, 2, \ldots)$, proving the lemma.

Lemma 6. The mapping

$$w = e^{i\varphi}\left(\frac{z - \beta}{1 - \bar{\beta}z}\right) \qquad (|\beta| < 1) \tag{5}$$

maps the unit circle onto itself.

Proof. Since w is regular in $|z| \leqq 1$ it is enough to show for $|z| = 1$ we have $|w| = 1$. But if $z = e^{i\theta}$,

$$\begin{aligned} |1 - \bar{\beta}z| &= |1 - \bar{\beta}e^{i\theta}| = |e^{-i\theta} - \bar{\beta}| \\ &= |e^{i\theta} - \beta| \\ &= |z - \beta|. \end{aligned}$$

An easy calculation shows that the mapping is univalent.

Lemma 7 (The Monodromy Theorem). Let $f(z)$ be regular in a circle $|z - z_0| \leqq \delta$. Let D be a simply connected domain containing z_0 and having the property that if z_1 is any point of D, then $f(z)$ can be analytically continued from z_0 to z_1, along every path joining z_0 to z_1 and lying in D. Then the value obtained for the continued function at z_1 is independent of the particular path chosen for the continuation, so that the original functional element centered at z_0 generates a uniquely defined analytic function throughout D.

Proof. We do not prove the monodromy theorem here, for this is done in most books on analytic function theory. We remark that the proof is quite deep, necessarily involving the topological properties of self-intersecting polygons quite intimately, and therefore many of the proofs given are not quite complete. A second comment is that sometimes the theorem is stated quite briefly as "a function regular in a simply connected domain is single-valued there," but this author takes the view that a function is by definition single valued, which forces the considerably longer statement given above.

We proceed now to the proof of the Riemann mapping theorem, supposing that a schlicht domain D is given and that it is desired to construct a function $f(z)$ mapping D 1-1 conformally onto $|w| < 1$.

Since D has at least two boundary points, we may suppose $z_1, z_2, z_1 \neq z_2$ are both boundary points of D.

Consider the function

$$h(z) = \sqrt{\frac{z - z_1}{z - z_2}} \tag{6}$$

defined for z in D. The meaning of the function (6) is that we choose any particular determination of the square root at one point of D, and define $h(z)$ throughout D by analytic continuation from that point. The monodromy theorem assures that we will not "step on our own toes" in the process and that $h(z)$ is thereby well defined and regular in D. It is trivial to see, by assuming the contrary, that $h(z)$ is univalent in D.

Next, we claim that if w is a point in $h(D)$ then $-w$ is not, for suppose $h(\zeta) = w$, $h(\eta) = -w$, then

$$w^2 = \frac{\zeta - z_1}{\zeta - z_2} = \frac{\eta - z_1}{\eta - z_2}$$

which implies $\zeta = \eta$.

Now, since $h(D)$ is a domain, it is in particular an open set. Hence, if w_0 is in $h(D)$, so is a neighborhood $|w - w_0| < \delta$ in $h(D)$. Thus the negatives of all the numbers in that neighborhood are not in $h(D)$, i.e., the numbers w in $|w + w_0| < \delta$ are not values of $h(z)$. We therefore have $|h(z) + w_0| \geqq \delta$ for all z in D. It follows that for the function

$$h_1(z) = \frac{\delta}{h(z) + w_0}$$

we have $|h_1(z)| \leqq 1$ for all z in D, so we have constructed a function which maps D 1-1 *into* (*not onto*) the unit circle $|w| < 1$.

Next, let F denote the family of all functions $f(z)$ regular and univalent in D and satisfying in D: (i) $|f(z)| \leqq 1$ and (ii) $|f'(\zeta)| \geqq |h_1'(\zeta)|$, ζ being a fixed point of D. By Lemma 5, F is a Montel family in D and is surely nonempty because $h_1(z)$ is in F.

It follows that the continuous functional

$$H[f] = |f'(\zeta)|$$

assumes its maximum value on a function $f_0(z)$ of F, that is, there is a function $f_0(z)$ in F such that

$$|f'(\zeta)| \leqq |f_0'(\zeta)|$$

for every $f(z)$ in F. We claim that this function $f_0(z)$ is the function we seek, and it maps D 1-1 conformally onto $|w| < 1$. Indeed, since $f_0(z)$ is in F, we already know that it is univalent in D and maps D *into* $|w| < 1$; we have only to show that $f_0(z)$ omits no value w in $|w| < 1$.

Thus, suppose $f_0(z)$ does omit the value w_0, where $|w_0| < 1$. Then we will construct a function $f_2(z)$ in F for which $|f_2'(\zeta)| > |f_0'(\zeta)|$, contradicting the extremal property of $f_0(z)$. In fact, if we write

$$f_1(z) = \sqrt{\frac{w_0 - f_0(z)}{1 - \overline{w}_0 f_0(z)}} \tag{7}$$

$$f_2(z) = \frac{f_1(z) - f_1(\zeta)}{1 - \overline{f_1(\zeta)} f_1(z)} \tag{8}$$

then we claim that $f_2(z)$ is such a function. For, first, $f_2(z)$ is in F since $f_0(z)$ was in F, and the transformations (7), (8) do not disturb the univalence or the boundedness, according to Lemma 6. Next, we calculate the derivative of $f_2(z)$ at $z = \zeta$. To do this let us first show that $f_0(\zeta) = 0$. For consider the function

$$\psi(z) = \frac{f_0(z) - f_0(\zeta)}{1 - \overline{f_0(\zeta)} f_0(z)}.$$

For this function, which belongs to F, we have

$$\psi'(z) = \frac{[1 - \overline{f_0(\zeta)} f_0(z)] f_0'(z) + [f_0(z) - f_0(\zeta)] \overline{f_0(\zeta)} f_0'(z)}{[1 - f_0(\zeta) f_0(z)]^2}$$

and therefore

$$\psi'(\zeta) = \frac{[1 - |f_0(\zeta)|^2] f_0'(\zeta)}{[1 - |f_0(\zeta)|^2]^2}$$

$$= \frac{f_0'(\zeta)}{1 - |f_0(\zeta)|^2}.$$

If $f_0(\zeta)$ were not zero, we would have $|\psi'(\zeta)| > |f_0'(\zeta)|$, contradicting the extremal property of $f_0(z)$. Hence $f_0(\zeta) = 0$. Now, from (8),

$$f_2'(\zeta) = \frac{f_1'(\zeta)}{1 - |f_1(\zeta)|^2} \tag{9}$$

whereas from (7),

$$f_1(\zeta) = \sqrt{w_0}$$

$$f_1'(\zeta) = -\frac{1 - |w_0|^2}{2\sqrt{w_0}} f_0'(\zeta). \tag{10}$$

Substituting (10) into (9), we find

$$f_2'(\zeta) = -\frac{1 + |w_0|}{2\sqrt{w_0}} f_0'(\zeta).$$

Hence

$$|f_2'(\zeta)| = \frac{1}{2}\frac{1+|w_0|}{\sqrt{|w_0|}}\,|f_0'(\zeta)|$$

$$= \frac{1}{2}\left\{\frac{1}{\sqrt{|w_0|}} + \sqrt{|w_0|}\right\}|f_0'(\zeta)|$$

$$> |f_0'(\zeta)|$$

where we have used the fact that $x + 1/x > 2$ for $0 < x < 1$. The assumption that $f_0(z)$ omits some value in $|w| < 1$ has therefore led to a contradiction and the theorem is proved.

6.6 A CONSTRUCTIVE APPROACH

The proof of the Riemann mapping theorem given in the preceding section was not constructive in that the existence of the desired mapping function was proved without exhibiting the function explicitly or even giving an algorithm for calculating it. Yet the argument does contain the germ of such a constructive procedure, which is readily adapted to automatic computation methods.

Indeed, the key step in arriving at a contradiction was the demonstration that if $f_0(z)$ omitted some value on $|w| < 1$, then the successive transformations (7), (8) would increase $|f'(\zeta)|$. One may suppose that the iteration of the process (7), (8) starting from any function $f_0(z)$ satisfying $|f_0(z)| \leqq 1$ and $f_0(\zeta) = 0$ will continually increase $|f'(\zeta)|$ and so will converge to the desired mapping function. The proof that this is in fact the case is sufficiently similar to what has already been shown to be omitted. We content ourselves with an explicit statement of the algorithm in question, the method of Koebe image-domains.†

Given a simply connected domain D with two boundary points (at least), z_1, z_2,

(*a*) take $f_0(z)$ to be the function

$$f_0(z) = \frac{\delta}{h(z) + w_0}$$

which maps D *into* $|w| < 1$.

(*b*) Having determined $f_0(z), f_1(z), \ldots, f_{2n}(z)$, let $z = a$ be the boundary point of $f_{2n}(D)$ which lies nearest the origin. Define

$$f_{2n+1}(z) = \sqrt{\frac{a - f_{2n}(z)}{1 - \bar{a}f_{2n}(z)}} \tag{11}$$

$$f_{2n+2}(z) = \frac{f_{2n+1}(z) - f_{2n+1}(\zeta)}{1 - \overline{f_{2n+1}(\zeta)}f_{2n+1}(z)} \qquad (n = 0, 1, 2, \ldots) \tag{12}$$

† Koebe [1].

then

$$\lim_{n\to\infty} f_{2n}(z) = f(z)$$

is the desired mapping function, mapping D 1-1 conformally *onto* $|w| < 1$.

6.7 THE SCHWARZ-CHRISTOFFEL MAPPING

In certain cases where the given domain D has a particularly simple shape, the desired mapping function can be written down at once. One of these situations arises when D is a polygon.

Let C be a Jordan arc in the z-plane and $f(z)$ a function which is regular in some domain containing C. Let ξ, η be vectors tangent to C and $f(C)$, respectively, at the points $P, f(P)$. Then we have

$$\xi = \lim_{\Delta z\to 0} \frac{\Delta z}{|\Delta z|} \tag{13}$$

$$\eta = \lim_{\Delta w\to 0} \frac{\Delta w}{|\Delta w|} \tag{14}$$

the limits being taken through points $P + \Delta z, f(P) + \Delta w$ which lie on C, $f(C)$, respectively. Further, if we regard ξ, η as the complex numbers (13), (14), then at P,

$$\begin{aligned} f'(z) &= \lim_{\Delta z\to 0} \frac{\Delta w}{\Delta z} \\ &= \lim_{\Delta z\to 0} \left(\frac{\Delta w}{|\Delta w|}\left|\frac{\Delta w}{\Delta z}\right| \Big/ \frac{\Delta z}{|\Delta z|}\right) \\ &= \eta\xi^{-1}|f'(z)| \end{aligned}$$

and we have, clearly,

$$\arg f'(z) = \arg \eta - \arg \xi. \tag{15}$$

In the particular case where C is a portion of the real axis, $\arg \xi = 0$, $\arg f'(z) = \arg \eta$, and we see that the slope of the image curve at a point $f(P)$ is simply $\arg f'(z)$.

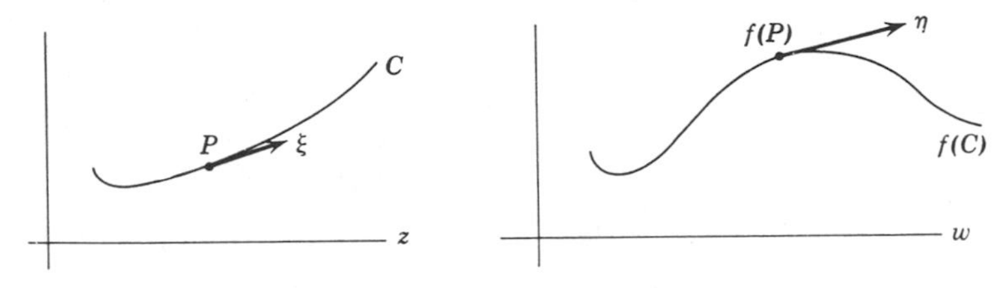

Figure 6.2

Now, let

$$a_1 < a_2 < \cdots < a_n$$

be points of the real axis, and consider the function $f(z)$ whose derivative is

$$(16) \qquad f'(z) = B(z - a_1)^{-k_1}(z - a_2)^{-k_2} \cdots (z - a_n)^{-k_n}.$$

For this function we have

$$(17) \quad \arg f'(z) = \arg B - k_1 \arg (z - a_1) - \cdots - k_n \arg (z - a_n).$$

Now, visualize the point z as moving from left to right along the real axis, starting to the left of the point a_1. When $z < a_1$, we have

$$\arg (z - a_1) = \arg (z - a_2) = \cdots = \arg (z - a_n) = \pi$$

whereas for $a_1 < z < a_2$, $\arg (z - a_1) = 0$, the others remaining at π. Hence, as z crosses a_1 from left to right, $\arg f'(z)$ increases by $k_1\pi$. It remains constant for $a_1 < z < a_2$ and increases by $k_2\pi$ as z crosses a_2, etc. In view of the remark following (15), then, the image of the segment $-\infty < z < a_1$ is a straight line, the image of $a_1 < z < a_2$ is another whose argument exceeds that of the first by $k_1\pi$, and so on.

If we constrain the numbers $k_1, k_2, \ldots, k_n$ to lie between -1 and 1, then the increments in the argument of $f'(z)$ will lie between $-\pi$ and π. Further, for $k_1 < 1, k_2 < 1, \ldots, k_n < 1$ it is obvious that the function $f(z)$ whose derivative is (16) is actually continuous at each of the points $a_1, a_2, \ldots, a_n$, and therefore the image of the moving point z will be a polygonal line.

The sum of the exterior angles of this polygonal line is

$$k_1\pi + k_2\pi + \cdots + k_n\pi.$$

If this is 2π, i.e., if

$$(18) \qquad k_1 + k_2 + \cdots + k_n = 2$$

the polygon will be closed. If all $k_j > 0$, the polygon will be convex.

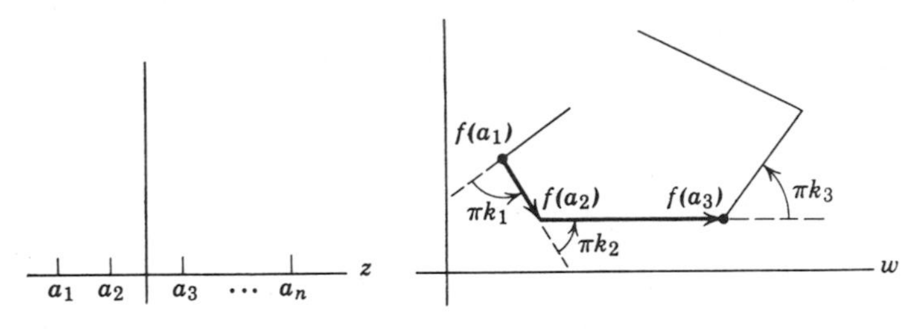

Figure 6.3

Integrating (16), we have the result that the function

$$f(z) = B\int^{z}(z' - a_1)^{-k_1}(z' - a_2)^{-k_2}\cdots(z' - a_n)^{-k_n}\,dz' + C \tag{19}$$

maps the x-axis onto a polygonal line which is closed if (18) holds. The mapping (19) is the Schwarz-Christoffel transformation.

6.8 APPLICATIONS OF CONFORMAL MAPPING

The utility of conformal mapping methods in the physical sciences results principally from the fact that, roughly speaking, a harmonic function of a harmonic function is harmonic.

More precisely, let $\varphi(u, v)$ be harmonic in a region D' of the u, v plane, that is

$$\nabla^2\varphi = \frac{\partial^2\varphi}{\partial u^2} + \frac{\partial^2\varphi}{\partial v^2} = 0$$

holds for all (u, v) in D'. Suppose the mapping

$$u = u(x, y)$$
$$v = v(x, y)$$

maps a region D in the x-y plane 1-1 onto D' and that $u(x, y)$, $v(x, y)$ are conjugate harmonic functions in D. We claim that the function

$$\Phi(x, y) = \varphi(u(x, y), v(x, y))$$

is harmonic in D. Indeed, for x, y, in D,

$$\begin{aligned}
\Phi_x &= \varphi_u u_x + \varphi_v v_x \\
\Phi_y &= \varphi_u u_y + \varphi_v v_y \\
\Phi_{xx} &= \varphi_u u_{xx} + u_x\varphi_{ux} + \varphi_v v_{xx} + v_x\varphi_{vx} \\
&= \varphi_u u_{xx} + u_x(\varphi_{uu}u_x + \varphi_{uv}v_x) + \varphi_v v_{xx} + v_x(\varphi_{vu}u_x + \varphi_{vv}v_x) \\
\Phi_{yy} &= \varphi_u u_{yy} + u_y(\varphi_{uu}u_y + \varphi_{uv}v_y) + \varphi_v v_{yy} + v_y(\varphi_{vu}u_y + \varphi_{vv}v_y).
\end{aligned}$$

Hence

$$\begin{aligned}
\nabla^2\Phi(x, y) &= \varphi_u(u_{xx} + u_{yy}) + \varphi_{uu}(u_x^2 + u_y^2) + \varphi_{uv}(u_x v_x + u_y v_y) \\
&\quad + \varphi_v(v_{xx} + v_{yy}) + \varphi_{vu}(u_x v_x + v_y u_y) + \varphi_{vv}(v_x^2 + v_y^2) \\
&= \varphi_{uu}(u_x^2 + u_y^2) + 2\varphi_{uv}(u_x v_x + u_y v_y) + \varphi_{vv}(v_x^2 + v_y^2).
\end{aligned}$$

However, by the Cauchy-Riemann equations,

$$u_x = v_y$$
$$u_y = -v_x$$

this becomes

$$\nabla^2\Phi(x, y) = (u_x^2 + u_y^2)(\varphi_{uu} + \varphi_{vv}) + 2\varphi_{uv}(v_x v_y - v_x v_y) = 0. \tag{20}$$

Theorem 6. *Let $\varphi(u, v)$ be harmonic in a region D' of the w-plane, and let $w = f(z)$ map a region D of the z-plane* 1-1 *onto D'. Then $\Phi(x, y) = \varphi\,(\mathbf{Re}\,f(z), \mathbf{Im}\,f(z))$ is harmonic in D.*

This simple result enables one to transform problems in difficult geometry into simple geometry, solve them, and transform the answer back again.

As an illustration of the method, as well as of the Schwarz-Christoffel transformation, we consider the following problem: Find a function $\varphi(x, y)$ which is harmonic in the half-strip

$$D\colon\ -\frac{\pi}{2} < x < \frac{\pi}{2} \qquad y > 0$$

of the x-y plane and satisfies on the boundary of D the conditions

$$\begin{cases} \varphi\left(-\dfrac{\pi}{2}, y\right) = 0 = \varphi\left(\dfrac{\pi}{2}, y\right) & (y > 0) \\ \varphi(x, 0) = 1 & \left(-\dfrac{\pi}{2} < x < \dfrac{\pi}{2}\right). \end{cases}$$

Now, suppose it has been shown that the successive mappings

$$w_1 = \sin z \tag{21}$$

$$w = \log \frac{w_1 - 1}{w_1 + 1} \tag{22}$$

map D onto the upper half of the w_1-plane, and then onto the horizontal strip

$$D''\colon\ 0 < \mathbf{Im}\, w < \pi \tag{23}$$

of the w-plane, in the manner shown in Figure 6.4.

If this be granted, for the moment, then we see that the side BCD which carries the boundary condition $\varphi = 1$ is mapped onto the line $\mathbf{Im}\ w = \pi$, while both of the sides carrying the condition $\varphi = 0$ are carried into $\mathbf{Im}\ w = 0$. Hence we need only find a function harmonic in the strip D'', vanishing on the lower boundary, having the value unity on the upper

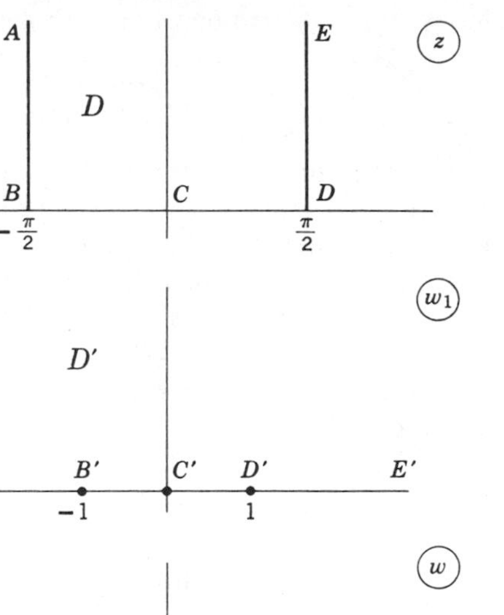

Figure 6.4

boundary, and then map it back to the z-plane. The required function is evidently

$$\frac{1}{\pi}\,\mathbf{Im}\, w \tag{24}$$

which is clearly harmonic and assumes the correct values on the boundary. We now proceed to unravel the mappings.

If (u, v), (u_1, v_1) are the real and imaginary parts of w, w_1, respectively, then from (22),

$$w = \log\left|\frac{w_1 - 1}{w_1 + 1}\right| + i\tan^{-1}\left\{\frac{2v_1}{u_1^2 + v_1^2 - 1}\right\}$$

and the function (24) is

$$\frac{1}{\pi}\tan^{-1}\left\{\frac{2v_1}{u_1^2 + v_1^2 - 1}\right\} \tag{25}$$

which is harmonic in D', by Theorem 6. Continuing, (21) gives

$$u_1 = \sin x \cosh y$$
$$v_1 = \cos x \ \sinh y$$

and substituting in (25), the solution of our problem is

$$\varphi(x, y) = \frac{2}{\pi} \tan^{-1} \left\{\frac{\cos x}{\sinh y}\right\} \tag{26}$$

It remains to show that the functions (21), (22) have the mapping properties claimed for them. We show this for (21), leaving (22) as an exercise. Rather than merely verifying that (21) does carry D into D', we will pretend not to know the answer and use the Schwarz-Christoffel formula to find the function which maps the upper half-plane onto the region D in the w-plane; and then the inverse of that function will be the one we seek.

Now, consider this strip D as a limiting form of the triangle ABC shown below as the point C tends to $\pi/2 + i\infty$. The exterior angles have the limiting values

$$\pi k_1 = \frac{\pi}{2} \quad \pi k_2 = \frac{\pi}{2} \quad \pi k_3 = \pi$$

and the transformation (19) has the form

$$f(z) = B \int^z (1 + z')^{-1/2}(z' - 1)^{-1/2}\, dz' + C_1$$

where, in accordance with the labeling of points in Figure 6-4, we have chosen $a_1 = -1$, $a_2 = 1$, $a_3 = \infty$. Performing the integration

$$\begin{aligned} f(z) &= iB \int_0^z \frac{dz'}{\sqrt{1 - z'^2}} + C_1 \\ &= iB \sin^{-1} z + C_1. \end{aligned}$$

If the constants are fixed so that $f(-1) = -\pi/2$, $f(1) = \pi/2$, we find

$$f(z) = \sin^{-1} z$$

as required.

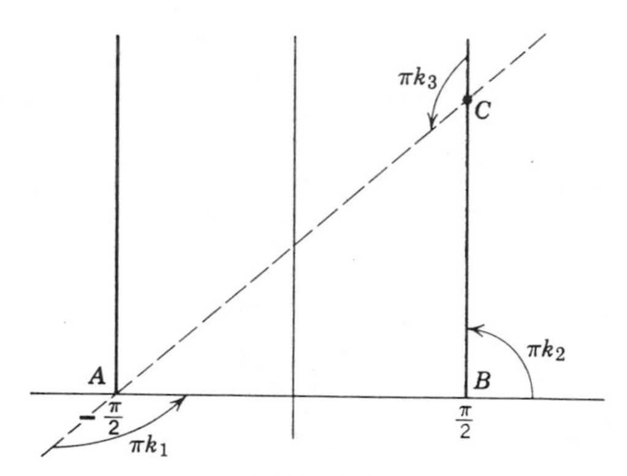

Figure 6.5

6.9 ANALYTICAL AND GEOMETRIC FUNCTION THEORY

We conclude this chapter with a short discussion of the interplay between the geometric and analytic properties of mapping functions in general.

Our first result concerns functions which map the unit circle into itself.

Theorem 7 (Schwarz' Lemma). *Let $f(z)$ be regular in $|z| < 1$ and let it satisfy the conditions*: (i) $f(0) = 0$ *and* (ii) $|f(z)| < 1$ *for* $|z| < 1$. *Then, actually*,

$$|f(z)| \leqq |z| \qquad (|z| < 1). \tag{27}$$

Proof. The function $g(z) = f(z)/z$ is also regular in $|z| < 1$. Hence, in the circle $|z| \leqq \rho$, $g(z)$ attains its maximum modulus on the circumference $|z| = \rho$. That is, for $|z| \leqq \rho$,

$$\begin{aligned} |g(z)| &\leqq \max_{|z|=\rho} |g(z)| \\ &= \frac{1}{\rho} \max_{|z|=\rho} |f(z)| \\ &\leqq \frac{1}{\rho} \end{aligned}$$

which says that for $|z| \leqq \rho$,

$$\left|\frac{f(z)}{z}\right| \leqq \frac{1}{\rho}$$

Making $\rho \to 1$, we have (27).

Next, we introduce the important idea of subordination. Roughly speaking, if we have two functions $f(z)$, $g(z)$ defined in a domain D, $g(z)$ is subordinate to $f(z)$ if $g(D)$ is contained entirely within $f(D)$. More precisely, suppose $f(z)$ is regular and univalent in D, that $g(z)$ is regular in D and that every value taken by $g(z)$ in D is also taken by $f(z)$. Then we say that $g(z)$ is *subordinate* to $f(z)$ in D.

Theorem 8.† *Let $f(z)$ and $g(z)$ be regular in $|z| < 1$, and suppose*

$$D = f(|z| < 1)$$

is a convex, schlicht domain. Further, suppose $g(z)$ is subordinate to $f(z)$ in $|z| < 1$, and that $f(z)$, $g(z)$ have the expansions

$$f(z) = \sum_{\nu=1}^{\infty} a_\nu z^\nu$$

$$g(z) = \sum_{\nu=1}^{\infty} b_\nu z^\nu.$$

† Löwner [1].

Then

$$|b_\nu| \leqq |a_1| \qquad (\nu = 1, 2, \ldots). \tag{28}$$

Proof. This result states, roughly, that geometric dominance is manifested in coefficient dominance in the power series expansions. Note, however, the key hypothesis that $f(|z| < 1)$ is convex and schlicht, without which the theorem is false.

Proceeding with the proof, since $g(|z| < 1)$ is contained in $f(|z| < 1)$, the inverse function of $f(z)$, $f^{-1}(z)$ (the function mapping D onto $|z| < 1$) is well defined at the point $g(z)$, for any $|z| < 1$. Hence the function

$$\begin{aligned} h(z) &= f^{-1}(g(z)) \\ &= \frac{b_1}{a_1} z + \cdots \end{aligned}$$

is regular in $|z| < 1$, satisfies $|h(z)| < 1$ there, and clearly $h(0) = 0$. According to Schwarz's lemma, $|h(z)| \leqq |z|$; but this obviously requires

$$\left|\frac{b_1}{a_1}\right| \leqq 1 \tag{29}$$

which is (28) with $\nu = 1$. Now, suppose that $\eta_1, \eta_2, \ldots, \eta_m$ are the m mth roots of unity, and consider the function

$$\begin{aligned} G(z) &= \frac{1}{m} \sum_{k=1}^{m} g(\eta_k z^{1/m}) \\ &= \frac{1}{m} \sum_{k=1}^{m} \sum_{\nu=1}^{\infty} b_\nu \eta_k{}^\nu z^{\nu/m} \\ &= \frac{1}{m} \sum_{\nu=1}^{\infty} b_\nu \left(\sum_{k=1}^{m} \eta_k{}^\nu \right) z^{\nu/m} \\ &= b_m z + b_{2m} z^2 + b_{3m} z^3 + \cdots. \end{aligned}$$

Now $G(z)$ is the center of gravity of the points $g(\eta_1 z^{1/m}), \ldots, g(\eta_m z^{1/m})$, all of which lie in $g(|z| < 1)$ and, therefore, also in $f(|z| < 1) = D$. Since D is convex, the center of gravity lies in D, that is, $G(z)$ is subordinate to $f(z)$. Then (29) applied to $G(z)$, $f(z)$ gives

$$|b_m| \leqq |a_1| \qquad (m = 1, 2, \ldots)$$

which was to be shown.

Two applications of this theorem are immediate.

Theorem 9.† *Let the function*

$$f(z) = z + a_2 z^2 + \cdots$$

† Gronwall [1], Löwner [1].

map the unit circle onto a convex, schlicht domain. Then

$$|a_n| \leqq 1 \qquad (n = 1, 2, \ldots). \tag{30}$$

Proof. Take $g(z) = f(z)$ in the previous theorem.

Theorem 10. *Let the function*

$$f(z) = \frac{1}{2} + a_1 z + a_2 z^2 + \cdots$$

map the unit circle into *the right half plane* (i.e., *suppose* $f(z)$ *has positive real part in the unit circle*). *Then*

$$|a_n| \leqq 1 \qquad (n = 1, 2, \ldots). \tag{31}$$

Proof. The function

$$f_0(z) = \frac{z}{1 - z} = z + z^2 + z^3 + \cdots$$

maps the unit circle onto the schlicht convex domain $\mathbf{Re}\, w > -\frac{1}{2}$. The function $f(z) - \frac{1}{2}$ is subordinate to $f_0(z)$, and the result follows by Theorem 8.

Theorem 11.† *Let the function*

$$f(z) = z + a_2 z^2 + \cdots \tag{32}$$

be regular in the unit circle, and suppose the coefficients $a_2, a_3, \ldots$ *are real. Suppose, further, that* $f(z)$ *is not real if* z *is not real. Then*

$$|a_n| \leq n \qquad (n = 2, 3, 4, \ldots). \tag{33}$$

Proof. We have for $r < 1$,

$$f(re^{i\theta}) = u(r, \theta) + iv(r, \theta)$$
$$= \sum_{n=1}^{\infty} a_n r^n e^{in\theta}$$

whence

$$v(r, \theta) = \sum_{n=1}^{\infty} a_n r^n \sin n\theta.$$

Solving for the coefficients of this Fourier series,

$$a_n r^n = \frac{2}{\pi} \int_0^{\pi} v(r, \theta) \sin n\theta \, d\theta \qquad (n = 1, 2, \ldots). \tag{34}$$

† Rogosinski [1]

Now our hypotheses state that $V(r, \theta)$ does not change sign when $0 < \theta < \pi$. Hence

$$
\begin{aligned}
|a_n r^n| &= \left| \frac{2}{\pi} \int_0^\pi v(r, \theta) \sin n\theta \, d\theta \right| \\
&= \left| \frac{2}{\pi} \int_0^\pi v(r, \theta) \sin \theta \frac{\sin n\theta}{\sin \theta} \, d\theta \right| \\
&\leqq \frac{2n}{\pi} \int_0^\pi v(r, \theta) \sin \theta \, d\theta \qquad (35) \\
&= n a_1 r \\
&= nr
\end{aligned}
$$

where we have used the inequality

$$\left| \frac{\sin n\theta}{\sin \theta} \right| \leqq n \qquad (0 < \theta < \pi). \tag{36}$$

The result follows by letting $r \to 1$ in (35).

Theorem 12. *Let the function*

$$f(z) = z + a_2 z^2 + \cdots$$

be regular and univalent in $|z| < 1$, *and suppose the coefficients* $a_2, a_3, \ldots$ *are real. Then*

$$|a_n| \leqq n \qquad (n = 1, 2, \ldots). \tag{37}$$

Proof. We claim that this function satisfies the conditions of the previous theorem. For, if not, $f(z_0)$ would be real for some z_0 which is not real, and we would have

$$f(z_0) = \overline{f(z_0)} = f(\bar{z}_0)$$

contradicting the univalence of $f(z)$ and proving the theorem.

Bibliography

For a well-balanced introduction to the subject of conformal mapping we recommend

1. Z. Nehari, *Conformal Mapping*, McGraw-Hill Book Co., New York, 1952.

The deeper mathematical aspects of the theory are well summarized in

2. W. K. Hayman, *Multivalent Functions*, Cambridge Tracts, No. 48, Cambridge University Press, 1958.

and are also discussed in

3. C. Caratheodory, *Conformal Representation*, Cambridge Tracts, No. 28, Cambridge University Press, 1932.

Applications to physical problems are, for instance, in

4. R. V. Churchill, *Introduction to Complex Variables and Applications*, McGraw-Hill Book Co., New York, 1948.

5. H. Jeffreys and B. Jeffreys, *Methods of Mathematical Physics*, Cambridge University Press, New York, 1950.

A complete proof of the convergence of the iterative procedure in Section 6.6 may be found in problems 88–96, Chapter IV, vol. 2 of

6. G. Pólya and G. Szegö, *Aufgaben und Lehrsätze aus der Analysis*, Springer, Berlin, 1954.

along with many other extremal problems in conformal mapping.

Finally, a rather lengthy list of the mapping properties of special functions is in

7. H. Kober, *A Dictionary of Conformal Representations*, Dover Publications, New York, 1952.

Exercises

1. What are the images of the unit circle $|z| < 1$ under the mappings

(*a*)
$$w = \frac{z}{(1 - z)^2}$$

(*b*)
$$w = \frac{1 + z}{1 - z}.$$

2. Show that the above functions are univalent in $|z| < 1$.
3. Let $g(z)$ be univalent in D and $f(z)$ univalent in $g(D)$. Then $f(g(z))$ is univalent in D.
4. The function
$$f(z) = (1 + z)^n - 1$$
is surely not univalent in $|z| < 1$ if $n \geqq 6$, even though $f'(z) \neq 0$ there. In fact, $f(z)$ takes the same value at some two points of the circle
$$|z| \leq \frac{4}{n - 1}.$$
(*Hint.* Apply (37) to $f(Rz)$).
5. Show that the function (22) has the mapping property claimed for it.
6. Find a function which maps the unit circle onto the strip $|\mathbf{Re}\, w| < \pi/2$.
7. Using the result of exercises 1, 2, show that (30), (31) would be false if the number 1 were replaced by any smaller number.
8. Similarly, show that (33), (34) would be false if n were replaced by anything smaller.
9. Does the function
$$f(z) = 3 + z + 7z^2 + 4z^3 + 4z^4 + \cdots$$
have positive real part in the unit circle?
10. The maximum modulus of a certain function
$$f(z) = z + a_2 z^2 + \cdots$$
on the circle $|z| = r < 1$ is $r/(1 - r)^{3/2}$, the a_j being real.
(*a*) Can $f(|z| < 1)$ be convex?
(*b*) Can $f(|z| < 1)$ be schlicht?
(*c*) Can $f(z)$ take a real value off the real axis in $|z| < 1$?

11. A function $\phi(x, y)$ is subharmonic in a region D of the x-y plane if

$$\nabla^2\varphi(x, y) > 0$$

for all points of D. Is this property invariant under a conformal mapping?

12. Among all functions regular and univalent in the upper half-plane and mapping it onto the unit circle, find the one for which $|f'(2i)|$ is largest.

chapter 7

Extremum problems

7.1 INTRODUCTION

In this chapter we study certain kinds of maximum-minimum problems which lend themselves to solution. In its most general setting (though we study only specific ones), an extremum problem consists of a set of objects together with a rule which assigns a number to each object in the set. It is then required to find the largest (smallest) number assigned by the rule to any member of the set and to describe the members, if any, for which the maximum is attained.

Example 1. Let n be given, and let $\mathscr{S}$ be the set of all polynomials of degree n whose highest coefficient is unity. To any $f(x)$ in $\mathscr{S}$, assign the number

$$H[f] = \max_{-1 \leqq x \leqq 1} |f(x)|.$$

What is the minimum value taken by $H[f]$? What polynomial gives $H[f]$ its minimum value?

Example 2. Down which curve joining the points (0, 1), (1, 0) will a ball roll in the shortest possible elapsed time? The set here is the set of all continuous curves joining the given points. Assigned to each such curve C is a number $T[C]$ which is the time of descent.

Example 3. Let $\mathscr{S}$ denote the set of all functions

$$f(z) = z + a_2 z^2 + \cdots$$

regular in $|z| < 1$, and univalent there. To each such $f(z)$ assign the number $|a_5|$. What is the maximum value of this functional?

Example 4. What is the shortest possible way, starting from Washington D.C., to visit each state capitol exactly once and return home? The reader may supply the set and rule in question.

Example 5.† How badly can one err by calculating the reciprocal of the average of a function instead of the average of the reciprocal? Precisely, in the set $\mathscr{S}$ of all decreasing, concave functions joining $(0, \alpha)$ to $(1, \beta)$ which one maximizes the functional

$$H[f] = \left\{\int_0^1 f(t)\, dt \int_0^1 \frac{dt}{f(t)}\right\}^{-1}?$$

These examples were chosen primarily to display the diversity of possible extremum problems. A rather wide range of difficulty is also shown. Indeed, in Example 3 the answer is unknown. In Example 4 there are only finitely many possibilities, so that, in principle, the problem is trivial; in fact, however, since 50! is rather a large number, the practical question of solvability turns on the efficiency of any proposed algorithm, and in that sense the problem is unsolved. The other three examples have yielded to more or less difficult analyses, and the answers are known.

The first question that arises concerns, of course, the existence of a maximum rather than merely a least upper bound (possibly infinite). In physical applications this is not ordinarily an over-riding consideration, the existence being intuitively clear. Yet, in the final analysis, settling the question of existence normally involves establishing the continuity of the functional and the compactness of the set $\mathscr{S}$. The first of these is ordinarily obvious, although the second can be quite troublesome, as we saw in the previous chapter. Since the question of compactness was considered at length in that chapter, we are here disposed to relegate it to the background and to concentrate on methods for finding the maxima, when they exist.

7.2 FUNCTIONS OF REAL VARIABLES

To start with a simple situation, suppose there is given a continuously differentiable function $f(x)$ defined on the finite, closed, real interval $[a, b]$. Suppose we know that at a certain point ξ of that interval we have $f'(\xi) = 0$. Then we may, with absolute confidence, assert that at the point ξ the function has either a local maximum, a local minimum, or neither, the three possibilities being illustrated, respectively, by $-x^2$, x^2, x^3 at $x = \xi = 0$. In other words, we know no more than previously.

To put a converse problem, suppose the point $x = \xi$ gives an absolute maximum to the function $f(x)$ on the interval $[a, b]$. Is $f'(\xi) = 0$? That the answer is again in the negative can be seen by considering $f(x) = x^{-1}$ on the

† Wilkins [1].

interval [2, 3]. By way of review we summarize the positive assertions which can be made in this case.

Theorem 1. *Let $f(x)$ be twice continuously differentiable on the finite interval $[a, b]$. Let $\xi_1, \ldots, \xi_n$ be the interior points of (a, b) at which $f'(x)$ vanishes. Then the absolute maximum (minimum) of $f(x)$ on the interval $[a, b]$ is among the numbers*

$$f(a), f(\xi_1), \ldots, f(\xi_n), f(b).$$

Further, if at ξ_i we have $f''(\xi_i) < 0\,(>0)$, then the point $x = \xi_i$ gives to the function $f(x)$ a local maximum (minimum). Finally, at any local minimum or maximum ξ of $f(x)$ which lies interior *to (a, b), we have $f'(\xi) = 0$.*

The situation in two variables (x, y) is more complicated. Indeed, let $f(x, y)$ be defined in a region G of the x-y plane, and suppose all second partial derivatives of $f(x, y)$ exist and are continuous throughout G. Now, suppose that at a certain *interior* point (ξ, η) of G we have

$$\left.\frac{\partial f(x, y)}{\partial x}\right|_{(\xi,\eta)} = \left.\frac{\partial f(x, y)}{\partial y}\right|_{(\xi,\eta)} = 0.$$

Then at the point (ξ, η), $f(x, y)$ may or may not have a local extremum. To distinguish the various possibilities one has in this case the matrix

$$F = \begin{pmatrix} f_{xx} & f_{xy} \\ f_{xy} & f_{yy} \end{pmatrix}$$

which plays a role analogous to the second derivative in the one-dimensional case. If F is strictly positive definite, (ξ, η) is a local minimum; if strictly negative definite, a local maximum. If F is non-negative or nonpositive definite there is a possibility of an extremum which must be explored by looking at higher derivatives. Finally, if F is strictly indefinite, we have a situation with no parallel in the one-dimensional situation, where, say $f(x, \eta)$ has a local maximum, while $f(\xi, y)$ has a local minimum, or conversely. In the last case, (ξ, η) is called a *saddle-point* of the function $f(x, y)$, the name arising in an obvious way from the geometrical picture.

The purpose of the foregoing remarks has been not so much to explain the theory of multidimensional extremum problems as to point out that if the theory is already so complex in n-dimensions, it will be even more so in some of the problems to be considered, and the answers even less clearcut.

7.3 THE METHOD OF LAGRANGE MULTIPLIERS

Let $f(x_1, x_2, \ldots, x_n)$ be a function of n variables defined throughout space, and possessing n continuous first partial derivatives. It is desired to find the maximum value of f, where, however, we do not have complete freedom in the choice of $x_1, \ldots, x_n$, but, instead, the variables $x_1, \ldots, x_n$ are "constrained" by side conditions of the form

$$\begin{cases} \phi_1(x_1, \ldots, x_n) = 0 \\ \phi_2(x_1, \ldots, x_n) = 0 \\ \quad\cdot \\ \quad\cdot \\ \quad\cdot \\ \phi_m(x_1, \ldots, x_n) = 0. \end{cases} \tag{1}$$

In other words, among all the points $(x_1, \ldots, x_n)$ which satisfy (1) we are to find the points that give to $f(x_1, \ldots, x_n)$ a maximum value.

The "straightforward" approach to such a problem would consist in eliminating m of the variables, say $x_1, \ldots, x_m$ by expressing them in terms of the remaining $n - m$ variables through (1). One could then maximize $f(x_1, \ldots, x_n)$ considered as a function of $n - m$ variables in the usual way.

The drawbacks of such a procedure are at once evident and to a certain extent, cannot be overcome with any "gadget" if the functions $\phi_i(x_1, \ldots, x_n)$ are very complicated. Nonetheless, the method of Lagrange, to be presented below, invariably can claim one distinction, which is that it does not disturb the symmetry of a given problem by making arbitrary choices of variables to be eliminated. In many cases this is quite important.

We illustrate the method in the case $n = 4$, $m = 2$. Thus, suppose it is desired to maximize

$$f(x_1, x_2, x_3, x_4)$$

subject to the conditions

$$\phi_1(x_1, x_2, x_3, x_4) = 0 \tag{2}$$

$$\phi_2(x_1, x_2, x_3, x_4) = 0. \tag{3}$$

If we suppose

$$\frac{\partial(\varphi_1, \varphi_2)}{\partial(x_3, x_4)} \neq 0$$

where the symbol denotes the usual Jacobian determinant, then we can solve (2), (3) for

$$x_3 = \alpha(x_1, x_2) \tag{4}$$

$$x_4 = \beta(x_1, x_2). \tag{5}$$

This being done, substitute (4), (5) into the function $f(x_1, x_2, x_3, x_4)$, and the problem is then to maximize the function of two variables

$$f(x_1, x_2, \alpha(x_1, x_2), \beta(x_1, x_2))$$

with no constraints. The conditions $\partial f/\partial x_1 = \partial f/\partial x_2 = 0$ yield

$$(6) \qquad f_{x_1} + f_{x_3}\alpha_{x_1} + f_{x_4}\beta_{x_1} = 0$$
$$(7) \qquad f_{x_2} + f_{x_3}\alpha_{x_2} + f_{x_4}\beta_{x_2} = 0.$$

On the other hand, since

$$\varphi_1(x_1, x_2, \alpha(x_1, x_2), \beta(x_1, x_2)) = 0$$
$$\phi_2(x_1, x_2, \alpha(x_1, x_2), \beta(x_1, x_2)) = 0$$

identically, we obtain by differentiation

$$(8) \qquad \varphi_{1,x_1} + \varphi_{1,x_3}\alpha_{x_1} + \varphi_{1,x_4}\beta_{x_1} = 0$$
$$(9) \qquad \varphi_{2,x_1} + \varphi_{2,x_3}\alpha_{x_1} + \varphi_{2,x_4}\beta_{x_1} = 0.$$

Regard (6), (8), and (9) as a system of linear homogeneous equations in $1, \alpha_{x_1}, \alpha_{x_4}$. Then we have

$$(10) \qquad \begin{vmatrix} f_{x_1} & f_{x_3} & f_{x_4} \\ \varphi_{1,x_1} & \varphi_{1,x_3} & \varphi_{1,x_4} \\ \varphi_{2,x_1} & \varphi_{2,x_3} & \varphi_{2,x_4} \end{vmatrix} = 0$$

at a nontrivial solution x_1, x_2, x_3, x_4. Similarly regarding (7), (8), and (9), we get

$$(11) \qquad \begin{vmatrix} f_{x_2} & f_{x_3} & f_{x_4} \\ \varphi_{1\,x_1} & \varphi_{1,x_3} & \varphi_{1,x_4} \\ \varphi_{2,x_1} & \varphi_{2,x_3} & \varphi_{2,x_4} \end{vmatrix} = 0.$$

Conditions (2), (3), (10), and (11) are four conditions which determine all the stationary points of the function $f(x_1, x_2, x_3, x_4)$.

With Lagrange's method we would introduce two numbers λ_1, λ_2, the *Lagrange multipliers*, and consider the function

$$(12) \quad F(x_1, x_2, x_3, x_4;\ \lambda_1, \lambda_2) = f(x_1, x_2, x_3, x_4) + \lambda_1\varphi_1(x_1, x_2, x_3, x_4) + \lambda_2\varphi_2(x_1, x_2, x_3, x_4)$$

as a function of six variables to be maximized with no constraints. Differentiation with respect to the space variables yields

$$f_{x_1} + \lambda_1\varphi_{1,x_1} + \lambda_2\varphi_{2,x_1} = 0$$
$$f_{x_2} + \lambda_1\varphi_{1,x_2} + \lambda_2\varphi_{2,x_2} = 0$$
$$f_{x_3} + \lambda_1\varphi_{1,x_3} + \lambda_2\varphi_{2,x_3} = 0$$
$$f_{x_4} + \lambda_1\varphi_{1,x_4} + \lambda_2\varphi_{2,x_4} = 0.$$

If λ_1, λ_2 are eliminated from the last two and the result substituted in the first two, one obtains the relations (10), (11). Finally, the conditions

$F_{\lambda_1} = F_{\lambda_2} = 0$ are just (2), (3), and therefore the unconstrained problem (12) has the same stationary points as the original problem with the constraints (2), (3). The general result is

Theorem 2 (Lagrange). *It is required to find the extrema of a function $f(x_1, \ldots, x_n)$ subject to the conditions (1). Suppose the functions $f, \phi_1, \ldots, \phi_m$ have continuous first partial derivatives in a region G of space, and that throughout G,*

$$\frac{\partial(\phi_1, \phi_2, \ldots, \phi_m)}{\partial(x_{k_1}, x_{k_2}, \ldots, x_{k_m})} \neq 0$$

where $x_{k_1}, \ldots, x_{k_m}$ is some fixed choice from the variables $x_1, \ldots, x_n$. Then the extrema in G of the original problem, and of the function

$$F(x_1, x_2, \ldots, x_n; \lambda_1, \ldots, \lambda_m) \equiv f(x_1, \ldots, x_n) + \sum_{\nu=1}^{m} \lambda_\nu \phi_\nu(x_1, \ldots, x_n) \tag{13}$$

of $n + m$ variables, are identical.

As an illustration, we find the extreme values of

$$f(x_1, x_2, \ldots, x_n) = \sum_{\mu,\nu=1}^{n} a_{\mu\nu} x_\mu x_\nu$$

subject to the condition

$$\varphi(x_1, x_2, \ldots, x_n) = \sum_{\mu=1}^{n} x_\mu^{\,2} - 1 = 0, \tag{14}$$

where $a_{\mu\nu} = a_{\nu\mu}$ $(\mu, \nu = 1, \ldots, n)$. Lagrange's function is

$$F(x_1, \ldots, x_n; \lambda) = \sum_{\mu,\nu=1}^{n} a_{\mu\nu} x_\mu x_\nu + \lambda\left(\sum_{\mu=1}^{n} x_\mu^{\,2} - 1\right).$$

Then

$$\frac{\partial F}{\partial x_\mu} = 2\sum_{\mu=1}^{n} a_{\mu\nu} x_\nu + 2\lambda x_\mu = 0 \qquad (\mu = 1, 2, \ldots, n)$$

or

$$\sum_{\nu=1}^{n} a_{\mu\nu} x_\nu = -\lambda x_\mu \qquad (\mu = 1, 2, \ldots, n)$$

which, together with (14), determine $\lambda, x_1, \ldots, x_n$. Naturally these equations tell us what we already knew—that any stationary point $(x_1, \ldots, x_n)$ is an eigenvector of the given matrix.

7.4 THE FIRST PROBLEM OF THE CALCULUS OF VARIATIONS

We are given a function $f(x, u, v)$ defined in a certain region G of (x, u, v) space. It is required to find the function $y(x)$ which gives to the integral

$$H[y] = \int_a^b f(x, y(x), y'(x))\, dx \tag{15}$$

a maximum value, in the class of all sufficiently smooth functions $y(x)$ which pass through two given points (a, α), (b, β).

The problem just posed is extremely difficult, and we will not be able to solve it here. All we can do is give *necessary* conditions for a maximum or, in other words, certain conditions which a maximizing function must satisfy.

Now, suppose that we have actually found the required maximizing arc $y(x)$. We suppose that $y(x)$ is twice continuously differentiable, that $f(x, u, v)$ is continuous, and that the first and second partial derivatives of f exist and are continuous.

Since $y(x)$ gives a local maximum to the integral (15), it follows that if we "disturb" the function $y(x)$ a bit, $H[y]$ will decrease. The disturbance, however, must not get us out of the class $\mathscr{S}$ of functions satisfying the given end conditions

$$y(a) = \alpha \qquad y(b) = \beta. \tag{16}$$

Hence consider a disturbed function of the form

$$y(x) + \varepsilon\eta(x) \tag{17}$$

where $\eta(x)$ is differentiable on (a, b) and vanishes at the end points, and ε is a small parameter. Such a function $\eta(x)$ is called a *variation*, and one sometimes writes

$$\delta y(x) = \varepsilon\eta(x).$$

Now, consider the function of a single real variable

$$\varphi(\varepsilon) = H[y + \varepsilon\eta] = \int_a^b f(x, y(x) + \varepsilon\eta(x), y'(x) + \varepsilon\eta'(x))\, dx. \tag{18}$$

This function has, by hypothesis, a local maximum at $\varepsilon = 0$, and therefore $\varphi'(0) = 0$. But,

$$\varphi'(\varepsilon) = \int_a^b \left\{\frac{\partial f}{\partial y}(x, y + \varepsilon\eta, y' + \varepsilon\eta')\eta(x) + \frac{\partial f}{\partial y'}(x, y + \varepsilon\eta, y' + \varepsilon\eta')\eta'(x)\right\} dx$$

and therefore,

$$\begin{aligned} 0 = \varphi'(0) &= \int_a^b \left\{\frac{\partial f(x, y, y')}{\partial y}\eta(x) + \frac{\partial f(x, y, y')}{\partial y'}\eta'(x)\right\} dx \\ &= \int_a^b \frac{\partial f(x, y, y')}{\partial y}\eta(x)\, dx + \left[\eta(x)\frac{\partial f(x, y, y')}{\partial y'}\right]_a^b \\ &\quad - \int_a^b \frac{d}{dx}\left\{\frac{\partial f(x, y, y')}{\partial y'}\right\}\eta(x)\, dx \\ &= \int_a^b \left\{\frac{\partial f}{\partial y} - \frac{d}{dx}\frac{\partial f}{\partial y'}\right\}\eta(x)\, dx \end{aligned} \tag{19}$$

where we have integrated by parts, and used the relations $\eta(a) = \eta(b) = 0$. Now, (19) holds for every variation $\eta(x)$, and it is therefore reasonable to suspect that the quantity in braces must vanish identically. We have, in fact,

Lemma 1. Let $h(x)$ be a continuous function, and suppose that

$$\int_a^b h(x)\eta(x)\,dx = 0$$

for every continuously differentiable function $\eta(x)$ which vanishes at a and b. Then $h(x) \equiv 0$ on (a, b).

Proof. Suppose $h(\xi) \neq 0$, where ξ is some point of (a, b). Since $h(x)$ is continuous, $h(x)$ must remain positive in some neighborhood of ξ, say $\alpha \leqq x \leqq \beta$. Then take

$$\eta(x) = \begin{cases} (x-\alpha)^2(x-\beta)^2 & \alpha \leqq x \leqq \beta \\ 0 & \text{otherwise} \end{cases}$$

This function is clearly continuously differentiable everywhere, vanishes at $x = a$, $x = b$, and is positive when $\alpha < x < \beta$. Thus

$$\int_a^b h(x)\eta(x)\,dx > 0$$

for this $\eta(x)$, contradicting the hypothesis. Hence $h(x)$ is never positive and, similarly, never negative, which was to be shown.

We have proved

Theorem 3. *If $y(x)$ is a twice continuously differentiable function which gives to the integral*

$$\int_a^b f(x, y(x), y'(x))\,dx \tag{20}$$

a local maximum or minimum value in the class of all such functions which assume given values at $x = a$, $x = b$, and if, in addition, the function $f(x, u, v)$ has continuous second partial derivatives, then we have

$$\frac{d}{dx}\frac{\partial f(x, y(x), y'(x))}{\partial y'} - \frac{\partial f(x, y(x), y'(x))}{\partial y} = 0 \tag{21}$$

for $a \leqq x \leqq b$.

Equation (21), normally a (nonlinear) differential equation of the second order, is called the *Euler equation* for the problem of finding the extreme values of (20) in the given class of functions. It gives only a necessary and by no means a sufficient condition for an extremum.

7.5 SOME EXAMPLES

The brachistochrone problem is that of determining the curve joining two given points down which a ball will roll in least time. More precisely, we are to find, in the class of all functions $y(x)$ which are twice differentiable on $[0, b]$, and which pass through the points $(0, h)$, $(b, 0)$, the function for which the elapsed time

$$T[y] = \frac{1}{\sqrt{2g}} \int_0^b \sqrt{(1 + y'^2)/y}\, dx$$

is least. Ignoring the constant,

$$f(x, y, y') = \left\{\frac{1 + y'^2}{y}\right\}^{1/2}$$

$$\frac{\partial f(x, y, y')}{\partial y} = -\tfrac{1}{2} y^{-3/2} \sqrt{1 + y'^2}$$

$$\frac{\partial f(x, y, y')}{\partial y'} = \frac{y'}{\sqrt{y(1 + y'^2)}}$$

and Euler's equation is

$$\frac{d}{dx}\left[\frac{y'}{\sqrt{y(1 + y'^2)}}\right] = -\tfrac{1}{2} y^{-3/2} \sqrt{1 + y'^2}.$$

Another example is that of finding the shortest distance between two given points. In Euclidean space of two dimensions, we are to minimize

$$\int_a^b \sqrt{1 + y'^2}\, dx$$

in the class of all twice differentiable arcs of joining $(a, y(a))$, $(b, y(b))$, where $y(a)$, $y(b)$ are given. In this case, $f_y = 0$, and Euler's equation is just

$$\frac{d}{dx}\left\{\frac{y'}{\sqrt{1 + y'^2}}\right\} = 0$$

which integrates at once to

$$y' = c_1 \sqrt{1 + y'^2}$$

whence

$$y'^2 = \frac{c_1^{\,2}}{1 - c_1^{\,2}}$$

is constant, and the required curve is a straight line. In this problem the existence of a minimizing curve was obvious (as was the answer). The method generalizes to finding the shortest distance between two points on an arbitrary surface, the Euler equations then representing the differential equations of geodesics on the surface.

Concerning the actual integration of Euler's equation, three important special cases arise rather frequently in practice.

(*a*) The function $f(x, y, y')$ is independent of y'. Then the equation reduces simply to

$$\frac{\partial f(x, y)}{\partial y} = 0$$

which determines the function $y(x)$ implicitly, though no arbitrary constants will be present. This means that the end conditions cannot be arbitrarily prescribed.

(*b*) The function $f(x, y, y')$ is independent of y. In this case (21) is

$$\frac{d}{dx}\frac{\partial f(x, y')}{\partial y'} = 0$$

which integrates at once to

$$\frac{\partial f(x, y')}{\partial y'} = c_1.$$

We may now solve for

$$y'(x) = g(x, c_1)$$

algebraically, and then

$$y(x) = \int g(x, c_1)\, dx.$$

(*c*) The function $f(x, y, y')$ is independent of x. In this case we have

$$\frac{d}{dx}\left(y'\frac{\partial f(y, y')}{\partial y'} - f(y, y')\right) = y''\frac{\partial f(y, y')}{\partial y'} + y'\frac{d}{dx}\frac{\partial f(y, y')}{\partial y'}$$
$$- \frac{\partial f(y, y')}{\partial y}y' - \frac{\partial f(y, y')}{\partial y'}y'' = y'\left\{\frac{d}{dx}\frac{\partial f}{\partial y'} - \frac{\partial f}{\partial y}\right\}.$$

Hence, for any solution of Euler's equation we have

$$\frac{d}{dx}\left\{y'\frac{\partial f}{\partial y'} - f\right\} = 0$$

and thus

$$y'\frac{\partial f(y, y')}{\partial y'} - f(y, y') = c_1.$$

We then find

$$y' = g(y, c_1)$$

by algebraic means and have

$$x + c_2 = \int \frac{dy}{g(y, c_1)}$$

as the solution.

7.6 DISTINGUISHING MAXIMA FROM MINIMA

In the preceding section we found an equation whose solution might be either a local maximum for the problem, a local minimum, or neither. Although we shall never be able to resolve these cases completely, we can, nonetheless, find an analogue of the conditions

$$f''(x) > 0, \quad f''(x) < 0$$

which distinguish maxima and minima for functions of a single real variable.

To do this, we return to the function $\varphi(\varepsilon)$ of (18) and calculate $\varphi''(0)$. We have

$$\begin{aligned}\varphi''(0) &= \int_a^b \left\{ f_{yy}\eta^2(x) + 2f_{yy'}\eta(x)\eta'(x) + f_{y'y'}\eta'^2(x) \right\} dx \\ &= \int_a^b (\eta(x), \eta'(x)) \begin{pmatrix} f_{yy} & f_{yy'} \\ f_{yy'} & f_{y'y'} \end{pmatrix} \begin{pmatrix} \eta(x) \\ \eta'(x) \end{pmatrix} dx.\end{aligned}$$

Thus, suppose the matrix

$$F = \begin{pmatrix} f_{yy} & f_{yy'} \\ f_{yy'} & f_{y'y'} \end{pmatrix} \tag{22}$$

is positive definite, the functions being evaluated at $(x, y(x), y'(x))$, where $y(x)$ is a solution of Euler's equation, and the definiteness holds for all $a \leqq x \leqq b$. Then, clearly, $\varphi''(0) > 0$, the function $\varphi(\varepsilon)$ having a local minimum at $\varepsilon = 0$. Hence the function $y(x)$ in question provides the integral with a local minimum with respect to differentiable perturbations $\eta(x)$.

In other words, sufficient conditions for a local minimum are

$$f_{y'y'} > 0 \tag{23a}$$

$$f_{yy}f_{y'y'} - f_{yy'}{}^2 > 0 \tag{23b}$$

while sufficient conditions for a local maximum are

$$f_{y'y'} < 0 \tag{24}$$

and (23b).

These conditions are very strong and are not necessary, only sufficient. To get an idea of what is necessary, suppose we have a function $y(x)$ which gives a local minimum. Then $\varphi''(0) \geqq 0$ for every choice of the variation $\eta(x)$. We choose a particular $\eta(x)$ which is zero except for a small interval around a point ξ in (a, b).

For instance,

$$\eta(x) = \begin{cases} \sqrt{h}\left\{1 - \dfrac{|x - \xi|}{h}\right\} & |x - \xi| \leq h \\ 0 & |x - \xi| > h \end{cases}$$

Then, for $|x - \xi| \leq h$, $\eta'^2(x) = 1/h$ and by substitution,

$$\begin{aligned} \varphi''(0) &= \int_a^b \{f_{yy}\eta^2(x) + 2f_{yy'}\eta(x)\eta'(x) + f_{y'y'}\eta'^2(x)\}\,dx \\ &= \int_{\xi-h}^{\xi+h}\left\{f_{yy}h\left\{1 - \frac{|x-\xi|}{h}\right\}^2 + 2f_{yy'}\left\{\pm\left[1 - \frac{|x-\xi|}{h}\right]\right\} + f_{y'y'}\frac{1}{h}\right\}dx \\ &= \left\{\max_{|x-\xi|\leq h} |f_{yy}|\right\}O(h^2) + 2\left\{\max_{|x-\xi|\leq h} |f_{yy'}|\right\}O(h) + \frac{1}{h}\int_{\xi-h}^{\xi+h} f_{y'y'}\,dx \quad (h \to 0) \\ &= o(1) + 2f_{y'y'}(\xi, y(\xi), y'(\xi)) \qquad (h \to 0). \end{aligned}$$

Since $\varphi''(0) \geqq 0$, we see by making $h \to 0$ that $f_{y'y'}(\xi, y(\xi), y'(\xi)) \geqq 0$. Since ξ was arbitrary, we must have

$$f_{y'y'} \geqq 0 \tag{25}$$

throughout (a, b).

Theorem 4. *Let $y(x)$ satisfy Euler's differential equation. For $y(x)$ to provide a local minimum with respect to differentiable variations, it is necessary that* (25) *hold, and sufficient that* (23) *hold, for $a \leq x \leq b$.*

The requirement (25) is called Legendre's condition.

7.7 PROBLEMS WITH SIDE CONDITIONS

In the problem already considered, the class of admissible functions consisted of all those differentiable functions which pass through two given points. We consider now problems with additional side conditions.

An example of such a problem is that of finding among all curves with given perimeter (arc length) the one which encloses the greatest area. This is the classical isoperimetric problem, and it is formulated by asking for the function $y(x)$ which maximizes the functional

$$H[y] = \int_a^b y(x)\,dx$$

in the class of all functions $y(x)$ satisfying $y(a) = y_1$, $y(b) = y_2$, with

$$\int_a^b \sqrt{1 + y'^2}\,dx = \text{given}.$$

Hence we pose the general question of finding the extreme values of

$$H[y] = \int_a^b f(x, y, y')\, dx \tag{26}$$

subject to the conditions $y(a) = y_1$, $y(b) = y_2$ and

$$\int_a^b g(x, y, y')\, dx = K \tag{27}$$

where K is given.

As before, suppose we have found a function $y(x)$ which gives a stationary value to $H[y]$. We wish to disturb $y(x)$ again so as to find the analytical conditions for the extremum. Yet the disturbance must be carried out in such a way that the perturbed functions remain always inside the class of functions being considered. This means that functions like

$$y(x) + \varepsilon\eta(x)$$

will no longer do, even if $\eta(a) = \eta(b) = 0$, for (27) will not, in general, be satisfied identically in ε. We consider then the function

$$y(x) + \varepsilon_1\eta_1(x) + \varepsilon_2\eta_2(x)$$

where $\eta_1(x)$, $\eta_2(x)$ each vanish at the endpoints, and the parameters ε_1, ε_2 are not independent, but are connected by the condition

$$\psi(\varepsilon_1, \varepsilon_2) = \int_a^b g(x, y + \varepsilon_1\eta_1 + \varepsilon_2\eta_2, y^1 + \varepsilon_1\eta_1{}^1 + \varepsilon_2\eta_2{}^1)\, dx = K. \tag{28}$$

Our problem then is that the function

$$\phi(\varepsilon_1, \varepsilon_2) = \int_a^b f(x, y + \varepsilon_1\eta_1 + \varepsilon_2\eta_2, y' + \varepsilon_1\eta_1' + \varepsilon_2\eta_2')\, dx \tag{29}$$

is to be stationary at $\varepsilon_1 = \varepsilon_2 = 0$ with respect to values of ε_1, ε_2 which satisfy (28).

The problem is now reduced to one of extremizing a function of two real variables with a side condition. We know from the theory of Lagrange multipliers (Theorem 2) that for some number λ, we have

$$\frac{\partial}{\partial\varepsilon_1}\{\varphi(\varepsilon_1, \varepsilon_2) + \lambda\psi(\varepsilon_1, \varepsilon_2)\}_{\varepsilon_1=\varepsilon_2=0} = 0$$

$$\frac{\partial}{\partial\varepsilon_2}\{\varphi(\varepsilon_1, \varepsilon_2) + \lambda\psi(\varepsilon_1, \varepsilon_2)\}_{\varepsilon_1=\varepsilon_2=0} = 0.$$

The first of these gives

$$\int_a^b \{f_y\eta(x) + f_{y'}\eta_1'(x) + \lambda g_y\eta_1(x) + \lambda g_{y'}\eta_1'(x)\}\, dx = 0$$

and the second gives

$$\int_a^b \{f_y\eta_2(x) + f_{y'}\eta_2'(x) + \lambda g_y\eta_2(x) + \lambda g_{y'}\eta_2'(x)\}\,dx = 0.$$

Integrating by parts,

$$\int_a^b \{[f_y - f_{y'}'] + \lambda[g_y - g_{y'}']\}\eta_1(x)\,dx = 0$$

$$\int_a^b \{[f_y - f_{y'}'] + \lambda[g_y - g_{y'}']\}\eta_2(x)\,dx = 0.$$

If $g_y \not\equiv g_{y'}'$, then the first of these states that λ is independent of $\eta_2(x)$, and the second shows that

$$f_y - f_{y'}' + \lambda(g_y - g_{y'}') = 0$$

or

$$(f + \lambda g)_y - \frac{d}{dx}(f + \lambda g)_{y'} = 0. \tag{30}$$

This is the Euler equation of the problem. The three constants c_1, c_2, λ introduced by the solution of the equation and the Lagrange multiplier are determined by the end conditions $y(a) = y_1$, $y(b) = y_2$, and (27).

As an illustration, we find the curve of length L which joins the points (0, 0), (1, 0), and encloses the maximum area between itself and the x-axis. Hence we wish to maximize

$$\int_0^1 y(x)\,dx$$

in the class of all differentiable arcs $y(x)$ satisfying $y(0) = y(1) = 0$ and

$$\int_0^1 \sqrt{1 + y'^2}\,dx = L.$$

Equation (30) becomes, after putting

$$f + \lambda g = y + \lambda\sqrt{1 + y'^2}$$

the differential equation

$$\frac{d}{dx}\left\{\frac{\lambda y'}{\sqrt{1 + y'^2}}\right\} = 1.$$

After differentiation this reads,

$$\frac{y''}{(1 + y'^2)^{3/2}} = \frac{1}{\lambda}$$

which says that the radius of curvature is constant, the required extremal is an arc of a circle, and λ is the radius.

7.8 SEVERAL UNKNOWN FUNCTIONS OR INDEPENDENT VARIABLES

Suppose it is desired to find functions $u_1(x), \ldots, u_n(x)$, which give to the integral

$$\int_a^b f(x, u_1(x), \ldots, u_n(x), u_1'(x), \ldots, u_n'(x))\, dx$$

extreme values, where $u_k(a)$, $u_k(b)$ $(k = 1, \ldots, n)$ are given. Just as before we introduce variations $\eta_1(x), \ldots, \eta_n(x)$ and consider

$$\varphi(\varepsilon_1, \varepsilon_2, \ldots, \varepsilon_n) = \int_a^b f(x, u_1 + \varepsilon_1\eta_1, \ldots, u_n + \varepsilon_n\eta_n, \\ u_1' + \varepsilon_1\eta_1', \ldots, u_n' + \varepsilon_n\eta_n')\, dx.$$

The requirement that this function be stationary at $\varepsilon_1 = \varepsilon_2 = \cdots = \varepsilon_n = 0$, regardless of the choice of $\eta_1(x), \ldots, \eta_n(x)$, leads in a straightforward way to the Euler equations

$$\frac{d}{dx}\frac{\partial f}{\partial u_i'} = \frac{\partial f}{\partial u_i} \qquad (i = 1, 2, \ldots, n). \tag{31}$$

In general, we now have to solve a *system* of n coupled ordinary differential equations of the second order, in n unknown functions.

In a similar way, if it is desired to find the extrema of

$$\iint_R f(x, y, u(x, y), v(x, y), u_x(x, y), v_x(x, y), u_y(x, y), v_y(x, y))\, dx\, dy$$

where $u(x, y)$, $v(x, y)$ have boundary values given around the boundary C of the region R of the x-y plane, the Euler equations take the form

$$\begin{aligned} \frac{\partial}{\partial x}\left(\frac{\partial f}{\partial u_x}\right) + \frac{\partial}{\partial y}\left(\frac{\partial f}{\partial u_y}\right) &= \frac{\partial f}{\partial u} \\ \frac{\partial}{\partial x}\left(\frac{\partial f}{\partial v_x}\right) + \frac{\partial}{\partial y}\left(\frac{\partial f}{\partial v_y}\right) &= \frac{\partial f}{\partial v}. \end{aligned} \tag{32}$$

These are two partial differential equations of the second order which, with the given boundary values, will normally determine the required functions. We omit the derivation of these equations, which is perfectly straightforward except for integration by parts in two dimensions, which results from an integral theorem of Gauss. Generalizations to more than two functions are obvious, and as usual, the conditions are only necessary, the question of sufficiency being quite difficult.

7.9 THE VARIATIONAL NOTATION

We have already remarked that if $y(x)$ is an arbitrary function, and $\eta(x)$ a variation, then we write, symbolically,

$$\begin{aligned} y(x) + \varepsilon\eta(x) &= y(x) + \delta y(x) \\ \delta y(x) &= \varepsilon\eta(x). \end{aligned} \tag{35}$$

Further, on replacing $y(x)$ by $y(x) + \varepsilon\eta(x)$, a given function $f(x\ y, y')$ changes to

$$\begin{aligned} f(x, y + \varepsilon\eta, y' + \varepsilon\eta') &= f(x, y, y') + \varepsilon \frac{\partial f}{\partial y}\eta + \varepsilon\frac{\partial f}{\partial y'}\eta' + O(\varepsilon^2) \\ &= f(x, y, y') + \frac{\partial f}{\partial y}\,\delta y + \frac{\partial f}{\partial y'}\,\delta y' + O(\varepsilon^2) \\ &= f + \delta f + O(\varepsilon^2) \end{aligned}$$

where

$$\delta f = \frac{\partial f}{\partial y}\,\delta y + \frac{\partial f}{\partial y'}\,\delta y'. \tag{36}$$

We note that if y changes to $y + \varepsilon\eta$, then y' changes to $y' + \varepsilon\eta'$, whence

$$\delta(y') = \varepsilon\eta' = (\varepsilon\eta)' = (\delta y)'$$

that is, the symbols d/dx, δ commute with each other. The usual formal rules of calculus have analogues here, such as

$$\begin{aligned} \delta(uv) &= u\,\delta v + v\,\delta u \\ \delta(u + v) &= \delta u + \delta v \\ \delta\left(\frac{u}{v}\right) &= \frac{v\,\delta u - u\,\delta v}{v^2} \end{aligned}$$

all of which can be verified by recourse to the definitions. In variational notation, a "proof" of the necessity of Euler's equation look like this:

$$\begin{aligned} \delta H[y] = 0 = \delta\int_a^b f(x, y, y')\,dx \\ = \int_a^b \delta f(x, y, y')\,dx \\ = \int_a^b (f_y\,\delta y + f_{y'}\,\delta y')\,dx \\ = \int_a^b (f_y - f'_{y'})\,\delta y\,dx. \end{aligned}$$

Such arguments are often quite useful as indicators of the direction in which the truth lies but are not in themselves proofs, for nothing was assumed and nothing proved.

As an illustration of a situation in which the variational notation considerably reduces the amount of labor involved, we consider a problem which is, in a sense, inverse to the ones we have been so far considering. We ask: Of what variational problem is the differential equation

$$\frac{d}{dx}\left(a(x)\frac{dy}{dx}\right) + b(x)y + \lambda c(x)y = 0 \tag{37}$$

$$y(x_0) = y(x_1) = 0$$

the Euler equation?

To answer this, multiply both sides of (37) by a variation $\delta y(x)$ and integrate over (x_0, x_1), getting

$$\int_{x_0}^{x_1} dx\left\{\delta y \frac{d}{dx}\left(a(x)\frac{dy}{dx}\right) + b(x)y\,\delta y + \lambda c(x)y\,\delta y\right\} = 0.$$

Now, using the relation

$$y\,\delta y = \tfrac{1}{2}\delta(y^2)$$

and integrating the first term by parts, we find

$$\int_{x_0}^{x_1} dx\left\{-a(x)\frac{dy}{dx}(\delta y)' + \frac{b(x)}{2}\,\delta y^2 + \frac{\lambda c(x)}{2}\,\delta y^2\right\} = 0.$$

Now, since

$$\frac{dy}{dx}(\delta y)' = y'(\delta y)' = y'\,\delta y' = \tfrac{1}{2}\delta y'^2$$

we have

$$\int_{x_0}^{x_1} \delta\{-a(x)y'^2(x) + [b(x) + \lambda c(x)]y^2(x)\}\,dx = 0$$

or equivalently,

$$\lambda\delta\int_{x_0}^{x_1} c(x)y^2(x)\,dx - \delta\int_{x_0}^{x_1}\{a(x)y'^2(x) - b(x)y^2(x)\}\,dx = 0.$$

Hence, if we write

$$H_1[y] = \int_{x_0}^{x_1} c(x)y^2(x)\,dx$$

$$H_2[y] = \int_{x_0}^{x_1}\{a(x)y'^2(x) - b(x)y^2(x)\}\,dx$$

then we have found that

$$\delta(H_2 - \lambda H_1) = 0$$

This would be the variational condition for finding the maximum value of H_2 subject to the side condition $H_1 = 1$. If we omit the normalization $H_1 = 1$, we can say, equivalently, that the function satisfying (37) is an extremal of the problem of finding the stationary values of the ratio

$$\lambda = \frac{H_2}{H_1} = \frac{\displaystyle\int_{x_0}^{x_1} \{a(x)y'^2(x) - b(x)y^2(x)\}\, dx}{\displaystyle\int_{x_0}^{x_1} c(x)y^2(x)\, dx}. \tag{38}$$

A rigorous proof can now be supplied by starting with (38) and working back to (37), the result being that the eigenvalues of the problem (37) are the stationary values of the ratio (38).

This technique is particularly useful in some branches of physics, where it is known as constructing a "variational principle." As an illustration of the utility of such a procedure, consider the question of determining the eigenvalues of the problem

$$\begin{cases} y'' + \lambda y = 0 \\ y(0) = y(1) = 0 \end{cases} \tag{39}$$

which are readily found, analytically, to be $\pi^2, 4\pi^2, 9\pi^2, \ldots$. According to (38) these eigenvalues are the stationary values of the ratio

$$\lambda = \frac{\displaystyle\int_0^1 y'^2\, dx}{\displaystyle\int_0^1 y^2\, dx}. \tag{40}$$

The smallest eigenvalue of the problem is then the least value of the ratio, so that if we choose any trial function satisfying the boundary conditions, say

$$y_0(x) = x(1 - x)(1 + \alpha x) \tag{41}$$

then we shall have

$$\lambda_{\min} \le \frac{\displaystyle\int_0^1 y_0'^2\, dx}{\displaystyle\int_0^1 y_0^{\,2}\, dx} = \frac{\frac{92}{15}\alpha^2 - \frac{5}{3}\alpha + \frac{1}{3}}{-\frac{58}{35}\alpha^2 + \frac{17}{10}\alpha + \frac{1}{30}} \tag{42}$$

for any value of α. The next step would be to differentiate this expression and find the optimal choice of α. We leave it simply with the remark that, even with $\alpha = 0$, the right side of (42) is 10, which is already fairly close to $\pi^2 = 9.87\ldots$.

7.10 THE MAXIMIZATION OF LINEAR FUNCTIONS WITH CONSTRAINTS

We begin with a simple example, the so-called "vitamin problem." Suppose there are two foods, say I and II. Food I contains one unit of vitamin A per pound, four units of vitamin B per pound, and costs three dollars per pound. For food II, the corresponding numbers are 1, 1, 2. If the minimum requirements of vitamins A and B are one unit and two units, respectively, the problem is to buy, at least cost, the required amount of each vitamin.

If the amounts of foods I and II bought are x and y, respectively, the total cost is

$$C = 3x + 2y. \tag{43}$$

This is to be minimized subject to the side conditions

$$x + y \geqq 1 \tag{44}$$

$$4x + y \geqq 2 \tag{45}$$

$$x \geqq 0, \quad y \geqq 0. \tag{46}$$

The restrictions (44) through (46) delineate a certain admissible portion of the x-y plane, which is sketched below.

Also shown in the figure is a line of constant cost $3x + 2y = C$. Geometrically, the problem is to find the line of constant cost with the least value of C, which still has a point in common with the shaded region. So stated, the answer is obvious: one simply draws the line $3x + 2y = C$ which passes through the vertex of the shaded region, since, clearly, no smaller value of C meets the requirements. This vertex is at $(\frac{1}{3}, \frac{2}{3})$, and the line is

$$3x + 2y = \frac{7}{3}$$

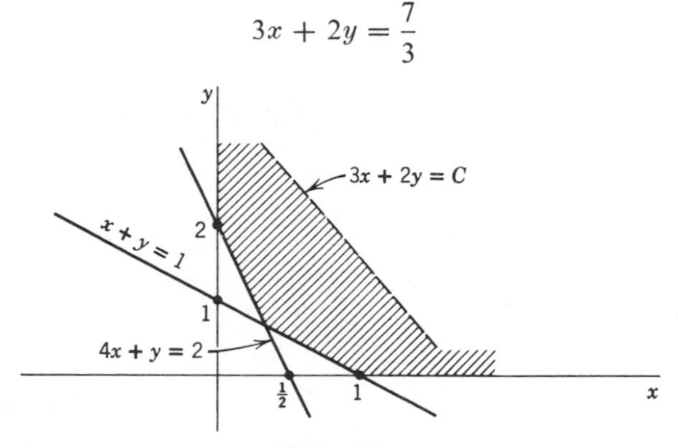

Figure 7.1

The solution of our problem, then, is to buy $\frac{1}{3}$ pound of food I and $\frac{2}{3}$ pound of food II, the minimum cost being \$2.33.

The most general problem of the type just considered is to maximize

$$P = C_1x_1 + \cdots + C_nx_n \tag{47}$$

subject to the conditions

$$\sum_{j=1}^{n} \alpha_{ij}x_j \leqq \beta_i \qquad (i = 1, 2, \ldots, m) \tag{48}$$

$$x_i \geqq 0 \qquad (i = 1, 2, \ldots, n). \tag{49}$$

In matrix form, we have to maximize $(\mathbf{c}, \mathbf{x})$, subject to

$$A\mathbf{x} \leqq \mathbf{b} \tag{50}$$

$$\mathbf{x} \geqq 0. \tag{51}$$

Geometrically, the constraints (50), (51) normally define a polyhedron K in n-space; the function to be maximized is a hyperplane in that space; and we wish to determine the "highest" position of the hyperplane which still intersects K.

We remark that K is invariably a convex polyhedron, for, suppose the points $x_1 = (x_{11}, \ldots, x_{1n})$, $x_2 = (x_{21}, \ldots, x_{2n})$ both belong to K (i.e., satisfy (50), (51)). If t is any number in $0 < t < 1$, we have

$$\begin{aligned} A(tx_1 + (1 + t)x_2) &= tAx_1 + (1 - t)Ax_2 \\ &\leqq tb + (1 - t)b \\ &= b \end{aligned}$$

and, clearly,

$$tx_1 + (1 - t)x_2 \geqq 0.$$

Hence, if the points x_1, x_2 belong to K, so does the line segment joining x_1 to x_2, and K is convex.

An *extreme point* of a convex set K is a point x of K such that x does not lie on any line segment of positive length which lies in K. More precisely, x is an extreme point of K if the conditions

(i) $x = tx_1 + (1 - t)x_2$
(ii) $0 < t < 1$
(iii) x_1, x_2, x in K

imply that

(iv) $x_1 = x_2 = x$.

If K is a polyhedron, the extreme points of K are just its vertices.

Now, suppose that the convex set K defined by (48), (49) is nonempty and bounded (i.e., contained in some sphere). Then, by Weierstrass' theorem, the linear function P attains its maximum at a point of K (or possibly at several).

We claim that among the points at which P attains its maximum value there is an extreme point of K, so that, in searching for the maximum, it is enough to restrict attention to the vertices of K.

To prove this, suppose K' is the set of all points of K at which $P(x)$ assumes its maximum value M. We have to show that in K' there is a vertex of K. Now K' is a convex set, for, if x_1, x_2 are in K', we have

$$\begin{aligned} P(tx_1 + (1 - t)x_2) &= tP(x_1) + (1 - t)P(x_2) \\ &= tM + (1 - t)M \\ &= M \end{aligned}$$

by linearity of $P(x)$, for every $0 \leqq t \leqq 1$. Hence, let x^* be any vertex of K'. We claim x^* is also a vertex of K, for, if not, we could write

$$x^* = tx_1 + (1 - t)x_2 \tag{52}$$

for some $0 < t < 1$, x_1, x_2 in K. Then we would have

$$\begin{aligned} P(x^*) &= M \\ &= tP(x_1) + (1 - t)P(x_2). \end{aligned}$$

Thus $P(x^*)$ lies between the smaller and larger of the numbers $P(x_1)$, $P(x_2)$. Since $P(x^*)$ is a maximum, it follows that $P(x_1) = P(x_2) = P(x^*) = M$. Hence x_1, x_2 are in K', but then (52) contradicts the assertion that x^* is a vertex of K', completing the proof.

Theorem 5. *The inequalities* (48), (49) *define a convex subset K of n-dimensional Euclidean space. If K is nonempty and bounded, then $P(x)$ of* (47) *assumes its maximum value in K at a vertex of K.*

7.11 THE SIMPLEX ALGORITHM

In order to solve the problem of maximizing a linear function subject to inequalities, we have seen that it is enough to examine only the vertices of the polyhedron in question. The Simplex method is, in essence, just a systematic procedure for examining these vertices in such a way that one continually passes from a given vertex to another at which the desired function has a larger value. Furthermore, the choice of the next vertex is, in a sense, optimal in that the function value increases as much as it could possibly by any single step to an adjacent vertex.

Given the problem

$$\text{maximize} \quad P(x) = C_1x_1 + C_2x_2 + \cdots + C_nx_n$$

subject to

$$(53) \qquad \sum_{j=1}^{n} \alpha_{ij}x_j \leqq \beta_i \qquad (i = 1, \ldots, m)$$

$$x_i \geqq 0 \qquad (i = 1, \ldots, n)$$

the first step is to introduce additional variables $x_{n+1}, \ldots, x_{n+m}$ which "take up the slack" in the inequalities and so are referred to as *slack variables*. In terms of these, the problem becomes

$$(54) \qquad \text{maximize} \quad C_1x_1 + \cdots + C_nx_n$$

subject to

$$(55) \qquad x_{n+i} + \sum_{j=1}^{n} \alpha_{ij}x_j = \beta_i \qquad (i = 1, \ldots, m)$$

$$(56) \qquad x_i \geqq 0 \qquad (i = 1, 2, \ldots, n + m).$$

It is clear that this new problem in $n + m$ variables, in which the inequalities (53) have been transformed into equalities, is entirely equivalent to the original problem. There is, however, no point in distinguishing between the "slack" variables and the original variables, and so we rephrase the problem in the symmetric form (with new names for the variables)

$$(57) \qquad \text{maximize} \quad C_1x_1 + C_2x_2 + \cdots + C_nx_n$$

subject to

$$(58) \qquad \sum_{j=1}^{n} \alpha_{ij}x_j = \beta_i \qquad (i = 1, \ldots, m)$$

$$(59) \qquad x_i \geqq 0 \qquad (i = 1, \ldots, n).$$

The next step is to find any vertex of the problem whatever. This means that we must find a point $(x_1, x_2, \ldots, x_n)$ in which exactly m of the x_i are positive, $n - m$ are zero, and (58), (59) are satisfied. This step can be mechanized, but we will not discuss the mechanization here, for quite often a choice of such a point will be obvious.

Let us now exhibit the matrix of coefficients of (58) in the form

$$(60) \qquad \begin{pmatrix} \alpha_{11} & \alpha_{12} & \cdots & \alpha_{1n} & \beta_1 \\ \alpha_{21} & \alpha_{22} & \cdots & \alpha_{2n} & \beta_2 \\ \cdot & \cdot & & \cdot & \cdot \\ \cdot & \cdot & & \cdot & \cdot \\ \cdot & \cdot & & \cdot & \cdot \\ \alpha_{m1} & \alpha_{m2} & & \alpha_{mn} & \beta_m \end{pmatrix}$$

and we note that the given constraints (58) are independent, so that the rank of the matrix in (60) is exactly m. Regarding the columns of this matrix as column vectors $\mathbf{P}_1, \ldots, \mathbf{P}_n, \mathbf{P}_0$,

$$P_j = \begin{pmatrix} \alpha_{1j} \\ \alpha_{2j} \\ \cdot \\ \cdot \\ \cdot \\ \alpha_{mj} \end{pmatrix}; \quad P_0 = \begin{pmatrix} \beta_1 \\ \beta_2 \\ \cdot \\ \cdot \\ \cdot \\ \beta_m \end{pmatrix}$$

suppose the variables are so numbered that the vertex which we have found is described by $x_1 > 0, \ldots, x_m > 0, x_{m+1} = \cdots = x_n = 0$. Then (58), in vector form, is

$$x_1\mathbf{P}_1 + x_2\mathbf{P}_2 + \cdots + x_m\mathbf{P}_m = \mathbf{P}_0. \tag{61}$$

If z_0 denotes the value of the function (57) to be maximized, at this point, we have

$$z_0 = C_1x_1 + \cdots + C_mx_m. \tag{62}$$

Since the vectors $\mathbf{P}_1, \ldots, \mathbf{P}_m$ are linearly independent, we can express each of the vectors $\mathbf{P}_0, \mathbf{P}_1, \ldots, \mathbf{P}_n$ as linear combinations of them,

$$\mathbf{P}_j = \gamma_{1j}\mathbf{P}_1 + \gamma_{2j}\mathbf{P}_2 + \cdots + \gamma_{mj}\mathbf{P}_m \qquad (j = 1, \ldots, n) \tag{63}$$

and we define, finally,

$$\phi_j = C_1\gamma_{1j} + C_2\gamma_{2j} + \cdots + C_m\gamma_{mj} \qquad (j = 1, 2, \ldots, n). \tag{64}$$

Now, suppose there is a particular value of j for which

$$C_j > \phi_j. \tag{65}$$

Then we shall find another vertex at which the value of z is larger than the value (62). Indeed, let us multiply (63) by a positive number θ and subtract from (61). There results

$$(x_1 - \theta\gamma_{1j})\mathbf{P}_1 + (x_2 - \theta\gamma_{2j})\mathbf{P}_2 + \cdots + (x_m - \theta\gamma_{mj})\mathbf{P}_m + \theta\mathbf{P}_j = \mathbf{P}_0. \tag{66}$$

Next, multiply (64) by θ and subtract from (62), getting

$$(x_1 - \theta\gamma_{1j})C_1 + \cdots + (x_m - \theta\gamma_{mj})C_m + \theta C_j = z_0 + \theta(C_j - \phi_j). \tag{67}$$

Equation (66) asserts that the constraints will be satisfied if we change the values $x_1, \ldots, x_m, 0, \ldots, 0$, to $x_1 - \theta\gamma_{1j}, \ldots, x_m - \theta\gamma_{mj}, 0, \ldots, 0, \theta, 0, \ldots, 0$ for any θ such that all these quantities are non-negative.

Equation (67) states that by so changing the values of the x_i, the value of the function we seek to maximize will *increase* (see (65)) from z_0 to

$$z_0 + \theta(C_j - \phi_j).$$

It follows that the optimum choice of θ is, for the given j, the largest possible consistent with the conditions

$$x_i - \theta\gamma_{ij} \geqq 0 \qquad (i = 1, \ldots, m)$$

that is, the value

$$\theta = \min_i \left(\frac{x_i}{\gamma_{ij}}\right) \qquad (\gamma_{ij} > 0) \tag{68}$$

which is a possible choice provided that at least one γ_{ij} is >0. Furthermore, since there may be several values of j for which (65) holds, the best choice of θ at each stage is the one which maximizes the increase in z_0, i.e., the one for which

$$\left\{\min_i \left(\frac{x_i}{\gamma_{ij}}\right)\right\}(C_j - \phi_j) \tag{69}$$

is as large as possible. Hence we first choose j to maximize (69), then θ according to (68). Having done so, it is clear that one, at least, of the numbers

$$x_1 - \theta\gamma_{1j}, x_2 - \theta\gamma_{2j}, \ldots, x_m - \theta\gamma_{mj}$$

is zero, and we have eliminated that variable and introduced the new variable x_j with the value θ. In other words, we have moved to the best possible adjacent vertex. We explicitly assume that no more than one of the above numbers is zero, referring the interested reader to the bibliography for the treatment of degenerate cases.

At this stage we are confronted with exactly the same situation with which we started, and the entire process can be repeated, with the result that one of the variables will be eliminated and another introduced, in such a way that z_0 will increase still more. The process must terminate after finitely many iterations because, since z always increases, no vertex will be encountered twice, and there are only a finite number of vertices.

The process terminates when either

(*a*) For some j, all $\gamma_{ij} \leqq 0$,

or

(*b*) for all j, $C_j \leqq \phi_j$.

Suppose the first alternative holds. By inspection of (66) it is clear that the variables remain non-negative for arbitrarily large values of θ. This means that the admissible polyhedron is unbounded, and the solution of our problem is $+\infty$. This case is exceptional in practice.

Next, suppose that alternative (*b*) holds, so that the process terminated because (65) was not fulfilled for any j. We claim that the current values of the variables at that time give the complete solution of the problem, i.e., for no other values of the variables can the current z_0 be increased.

To prove this, suppose $y_1, y_2, \ldots, y_n$ denote any other values of the variables which satisfy the constraints of the problem

$$y_1\mathbf{P}_1 + y_2\mathbf{P}_2 + \cdots + y_n\mathbf{P}_n = \mathbf{P}_0 \tag{70a}$$

$$C_1y_1 + C_2y_2 + \cdots + C_ny_n = z_1 \tag{70b}$$

$$y_1 \geqq 0, \ldots, y_n \geqq 0. \tag{70c}$$

We will show that z_0, the value of z for the point already found, is not less than z_1. Since $C_j \leqq \phi_j$ for all j (this is why the algorithm stopped), we see from (70b), (70c) that

$$z_1 \leqq \varphi_1y_1 + \varphi_2y_2 + \cdots + \varphi_ny_n. \tag{71}$$

Now, substitute the definition (64) of ϕ_j in (71), getting

$$\begin{aligned} z_1 &\leq \sum_{k=1}^{n} y_k\varphi_k \\ &= \sum_{k=1}^{n} y_k \sum_{l=1}^{m} C_l\gamma_{lk} \\ &= \sum_{l=1}^{m} \left\{ \sum_{k=1}^{n} \gamma_{lk}y_k \right\} C_l. \end{aligned} \tag{72}$$

Next, we substitute the expansion (63) of P_j into (70a) and find

$$\begin{aligned} \mathbf{P}_0 &= \sum_{k=1}^{n} \mathbf{P}_k y_k \\ &= \sum_{k=1}^{n} y_k \left\{ \sum_{l=1}^{m} \gamma_{lk}\mathbf{P}_l \right\} \\ &= \sum_{l=1}^{m} \left\{ \sum_{k=1}^{n} \gamma_{lk}y_k \right\} \mathbf{P}_l. \end{aligned} \tag{73}$$

Since the vectors $\mathbf{P}_1, \ldots, \mathbf{P}_m$ were assumed independent, the coefficients of the expansions (73) and (61) must agree, whence

$$x_l = \sum_{k=1}^{n} \gamma_{lk}y_k \qquad (l = 1, \ldots, m)$$

and substituting in (72),

$$z_1 \leq \sum_{l=1}^{m} x_lc_l = z_0$$

which was to be shown.

We have proved

Theorem 6.† *The Simplex algorithm will halt after a certain stage if and only if (a) the given inequalities were nonrestrictive and the solution is* $z = +\infty$ *or (b) the desired maximum is finite and has been found.*

7.12 ON BEST APPROXIMATION BY POLYNOMIALS

We conclude our study of extremum problems with some remarks about the approximation of given functions by polynomials in the "best possible" way.

Suppose there is given a continuous function $f(x)$ defined on a closed interval, which we suppose to be [0, 1]. We ask, first, whether such a function can be approximated arbitrarily well by polynomials, and, second, whether, if the degree n of the polynomial is specified, we can find a polynomial which does a better job than any other, in a sense to be specified.

The first question is completely settled by

Theorem 7. (*Weierstrass Approximation Theorem*). *Let* $f(x)$ *be continuous on* [0, 1]. *For any* $\varepsilon > 0$ *there is a polynomial* $P(x)$ *such that*

$$|f(x) - P(x)| < \varepsilon \qquad (0 \leqq x \leqq 1). \tag{74}$$

Proof. First, we claim that it is enough to prove the theorem when $f(0) = f(1) = 0$, for if that has been proved, then for any $f(x)$, define

$$g(x) = f(x) - f(0) - x\{f(1) - f(0)\}.$$

Then $g(0) = g(1) = 0$, whence $g(x)$ can be approximated as required, hence so can $f(x)$.

Supposing, then, that $f(0) = f(1) = 0$, we define $f(x)$ outside [0, 1] to be identically zero. Then $f(x)$ is everywhere continuous.

Consider the polynomials

$$\varphi_n(x) = \left\{2^{-(2n+1)} \frac{(2n+1)!}{n!^2}\right\}(1 - x^2)^n \qquad (n = 1, 2, \ldots) \tag{75}$$

which are so normalized that

$$\int_{-1}^{1} \varphi_n(x)\, dx = 1 \qquad (n = 1, 2, \ldots). \tag{76}$$

Now, by Stirling's formula,

$$2^{-(2n+1)} \frac{(2n+1)!}{(n!)^2} \sim \sqrt{\frac{n}{\pi}} \qquad (n \to \infty)$$

† Dantzig [1].

whence, on the interval $\delta \leqq |x| \leqq 1$

$$\varphi_n(x) = O(\sqrt{n}\,(1 - \delta^2)^n) \qquad (n \to \infty)$$

and therefore $\varphi_n(x)$ approaches zero uniformly on $\delta \leqq |x| \leqq 1$.

We claim that the polynomial

$$P(x) = \int_{-1}^{1} f(x + y)\varphi_n(y)\, dy \tag{77}$$

meets the conditions of the problem, if n is large enough. First, $P(x)$ is a polynomial, since, recalling that $f(x)$ vanishes outside [0, 1],

$$\begin{aligned} P(x) &= \int_{-x}^{1-x} f(x + y)\varphi_n(y)\, dy \\ &= \int_0^1 f(\xi)\varphi_n(\xi - x)\, d\xi \end{aligned}$$

the last expression being obviously a polynomial.

Since $f(x)$ is uniformly continuous, we can, for the given ε, find δ such that for $|x_1 - x_2| < \delta$,

$$|f(x_1) - f(x_2)| < \frac{\varepsilon}{2}.$$

Further, since a continuous function on a compact set is bounded, we have for all x,

$$|f(x)| \leqq A.$$

Then, for all x in [0, 1],

$$\begin{aligned} |P(x) - f(x)| &= \left| \int_{-1}^{1} f(x + y)\varphi_n(y)\, dy - f(x) \right| \\ &= \left| \int_{-1}^{1} f(x + y)\varphi_n(y)\, dy - \int_{-1}^{1} f(x)\varphi_n(y)\, dy \right| \\ &= \left| \int_{-1}^{1} [f(x + y) - f(x)]\varphi_n(y)\, dy \right| \\ &\leqq \int_{-1}^{1} |f(x + y) - f(x)|\, \varphi_n(y)\, dy \\ &= \int_{-1}^{-\delta} + \int_{-\delta}^{\delta} + \int_{\delta}^{1} \\ &\leqq 2A \int_{-1}^{-\delta} \varphi_n(y)\, dy + \frac{\varepsilon}{2} \int_{-\delta}^{\delta} \varphi_n(y)\, dy + 2A \int_{\delta}^{1} \varphi_n(y)\, dy \\ &= 2AO(\sqrt{n}(1 - \delta^2)^n) + \frac{\varepsilon}{2} \\ &< \varepsilon \end{aligned}$$

if n is sufficiently large, proving the theorem.

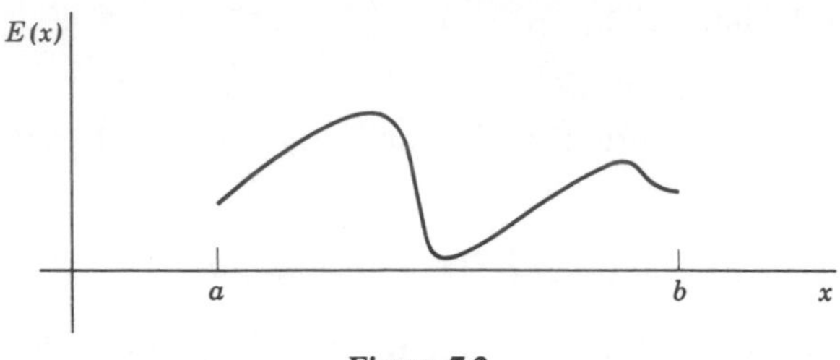

Figure 7.2

Having shown that approximating polynomials exist (indeed, the proof actually constructed them), we propose next to search for the best one of given degree. A natural criterion for the "best" one consists in looking at the error curve

$$E(x) = f(x) - P_n(x) \tag{78}$$

where $f(x)$ is given, and $P_n(x)$ is a polynomial, on the relevant interval $[a, b]$, and asking that the maximum error be as small as possible.

Our problem, then, is precisely this: *A continuous function $f(x)$ is given on an interval $[a, b]$, and an integer n is fixed. It is required to find a polynomial $P_n(x)$ of degree n for which*

$$\max_{a \leq x \leq b} |f(x) - P_n(x)|$$

is least.

A short intuitive argument will point the way to the solution. Indeed, suppose for a certain polynomial $P_n(x)$, the error curve $E(x)$ had the appearance shown above. One would then have a feeling that a better choice of $P_n(x)$ could be made, which would "push down" the maxima and the minimum so that they were more nearly equal. Carrying this feeling to its logical conclusion, one might conjecture that the best possible approximation $P_n(x)$ would result in an error curve in which the maxima and minima all had the same absolute value, and alternated with each other, as is indeed the case.

Theorem 8. *Let $f(x)$ be continuous on $[a, b]$, and let the integer n be given. Then*

(*a*) *there is a polynomial $P_n(x)$ of degree n for which*

$$\max_{a \leq x \leq b} |f(x) - P_n(x)| \tag{79}$$

is least.

(*b*) *for $P_n(x)$ to have this property, it is necessary and sufficient that $E(x) = f(x) - P_n(x)$ attain its maximum absolute value M at at least $n + 2$ points of $[a, b]$, and that the maxima of $E(x)$ alternate with its minima.*

(*c*) *the polynomial $P_n(x)$ is unique.*

Proof. To show the existence of a minimizing polynomial, first let $g_n(x)$ be any fixed polynomial of degree n, and put

$$M_0 = \max_{a \leqq x \leqq b} |f(x) - g_n(x)|.$$

Then, in seeking the minimum of (79), we may confine attention to the class of polynomials of degree n for which (79) is $\leqq M_0$. Since $f(x)$ is bounded on $[a, b]$, say $|f(x)| \leqq A$ there, we have for any polynomial $P(x)$ in this class, and x in $[a, b]$,

$$\begin{aligned} |P(x)| - |f(x)| &\leqq |f(x) - P(x)| \\ &\leqq \max_{a \leqq x \leqq b} |f(x) - P(x)| \\ &\leqq M_0. \end{aligned}$$

Hence

$$|P(x)| \leqq M_0 + A$$

for all $P(x)$ in the class considered.

Therefore, if

$$P(x) = c_0 + c_1 x + \cdots + c_n x^n$$

is one of these, choose $n + 1$ points $x_0, x_1, \ldots, x_n$ which are distinct and lie in $[a, b]$. Regard the equations

$$c_0 + c_1 x_i + c_2 x_i^2 + \cdots + c_n x_i^n = P(x_i) \qquad (i = 0, 1, \ldots, n)$$

as linear equations in $c_0, \ldots, c_n$. Since the matrix of coefficients is non-singular, we may solve for the c_i as linear combinations of the values $P(x_i)$,

$$c_i = \sum_{j=0}^{n} \gamma_{ij} P(x_j) \qquad (i = 0, 1, \ldots, n)$$

and

$$\begin{aligned} |c_i| &\leqq \sum_{j=0}^{n} |\gamma_{ij}| \, |P(x_j)| \\ &\leqq (M_0 + A) \sum_{j=0}^{n} |\gamma_{ij}| \qquad (i = 0, 1, \ldots, n) \end{aligned}$$

the bound on the right being independent of the particular polynomial chosen. Hence we are seeking the minimum of a continuous function

$$\varphi(c_0, c_1, \ldots, c_n) = \max_{a \leqq x \leqq b} |f(x) - c_0 - c_1 x - \cdots - c_n x^n|$$

of $n + 1$ variables in a *compact* subset of $n + 1$ dimensional space, and therefore the minimum is surely attained, and $P_n(x)$ exists.

Next we prove the sufficiency of condition (*b*), for, suppose $P_n(x)$ is a polynomial such that $f(x) - P_n(x)$ attains its maximum modulus M, with alternating signs, at $n + 2$ points of $[a, b]$. If $Q_n(x)$ is any other polynomial of degree n, we cannot have $|f(x) - Q_n(x)| < M$ throughout $[a, b]$ because then the polynomial

$$Q_n(x) - P_n(x) = [f(x) - P_n(x)] - [f(x) - Q_n(x)]$$

would be of alternating sign at the $n + 2$ points in question, and therefore would vanish at $n + 1$ points, which is impossible.

Next we show that the condition is necessary, for suppose the maximum error M is attained at fewer than $n + 2$ points. Then the interval $[a, b]$ can be divided into $n + 1$ subintervals, in each of which we have one or the other of the inequalities

$$-M \leqq f(x) - P_n(x) < M - \varepsilon \quad \text{or} \quad -M + \varepsilon < f(x) - P_n(x) \leqq M$$

where ε is a positive number. (This can be done by taking each subinterval to include one extremum of $f(x) - P_n(x)$.) Let $Q_n(x)$ be a polynomial which vanishes only at the n points common to two of these subintervals. Then $Q_n(x)$ is of constant sign inside each subinterval, and therefore, for some choice of the parameter η, we will have

$$|f(x) - P_n(x) - \eta Q_n(x)| < M$$

contradicting the extremal property of $P_n(x)$.

Finally, concerning uniqueness, suppose $P_n(x)$, $Q_n(x)$, $P_n(x) \not\equiv Q_n(x)$ are both extremals of our problem. Then, so is

$$R_n(x) = \frac{1}{2}\{P_n(x) + Q_n(x)\}$$

but $f(x) - R_n(x)$ attains its extrema at fewer than $n + 2$ points, which is impossible.

We conclude with a remark about the "normal" method of application of this theorem. If it is required to approximate a *differentiable* function $f(x)$ on $[a, b]$ by a polynomial of degree n in the best way, let

$$a = \xi_0 < \xi_1 < \cdots < \xi_n < \xi_{n+1} = b \tag{80}$$

be the $n + 2$ points referred to in the theorem. The equations

$$M = P(\xi_0) - f(\xi_0) = -[P(\xi_1) - f(\xi_1)] = \cdots = (-1)^{n+1}[P(\xi_{n+1}) - f(\xi_{n+1})] \tag{81}$$

$$P'(\xi_i) - f'(\xi_i) = 0 \qquad (i = 1, 2, \ldots, n) \tag{82}$$

constitute $2n + 2$ equations in the $n + 1$ coefficients of $P(x)$, n points $\xi_1, \ldots, \xi_n$ and M, which normally determine them uniquely if (80) is taken into account.

As an example, we approximate $f(x) = x^2$ on $[0, 1]$ by a linear function

$$P(x) = a + bx.$$

Here (81), (82) have the form

$$M = a = -[a + b\xi_1 - \xi_1^2] = a + b - 1$$

$$b - 2\xi_1 = 0$$

$$\xi_0 = 0, \quad \xi_2 = 1.$$

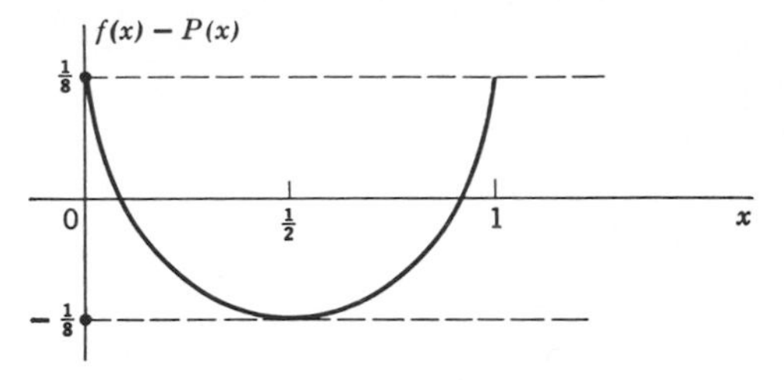

Figure 7.3

The solution

$$a = -\frac{1}{8}, \quad b = 1, \quad \xi_1 = \frac{1}{2}, \quad |M| = \frac{1}{8}$$

is readily found, the desired function being

$$P(x) = x - \tfrac{1}{8}.$$

The error curve has the appearance shown above.

Bibliography

The calculus of variations has been well treated in several references. For other introductory discussions, more complete than our own, see

1. R. Courant and D. Hilbert, *Methods of Mathematical Physics*, Interscience Publishers, New York, 1953,

as well as Chapter 2 in

2. F. B. Hildebrand, *Methods of Applied Mathematics*, Prentice-Hall, New York, 1952.

A thorough and rigorous development of the foundations of the subject is in

3. G. A. Bliss, *Lectures on the Calculus of Variations*, University of Chicago Press, 1946.

When confronted with an unusual or difficult problem, perhaps with differential side conditions, inequalities for side conditions, and so on, before consulting a digital computer consult

4. *Contributions to the Calculus of Variations*, University of Chicago Press, 1931.

The Simplex method was first enunciated in

5. G. Dantzig, *The Maximization of a Linear Function of Variables Subject to Linear Inequalities; Activity Analysis of Production and Allocation*, Cowles Commission Monograph 13, John Wiley and Sons, New York, 1951,

which remains an excellent reference. To see how the method actually appears on a computer, refer to

6. A. Ralston and H. Wilf, *Mathematical Methods for Digital Computers*, John Wiley and Sons, New York, 1960.

The subject of optimum polynomial approximation has an enormous literature. The early developments in the field are beautifully described in

7. S. Bernstein, *Leçons sur les Propriétés Extrémales et la Meilleure Approximation des Fonctions Analytiques d'une Variable Réele*, Gauthier-Villars, Paris, 1926,

which contains a seemingly endless parade of elegant problems and solutions.

Exercises

1. Find the maximum value of the function $x + y^2$ on the closed triangle whose vertices are (0, 0), (1, 0), (0, 1).
2. Among all polynomials of degree n, find the one which best approximates $f(x)$ in the least squares sense
$$\int_{-1}^{1} [f(x) - P(x)]^2\, dx = \text{minimum}$$
on the interval $[-1, 1]$.
3. Among all polynomials
$$z + a_2 z^2 + \cdots + a_n z^n$$
of given degree n which are univalent in $|z| < 1$, find one which has the largest possible $|a_n|$.
4. Let $\mathbf{a}$ be a given n-vector. Among all vectors $\mathbf{x}$ satisfying $(\mathbf{a}, \mathbf{x}) = 1$, minimize $(\mathbf{x}, \mathbf{x})$.
5. Discuss the following problem:
   Find a curve $y(x)$ connecting a point P and a given curve C such that the mean value of its slope is least.
6. How could you solve
$$\int F(x, y, y')\, dx \int G(x, y, y')\, dx = \text{minimum?}$$
7. Write down the Euler equation for the following problem:
   A ball rolls down a path of given length L joining P_1 and P_2. Minimize the time of descent. Find a first integral of the equations.
8. Find the Euler equations for the extreme values of
$$\int_a^b F(x, y(x), \quad y(x + h))\, dx$$
in the class of continuous functions $y(x)$ defined on $[a, b + h]$.
9. Maximize the function $x + 2y$ subject to the conditions
   (a) $y - x \leqq 1$
   (b) $y + \frac{1}{2}x \geqq 2$
   (c) $y - \frac{1}{2}x \leqq 3$
   (d) $x \leqq 5$
   (e) $x \geqq 0, \quad y \geqq 0$
   Do this first geometrically, then by the Simplex method starting from the vertex $(\frac{2}{3}, \frac{5}{3})$.
10. Find the best linear approximation to e^x on $[0, 1]$.

Solutions of the exercises

CHAPTER 1

1.
$$\begin{aligned}\|\mathbf{x}+\mathbf{y}\|^2 &= (\mathbf{x}+\mathbf{y}, \mathbf{x}+\mathbf{y}) = (\mathbf{x},\mathbf{x}) + (\mathbf{x},\mathbf{y}) + (\mathbf{y},\mathbf{x}) + (\mathbf{y},\mathbf{y}) \\ &= (\mathbf{x},\mathbf{x}) + 2\mathrm{Re}\,(\mathbf{x},\mathbf{y}) + (\mathbf{y},\mathbf{y}) \\ &\leqq (\mathbf{x},\mathbf{x}) + 2\,|(\mathbf{x},\mathbf{y})| + (\mathbf{y},\mathbf{y}) \\ &\leqq (\mathbf{x},\mathbf{x}) + 2\sqrt{(\mathbf{x},\mathbf{x})(\mathbf{y},\mathbf{y})} + (\mathbf{y},\mathbf{y}) \\ &= (\sqrt{(\mathbf{x},\mathbf{x})} + \sqrt{(\mathbf{y},\mathbf{y})})^2 \\ &= (\|\mathbf{x}\| + \|\mathbf{y}\|)^2\end{aligned}$$

2. Suppose $\Sigma c_\nu \mathbf{x}_\nu = 0$. Take the inner product of both sides with $\mathbf{x}_\mu$, and find $c_\mu(\mathbf{x}_\mu, \mathbf{x}_\mu) = 0$, whence $c_\mu = 0$ for each $\mu = 1, 2, \ldots$.

3. $P_0(x) =$ const., $P_1(x) =$ (const.) x, $P_2(x) =$ (const.) $(3x^2 - 1)$

4. (*a*) $(T^{-1})^* T^* = (TT^{-1})^* = I^* = I$

(*b*) obvious from definition of adjoint operator.

5. (*a*) For any $\mathbf{x}$,

$$(\mathbf{x}, T\mathbf{x}) = (T^*\mathbf{x}, \mathbf{x}) = -(T\mathbf{x}, \mathbf{x}) = -\overline{(\mathbf{x}, T\mathbf{x})}$$

hence $(\mathbf{x}, T\mathbf{x})$ is purely imaginary. If $T\mathbf{x} = \lambda\mathbf{x}$, $(\mathbf{x}, T\mathbf{x}) = \lambda(\mathbf{x}, \mathbf{x})$ shows that λ is purely imaginary.

(*b*) $S = \frac{1}{2}(S + S^*) + \frac{1}{2}(S - S^*)$

6. Not necessarily, unless A is nonsingular, for example

$$\begin{pmatrix}0 & 1\\ 0 & 0\end{pmatrix}\begin{pmatrix}1 & 2\\ 3 & 4\end{pmatrix} = \begin{pmatrix}0 & 1\\ 0 & 0\end{pmatrix}\begin{pmatrix}5 & 6\\ 3 & 4\end{pmatrix}$$

7. B is clearly $m \times m$. Further,

$$B^* = (AA^*)^* = A^{**}A^* = AA^* = B.$$

8. The eigenvalues are $\lambda_1 = \lambda_2 = 0$. The only eigenvector is (const.) (1, 0), so A is not diagonalizable.

9. From $A^* = A$ we find $b = c$. From $A^2 = A$ we get

$$a^2 + b^2 = a,\ b(a + d) = b,\ b^2 + d^2 = d.$$

With the exceptional case $b = 0$, we find $a = 1$ or 0, $d = 1$ or 0. Otherwise, $a + d = 1$, $a = 1 - d$, $b = \pm\sqrt{d(1 - d)}$. Hence $0 \leqq d \leqq 1$, and we may write $d = \cos^2\theta$ for some θ. Then $b = \pm \sin\theta\cos\theta$, $a = \sin^2\theta$. Finally, if $m = \tan\theta$, we have found that all such matrices are

$$A = \begin{pmatrix} 0 & 0 \\ 0 & 0 \end{pmatrix}; \quad A = \begin{pmatrix} 1 & 0 \\ 0 & 0 \end{pmatrix}; \quad A = \begin{pmatrix} 0 & 0 \\ 0 & 1 \end{pmatrix};$$

$$A = \begin{pmatrix} 1 & 0 \\ 0 & 1 \end{pmatrix}; \quad A = \frac{1}{1 + m^2}\begin{pmatrix} m^2 & \pm m \\ \pm m & 1 \end{pmatrix}$$

The first projects on the null vector, the second on the x-axis, the third on the y-axis, the fourth on the whole space, the last on the line $y = mx$.

10. From $AA^* = I$ we find, as in (9), $a^2 + b^2 = 1$, $ac + bd = 0$, $c^2 + d^2 = 1$. Taking $a = \sin\theta$. $b = \cos\theta$, $c = \sin\psi$, $d = \cos\psi$, the second equation gives $\cos(\theta - \psi) = 0$, $\psi = \theta + \pi/2$, and the matrix is

$$A = \begin{pmatrix} \sin\theta & \cos\theta \\ -\cos\theta & \sin\theta \end{pmatrix}$$

11. (*i*) $A = IAI^{-1}$ (*ii*) If $A = PBP^{-1}$, then $B = (P^{-1})A(P^{-1})^{-1}$
(*iii*) If $A = PBP^{-1}$, $B = QCQ^{-1}$, $A = (PQ)C(PQ)^{-1}$.

12. $A \sim \Lambda,\ B \sim \Lambda \Rightarrow A \sim \Lambda,\ \Lambda \sim B \Rightarrow A \sim B.$

13.
$$\text{Tr}[A, B] = \sum_{i=1}^{n} [A, B]_{ii} = \sum_{i=1}^{n}\left\{\sum_{k=1}^{n} a_{ik}b_{ki} - \sum_{k=1}^{n} b_{ik}a_{ki}\right\}$$
$$= \sum_{i,k} a_{ik}b_{ki} - a_{ki}b_{ik} = 0.$$

14.
$$A^2 - 6A + 8I = 0, \text{ or } 8A^{-1} = 6I - A = \begin{pmatrix} 3 & -1 \\ -1 & 3 \end{pmatrix}$$

and
$$A^{-1} = \begin{pmatrix} \frac{3}{8} & -\frac{1}{8} \\ -\frac{1}{8} & \frac{3}{8} \end{pmatrix}$$

15. The companion matrix is

$$A = \begin{pmatrix} -11 & 2 & -1 \\ 1 & 0 & 0 \\ 0 & 1 & 0 \end{pmatrix}$$

The matrix

$$A^2 = \begin{pmatrix} 123 & -23 & 11 \\ -11 & 2 & -1 \\ 1 & 0 & 0 \end{pmatrix}$$

clearly has the required property.

16. If $A\mathbf{x} = \lambda\mathbf{x}$ then $A^2\mathbf{x} = \lambda A\mathbf{x} = \lambda^2\mathbf{x}, \cdots, A^n\mathbf{x} = \lambda^n\mathbf{x}, \cdots$. Hence

$$f(A)\mathbf{x} = \{c_0 I + c_1 A + \cdots + c_m A^m\}\mathbf{x} = (c_0 + c_1\lambda + \cdots + c_m\lambda^m)\mathbf{x} = f(\lambda)\mathbf{x}$$

17. Since the eigenvalues of $\tilde{A}$ separate those of A at most one negative eigenvalue is introduced by the bordering process. Writing (108) in the form

$$\phi(\lambda) = \sum_{i=1}^{n} \frac{|(\mathbf{u}, \mathbf{x}_i)|^2}{\lambda - \lambda_i} - \lambda + \alpha = 0$$

we see that just to the left of the smallest eigenvalue λ_1, of A, $\phi(\lambda)$ is large and negative. To avoid introducing a negative eigenvalue the graph of $\phi(\lambda)$ must cross the λ-axis again between $\lambda = 0$, and $\lambda = \lambda_1$. The condition for this is simply $\phi(0) \geqq 0$, or

$$\sum_{i=1}^{n} \frac{|(\mathbf{u}, \mathbf{x}_i)|^2}{\lambda_i} \leqq \alpha.$$

To put this in more manageable form, expand $\mathbf{u}$ in the $\mathbf{x}_i$,

$$\mathbf{u} = (\mathbf{u}, \mathbf{x}_1)\mathbf{x}_1 + \cdots + (\mathbf{u}, \mathbf{x}_n)\mathbf{x}_n.$$

Then

$$A^{-1}\mathbf{u} = \frac{(\mathbf{u}, \mathbf{x}_1)}{\lambda_1}\mathbf{x}_1 + \cdots + \frac{(\mathbf{u}, \mathbf{x}_n)}{\lambda_n}\mathbf{x}_n$$

and therefore $(\mathbf{u}, A^{-1}\mathbf{u}) = \sum_{i=1}^{n} \frac{|(\mathbf{u}, \mathbf{x}_i)|^2}{\lambda_i}$. The necessary and sufficient condition is thus

$$(\mathbf{u}, A^{-1}\mathbf{u}) \leqq \alpha.$$

18. (*a*) $(\mathbf{x}, (A + B)\mathbf{x}) = (\mathbf{x}, A\mathbf{x}) + (\mathbf{x}, B\mathbf{x}) > 0$. Yes.
(*b*) The eigenvalues of A^2 are $\lambda_1^2, \ldots, \lambda_n^2$, all > 0. Yes.
(*c*) As in (*b*), if and only if the numbers $f(\lambda_1), \ldots, f(\lambda_n)$ are positive.

19. Let $\mathbf{x}$ be a vector of E_m. Then

$$(\mathbf{x}, AA^*\mathbf{x}) = (A^*\mathbf{x}, A^*\mathbf{x}) > 0.$$

20. A is negative definite if and only if $-A$ is positive definite. Watching the signs of the discriminants, the conditions are

$$\Delta_1 < 0, \Delta_2 > 0, \Delta_3 < 0, \ldots$$

21. $\Delta_m = \det A_m =$ product of eigenvalues of $A_m =$ real.

22. If $A = U\Lambda U^*$, $\lambda_i > 0$, define $\sqrt{A} = U\sqrt{\Lambda}U^*$. Then

$$\sqrt{A}\sqrt{A} = U\sqrt{\Lambda}U^*U\sqrt{\Lambda}U^* = U\Lambda U^* = A.$$

For each of the n elements of Λ we can choose either sign for the square root. We find 2^n square roots in this way, one of which is positive definite.

23. (*a*)
$$\left|\sum_{n=1}^{N} n^{-s}\right|^2 = \sum_{n=1}^{N} n^{-s} \sum_{m=1}^{N} \overline{m^{-s}}$$
$$= \sum_{m,n=1}^{N} n^{-s} m^{-\bar{s}}$$
$$= \sum_{m,n=1}^{n} n^{-\sigma}(\cos t \log n$$
$$+ i \sin t \log n) m^{-\sigma}(\cos t \log m + i \sin t \log m)$$
$$= \sum_{m,n=1}^{N} m^{-\sigma} a_{mn} n^{-\sigma}.$$

(*b*)
$$a_{mn} = \cos(t \log m - t \log n) = \cos(t \log m)\cos(t \log n)$$
$$+ \sin(t \log m)\sin(t \log n) = f_m f_n + g_m g_n.$$

Thus
$$A = \mathbf{f}\mathbf{f}^T + \mathbf{g}\mathbf{g}^T$$
and is plainly of rank two.

(*c*)
$$A^T = (\mathbf{f}\mathbf{f}^T + \mathbf{g}\mathbf{g}^T)^T = A$$
$$(\mathbf{x}, A\mathbf{x}) = (\mathbf{x}, (\mathbf{f}\mathbf{f}^T + \mathbf{g}\mathbf{g}^T)\mathbf{x})$$
$$= (\mathbf{x}, \mathbf{f}\mathbf{f}^T\mathbf{x}) + (\mathbf{x}, \mathbf{g}\mathbf{g}^T\mathbf{x})$$
$$= (\mathbf{f}^T\mathbf{x}, \mathbf{f}^T\mathbf{x}) + (\mathbf{g}^T\mathbf{x}, \mathbf{g}^T\mathbf{x}) = |(\mathbf{f}, \mathbf{x})|^2 + |(\mathbf{g}, \mathbf{x})|^2 \geqq 0.$$

(*d*) If $A\mathbf{x} = \lambda\mathbf{x}$, then
$$\lambda\mathbf{x} = A\mathbf{x} = (\mathbf{f}\mathbf{f}^T + \mathbf{g}\mathbf{g}^T)\mathbf{x} = (\mathbf{f}, \mathbf{x})\mathbf{f} + (\mathbf{g}, \mathbf{x})\mathbf{g}.$$

Thus, if $\mathbf{x}$ is perpendicular to $\mathbf{f}$ and $\mathbf{g}$ and $\lambda = 0$, $\mathbf{x}$ is an eigenvector. This accounts for $N - 2$ eigenvalues and eigenvectors. If $\lambda \neq 0$, then $\mathbf{x}$ is a linear combination of $\mathbf{f}$ and $\mathbf{g}$. $\mathbf{x} = c_1\mathbf{f} + c_2\mathbf{g}$. Then
$$\lambda\mathbf{x} = (\lambda c_1)\mathbf{f} + (\lambda c_2)\mathbf{g} = \{c_1(\mathbf{f}, \mathbf{f}) + c_2(\mathbf{f}, \mathbf{g})\}\mathbf{f} + \{c_1(\mathbf{g}, \mathbf{f}) + c_2(\mathbf{g}, \mathbf{g})\}\mathbf{g}.$$

Hence
$$(\mathbf{f}, \mathbf{f})c_1 + (\mathbf{f}, \mathbf{g})c_2 = \lambda c_1$$
$$(\mathbf{f}, \mathbf{g})c_1 + (\mathbf{g}, \mathbf{g})c_2 = \lambda c_2$$
and the other two eigenvalues of A are the roots of
$$\begin{vmatrix} (\mathbf{f}, \mathbf{f}) - \lambda & (\mathbf{f}, \mathbf{g}) \\ (\mathbf{f}, \mathbf{g}) & (\mathbf{g}, \mathbf{g}) - \lambda \end{vmatrix} = 0.$$

24. Let A be the given matrix. Express A in the form
$$A = D + A_1 + A_2 + \cdots + A_p$$
where D is diagonal, and each A_i is of rank one. One way of doing this is to put the diagonal elements of A in D, the rest of the first column of A in A_1, the second

column of A in A_2, and so on. Since D^{-1} is known, find $(D + A_1)^{-1}$, then $(D + A_1 + A_2)^{-1}$, etc. from (118) terminating with A^{-1}. Each step uses about $2n^2$, multiplications, so about $2n^3$ are needed altogether.

25.
$$A^{-1} = A^*(AA^*)^{-1}.$$

26. We have to show that there is a vector $\mathbf{e}_{\nu+1}$ independent of $\mathbf{e}_1, \ldots, \mathbf{e}_\nu$. Suppose not. Then for every vector $\mathbf{x}$ the set $\mathbf{e}_1, \ldots, \mathbf{e}_\nu, \mathbf{x}$ is dependent. Hence every $\mathbf{x}$ is a linear combination of $\mathbf{e}_1, \ldots, \mathbf{e}_\nu$, i.e., $\mathbf{e}_1, \ldots, \mathbf{e}_\nu$ is a basis for E_n, which is absurd unless $\nu = n$. Thus vectors can be adjoined until there are n of them.

27. We know that for AB to be Hermitian we need $[A, B] = 0$. We will show that this also insures the definiteness of AB. Indeed, A and B are simultaneously diagonalizable:

$$A = U\Lambda_a U^*; \quad B = U\Lambda_b U^*; \quad \Lambda_a > 0; \quad \Lambda_b > 0.$$

Hence, for any $\mathbf{x} \neq \mathbf{0}$,

$$\begin{aligned}(\mathbf{x}, AB\mathbf{x}) &= (\mathbf{x}, U\Lambda_a\Lambda_b U^*\mathbf{x}) = (U^*\mathbf{x}, \Lambda_a\Lambda_b U^*\mathbf{x}) \\ &= (U^*\mathbf{x}, \sqrt{\Lambda_a\Lambda_b}\, \sqrt{\Lambda_a\Lambda_b}\, U^*\mathbf{x}) \\ &= (\sqrt{\Lambda_a\Lambda_b}\, U^*\mathbf{x}, \sqrt{\Lambda_a\Lambda_b}\, U^*\mathbf{x}) \\ &> 0.\end{aligned}$$

28. No. For example, if A and B are arbitrary, $A \smile I$, $I \smile B$ but $A \smile B$ may be false.

29. First if A is diagonal,

$$f(A) = \frac{1}{2\pi i} \oint_c \frac{f(z)}{(zI - A)}\, dz$$

is just n independent statements of the usual scalar Cauchy integral formula, since $zI - A$ is also diagonal, and $(zI - A)_{ii}^{-1} = (z - A_{ii})^{-1}$. Generally, if $A = PDP^{-1}$,

$$\begin{aligned}\frac{1}{2\pi i} \oint_c [(zI - PDP^{-1})^{-1} f(z)\, dz &= P \frac{1}{2\pi i} \oint_c (zI - D)^{-1} f(z)\, dz P^{-1} \\ &= Pf(D)P^{-1} = f(A).\end{aligned}$$

30.
$$(zI - A)^{-1} = \begin{pmatrix} \dfrac{1}{z - 1} & \dfrac{2}{(z - 1)(z - 2)} \\ 0 & \dfrac{1}{z - 2} \end{pmatrix}$$

Hence

$$\begin{aligned} f(A)_{11} &= \frac{1}{2\pi i} \oint_c \frac{f(z)}{z - 1}\, dz = f(1) \\ f(A)_{12} &= \frac{1}{2\pi i} \oint_c \frac{2f(z)\, dz}{(z - 1)(z - 2)} = 2[f(2) - f(1)] \\ f(A)_{21} &= 0 \\ f(A)_{22} &= \frac{1}{2\pi i} \oint_c \frac{f(z)}{z - 2}\, dz = f(2). \end{aligned}$$

In particular

$$e^{At} = \begin{pmatrix} e^t & 2(e^{2t} - e^t) \\ 0 & e^{2t} \end{pmatrix}$$

31.
$$\begin{aligned}(\mathbf{x}, (A - C)\mathbf{x}) &= (\mathbf{x}, (A - B + B - C)\mathbf{x}) \\ &= (\mathbf{x}, (A - B)x) + (x, (B - C)\mathbf{x}) \\ &\geqq 0.\end{aligned}$$

32. Suppose $r(B) = r$, $A = PBP^{-1}$. Let $\mathbf{e}_1, \ldots, \mathbf{e}_r$ be a basis for $\mathscr{R}(B)$. It is enough to show that $P\mathbf{e}_1, \ldots, P\mathbf{e}_r$ form a basis for $\mathscr{R}(A)$. First, these are independent since if

$$c_1 P\mathbf{e}_1 + \cdots + c_r P\mathbf{e}_r = 0$$

there follows $P(c_1\mathbf{e}_1 + \cdots + c_r\mathbf{e}_r) = 0$, but P is nonsingular, hence $c_1\mathbf{e}_1 + \cdots + c_r\mathbf{e}_r = 0$, and $c_1 = c_2 = \cdots = c_r = 0$ since the $\mathbf{e}_j$ are independent. Next, these vectors span $\mathscr{R}(A)$, for let $\mathbf{y}$ be any vector in $\mathscr{R}(A)$. Then $\mathbf{y} = A\mathbf{x}$ for some $\mathbf{x}$, hence $\mathbf{y} = PBP^{-1}\mathbf{x} = P(BP^{-1}\mathbf{x})$. But $BP^{-1}\mathbf{x}$ is clearly in $\mathscr{R}(B)$, and is therefore a linear combination of $\mathbf{e}_1, \ldots, \mathbf{e}_r$. Thus $y = P(BP^{-1}\mathbf{x})$ is a linear combination of $P\mathbf{e}_1, \ldots, P\mathbf{e}_r$, which was to be shown.

33. The matrix of exercise 8 has nullity one, but a zero eigenvalue of multiplicity two. If A is diagonalizable, $A \sim D$ for some diagonal D. By exercise (32), $r(A) = r(D)$; by theorem 26, $\nu(A) = \nu(D)$, which is clearly the multiplicity of the zero eigenvalue.

34. The condition states that if $d_1, d_2, \ldots, d_n$ are arbitrary,

$$a_{ij}(d_i - d_j) = 0 \qquad (i, j = 1, 2, \ldots, n).$$

Choosing $d_i \neq d_j$, $a_{ij} = 0$ for $i \neq j$.

35. They are $\mathbf{uu}^*$, $\mathbf{u}$ being an arbitrary column vector.

36. Assume $(I + \mathbf{uv}^*)^n = I + \sigma_n \mathbf{uv}^*$, σ_n to be determined. Then $(I + \mathbf{uv}^*)^{n+1} = (I + \mathbf{uv}^*)(I + \sigma_n\mathbf{uv}^*) = I + [\sigma_n + (\mathbf{v}, \mathbf{u})\sigma_n + 1]\mathbf{uv}^*$ and $\sigma_{n+1} = [1 + (\mathbf{v}, \mathbf{u})]\sigma_n + 1$, $\sigma_1 = 1$. Hence, $\sigma_n = \dfrac{1}{(\mathbf{v}, \mathbf{u})}\{[1 + (\mathbf{v}, \mathbf{u})]^n - 1\}$.

37. If $\mathbf{e}_1, \ldots, \mathbf{e}_r$ are a basis for $\mathscr{R}(B)$, $A\mathbf{e}_1, \ldots, A\mathbf{e}_r$ surely span $\mathscr{R}(AB)$, though they need not be independent. Hence $r(AB) \leqq r(B)$. Next

$$r(AB) = r((AB)^T) = r(B^T A^T) \leqq r(A^T) = r(A) \qquad QED.$$

38. No. If $[A\ \ B] = 0$, for in that case,

$$(A + B)^n = \sum_{k=0}^{n} \binom{n}{k} A^{n-k} B^k$$

and one finds $e^A e^B = e^{A+B}$ by comparing the power series developments of both sides.

39.
$$I = \{p + 1, p + 2, \ldots, n\}, J = \{1, 2, \ldots, p\}.$$

40. It is enough to show that if rows i and j are interchanged and columns i and j are also interchanged, the reducibility of A is unaffected. Let I and J be the sets of integers which show that A is reducible. It is easy to see that interchanging i and j in these two sets shows that the transformed matrix is still reducible.

CHAPTER 2

1. (*a*)
$$\mu_r = \int_0^\infty x^r e^{-x}\,dx = r! \qquad (r = 0, 1, 2, \ldots).$$

Thus the moments matrix is
$$M_{ij} = \mu_{i+j} = (i + j)! \qquad (i, j = 0, 1, 2, \ldots)$$

(*b*) $\phi_0(x) = 1$; for $n = 1$, equation (8) is
$$\mu_0 \alpha_0 = -\mu_1$$
and therefore $\alpha_0 = -\dfrac{\mu_1}{\mu_0} = -1$. Thus $\phi_1(x) = -1 + x$; for $n = 2$, (8) reads
$$\begin{pmatrix} \mu_0 & \mu_1 \\ \mu_1 & \mu_2 \end{pmatrix} \begin{pmatrix} \alpha_0 \\ \alpha_1 \end{pmatrix} = - \begin{pmatrix} \mu_2 \\ \mu_3 \end{pmatrix}$$
or
$$\begin{pmatrix} 1 & 1 \\ 1 & 2 \end{pmatrix} \begin{pmatrix} \alpha_0 \\ \alpha_1 \end{pmatrix} = - \begin{pmatrix} 2 \\ 6 \end{pmatrix}$$

Solving, $\alpha_0 = 2$, $\alpha_1 = -4$ and $\phi_2(x) = 2 - 4x + x^2$

(*c*) $\phi_1(x) = 0$ at $x = 1$. $\phi_2(x) = 0$ at $x = 2 \pm \sqrt{2}$. These are real, distinct, and lie in $(0, \infty)$ as required.

2. This is merely the statement that the eigenvectors of the symmetric matrix J are orthogonal (see Theorem 5 and (69)).

3. (*a*) $\det(xI - J)$ is zero when x is an eigenvalue of J, i.e., when x is a zero of $\phi_N(x)$. Hence the two polynomials of degree N, $\phi_N(x)$ and $\det(xI - J)$ have the same zeros and therefore differ by at most a multiplicative constant, which must be k_N, as can be seen by matching the coefficients of the highest power of x.

(*b*) This is the Cayley-Hamilton Theorem (Theorem 19 of Chapter 1) for the matrix J.

4. The zeros of $\phi_{n+1}(x)$ are the eigenvalues of the $n \times n$ symmetric matrix J_n bordered by a row and column to make J_{n+1}. The result follows at once from Theorem 21 of Chapter 1.

5. Putting $x = x_i$, a zero of $\phi_{n+1}(x)$, in the recurrence relation (24), we find
$$\phi_n(x_i) = -\frac{k_n k_{n-2}}{k_{n-1}^2}\,\phi_{n-2}(x_i)$$
and $\phi_n(x_i)$, $\phi_{n-2}(x_i)$ must have opposite sign, as required.

6. (*a*) If $L_n(x) = \lambda_n \tilde{L}_n(x)$, where $\tilde{L}_n(x)$ are normalized, we find, after substituting, that
$$x\tilde{L}_n(x) = (2n + 1)\tilde{L}_n(x) - \frac{\lambda_{n+1}}{\lambda_n}\,\tilde{L}_{n+1}(x) - n^2\,\frac{\lambda_{n-1}}{\lambda_n}\,\tilde{L}_{n-1}(x).$$

The normalization condition is

$$\frac{\lambda_n}{\lambda_{n-1}} = n^2 \frac{\lambda_{n+1}}{\lambda_n}$$

or

$$\frac{\lambda_n}{\lambda_{n-1}} = n.$$

Thus $\lambda_n = \lambda_0 n!$, and the recurrence is

$$x\tilde{L}_n(x) = (2n+1)\tilde{L}_n(x) - (n+1)\tilde{L}_{n+1}(x) - n\tilde{L}_{n-1}(x)$$

(*b*)
$$J = \begin{pmatrix} 1 & -1 \\ -1 & 3 \end{pmatrix}$$

Eigenvalues are roots of $\lambda^2 - 4\lambda + 2 = 0$, as in problem 1*c*.

(*c*) Here $k_n/k_{n+1} = n + 1$, as can be seen by comparing the recurrence found in (*a*) with (24) and changing the polynomials so that their highest coefficient is positive. Along the right-hand sides of (47) the largest occurs when $i = N - 2$; hence the zeros of $L_N(x)$ are in the interval $\left(0, \beta_{N-2} + \frac{k_{N-3}}{k_{N-2}} + \frac{k_{N-2}}{k_{N-1}}\right)$, namely, the interval $(0, 4N - 6)$.

(*d*) From problem 1, part (*c*), $x_1 = 2 - \sqrt{2}$, $x_2 = 2 + \sqrt{2}$. From (68),

$$H_1 = -\frac{k_2}{k_1}\frac{1}{L_1(x_1)L_2'(x_1)}$$

$$H_2 = -\frac{k_2}{k_1}\frac{1}{L_1(x_2)L_2'(x_2)}.$$

Instead of normalizing the polynomials, as is required in these formulas, we calculate

$$\frac{H_1}{H_2} = \frac{L_1(x_2)L_2'(x_2)}{L_1(x_1)L_2'(x_1)} = \frac{(1-x_2)(2x_2-4)}{(1-x_1)(2x_1-4)} = \frac{(1+\sqrt{2})\,2\sqrt{2}}{(\sqrt{2}-1)(2\sqrt{2})} = 3 + 2\sqrt{2}$$

and $H_1 + H_2 = \int_0^\infty e^{-x}\,dx = 1$. Solving, $H_1 = \frac{1}{2} + \frac{\sqrt{2}}{4}$, $H_2 = \frac{1}{2} - \frac{\sqrt{2}}{4}$. The complete formula is

$$\int_0^\infty e^{-x} f(x)\,dx = \left(\frac{1}{2} + \frac{\sqrt{2}}{4}\right) f(2 - \sqrt{2}) + \left(\frac{1}{2} - \frac{\sqrt{2}}{4}\right) f(2 + \sqrt{2})$$

which is exact for polynomials $f(x)$ of degree $\leq 2 \cdot 2 - 1 = 3$.

(*e*) The exact answer is $4! = 24$. Approximately, we find

$$\int_0^\infty e^{-x}x^4\,dx \simeq \left(\frac{1}{2} + \frac{\sqrt{2}}{4}\right)(2 - \sqrt{2})^4 + \left(\frac{1}{2} - \frac{\sqrt{2}}{4}\right)(2 + \sqrt{2})^4$$

$$= 20$$

the error being 16%, in this case.

7. (*a*) The recurrence is $T_{n+1}(x) = 2xT_n(x) - T_{n-1}(x)$. The relation $T_n(x) = \cos(n\cos^{-1}x)$ is plainly true for $n = 0$ and $n = 1$. If true, inductively, for $n = 0, 1, \ldots, N$, then, writing $x = \cos\theta$, what we have to show is that

$$\cos(N+1)\theta = 2\cos\theta\cos N\theta - \cos(N-1)\theta,$$

an easy identity.

(*b*) $T_N(x) = 0$ when $\cos(N\cos^{-1}x) = 0$, i.e., when

$$N\cos^{-1}x = (k+\tfrac{1}{2})\pi \qquad (k = 0, 1, \ldots).$$

Thus the zeros are

$$x_{Nk} = \cos(k+\tfrac{1}{2})\frac{\pi}{N} \qquad (k = 0, 1, \ldots, N-1).$$

(*c*) First,

$$\begin{aligned} T_{N-1}(x_{Nk}) &= \cos\left((N-1)(k+\tfrac{1}{2})\frac{\pi}{N}\right) \\ &= \cos(k+\tfrac{1}{2})\pi\cos(k+\tfrac{1}{2})\frac{\pi}{N} + \sin(k+\tfrac{1}{2})\pi\sin\left((k+\tfrac{1}{2})\frac{\pi}{N}\right) \\ &= (-1)^k\sin(k+\tfrac{1}{2})\frac{\pi}{N} \end{aligned}$$

and

$$\begin{aligned} T_N'(x_{Nk}) &= \frac{N\sin(N\cos^{-1}x_{Nk})}{\sqrt{1-x_{nk}^2}} \\ &= \frac{N\sin(k+\tfrac{1}{2})\pi}{\sin(k+\tfrac{1}{2})\pi/N} = N(-1)^k\frac{1}{\sin(k+\tfrac{1}{2})\pi/N}. \end{aligned}$$

From (68), since $k_n = 2^n$, $\gamma_n = \dfrac{\pi}{2}$,

$$\begin{aligned} H_k &= \frac{2^N}{2^{N-1}}\frac{\pi}{2}\frac{1}{T_{N-1}(x_{Nk})T_N'(x_{Nk})} \\ &= \pi\frac{\sin(k+\tfrac{1}{2})\pi/N}{(-1)^k[\sin(k+\tfrac{1}{2})\pi/N](N)(-1)^k} \\ &= \frac{\pi}{N} \qquad (k = 1, 2, \ldots, N). \end{aligned}$$

(*d*)

$$\begin{aligned} \int_{-1}^{1} T_m(x)T_n(x)\frac{dx}{\sqrt{1-x^2}} &= \int_{-\pi}^{\pi} T_m(\cos\theta)\,T_n(\cos\theta)\,d\theta \\ &= \int_{-\pi}^{\pi}\cos m\theta\cos n\theta\,d\theta \\ &= 0 \qquad (m \neq n). \end{aligned}$$

(*e*) The identity is

$$\frac{1}{2} + \sum_{n=1}^{N}\cos\theta\cos n\psi = \frac{\cos(N+1)\theta\cos N\psi - \cos N\theta\cos(N+1)\psi}{\cos\theta - \cos\psi}.$$

8. (*a*) The formula is

$$\int_0^1 f(x)\,dx = H_0(f(x_1) + f(x_2)).$$

Putting $f(x) = 1, x, x^2$, we find, successively,

$$H_0 = \tfrac{1}{2}$$
$$\tfrac{1}{2} = \tfrac{1}{2}(x_1 + x_2)$$
$$\tfrac{1}{3} = \tfrac{1}{2}({x_1}^2 + {x_2}^2)$$

with solution $x_1 = \frac{1}{2}(1 - \sqrt{\frac{1}{3}})$, $x_2 = \frac{1}{2}(1 + \sqrt{\frac{1}{3}})$, which are in (0, 1).

(*b*) Proceeding as above, we get for $N = 2$, $H_0 = \frac{1}{2}$, $x_1 = 0$, $x_2 = 2$. For $H = 3$, however, the points x_1, x_2, x_3 turn out to be the three roots of the cubic equation

$$x^3 - 3x^2 + \tfrac{3}{2}x - \tfrac{3}{2} = 0$$

which, as can be seen from a graph, has only one real root, and no such formula exists.

9. This is merely a restatement of (69), using Theorem 5.

10. (*a*) The nth partial sum is

$$S_n(x) = \frac{1}{2} + \sum_{k=1}^{n} \cos kx = \frac{\sin (2n + 1)x/2}{2 \sin x/2}$$

which clearly does not approach a limit as $n \to \infty$, no matter what x is.

(*b*) The nth Féjér mean is

$$\sigma_{n-1}(x) = \frac{1}{n}\sum_{k=0}^{n-1} S_k(x) = \frac{1}{2n \sin x/2}\sum_{k=0}^{n-1} \sin (2k + 1)\frac{x}{2}$$
$$= \frac{1}{2n}\left(\frac{\sin nx/2}{\sin x/2}\right)^2$$

If $\sin x/2 \neq 0$, then, as $n \to \infty$, the quantity in the parentheses remains bounded, and hence $\sigma_{n-1}(x) \to 0$. At any point x where $\sin x/2 = 0$, the quantity in parentheses is n^2, and $\sigma_{n-1}(x)$ diverges to $+\infty$ as $n \to \infty$. Hence the series is summable to zero except at the points $x_k = \pm 2k\pi$ ($k = 0, 1, \ldots$).

(*c*) For the given a, we have

$$\lim_{n\to\infty} \int_{-a}^{a} f(t)\sigma_n(t)\,dt = \lim_{n\to\infty}\int_{-\delta}^{\delta} f(t)\sigma_n(t)\,dt + \lim_{n\to\infty}\left(\int_{-a}^{-\delta} + \int_{\delta}^{a}\right) f(t)\sigma_n(t)\,dt.$$

The second limit vanishes, according to part (*b*), since $\sigma_n(t) \to 0$ uniformly on $[-a, -\delta]$ and $[\delta, a]$. Hence the limit we seek is independent of the choice of a. Next, since $\sigma_n(t) \geqq 0$, we have

$$\min_{(-\delta,\delta)} f(t)\int_{-\delta}^{\delta} \sigma_n(t)\,dt \leqq \int_{-\delta}^{\delta} f(t)\,\sigma_n(t)\,dt \leqq \max_{(-\delta,\delta)} f(t)\int_{-\delta}^{\delta} \sigma_n(t)\,dt.$$

But, by (115),

$$\int_{-\pi/2}^{\pi/2} \sigma_n(t)\,dt = \frac{\pi}{2} = \int_{-\delta}^{\delta} \sigma_n(t)\,dt + \left\{\int_{-\pi/2}^{-\delta} + \int_{\delta}^{\pi/2}\right\}\sigma_n(t)\,dt.$$

Taking the limit, $n \to \infty$,

$$\lim_{n\to\infty} \int_{-\delta}^{\delta} \sigma_n(t)\,dt = \frac{\pi}{2}$$

independent of δ. Hence, taking the limit in the preceding equation,

$$\frac{\pi}{2} \min_{(-\delta,\delta)} f(t) \leqq \lim_{n\to\infty} \int_{-a}^{a} f(t)\sigma_n(t)\,dt \leqq \max_{(-\delta,\delta)} f(t)\frac{\pi}{2}.$$

Since this is valid for arbitrary $\delta > 0$, we can make $\delta \to 0$, and find

$$\lim_{n\to\infty} \int_{-a}^{a} f(t)\sigma_n(t)\,dt = \frac{\pi}{2} f(0)$$

as required.

CHAPTER 3

1. This matrix has the properties (*i*) $a_{1j} \neq 0$ ($j = 1, \ldots, n$) and (*ii*) $a_{j+1,j} \neq 0$ ($j = 1, \ldots, n - 1$). Suppose A is reducible, and let I, J be the sets of integers which show this. If i_0 is the largest integer in I, then by (*ii*), $i_0 - 1$ is not in J; hence it is in I. Thus I consists of $1, 2, \ldots, i_0$ and J consists of $i_0 + 1, \ldots, n$, which is impossible, by (*i*). Hence A is irreducible.

2. (*a*) $g(z) = a_n + a_{n-1}z + \cdots + a_0 z^n$

(*b*) $\dfrac{1}{z_1}, \dfrac{1}{z_2}, \ldots, \dfrac{1}{z_n}$

(*c*) If $R(a_0, \ldots, a_n)$ is any of the upper bounds for the moduli of the zeros of $f(z)$, then $R(a_n, \ldots, a_0)$ is an upper bound for the moduli of the zeros of $g(z)$, i.e.

$$\left|\frac{1}{z_\nu}\right|_{\max} = \frac{1}{|z_\nu|_{\min}} \leqq R(a_n, \ldots, a_0)$$

or

$$|z_\nu|_{\min} \geqq \{R(a_n, \ldots, a_0)\}^{-1}.$$

Hence all the zeros of $f(z)$ lie in the ring

$$\{R(a_n, \ldots, a_0)\}^{-1} \leqq |z| \leqq R(a_0, \ldots, a_n).$$

(*d*) By using Kojima's bounds (21), all zeros of $f(z)$ are in the ring

$$\tfrac{1}{2} \leqq |z| \leqq 2$$

The zeros of

$$1 + z + \cdots + z^n = \frac{1 - z^{n+1}}{1 - z}$$

are clearly the $(n + 1)$st roots of unity, and all lie on the circle $|z| = 1$.

3. Here

$$f_{2n}(z) = \sum_{\nu=0}^{n} (-1)^\nu \frac{z^{2\nu}}{(2\nu)!}$$

and

$$f_{2n}(\sqrt{z}) = \sum_{\nu=0}^{n} (-1)^\nu \frac{z^\nu}{(2\nu)!}.$$

By using Fujiwara's bound (20), the zeros of $f_{2n}(\sqrt{z})$ lie in

$$|z| \leqq 2 \max \{(2n)(2n-1), \sqrt{2n(2n-1)(2n-2)(2n-3)}, \ldots, (2n)!^{1/n}\}$$
$$= 4n(2n-1)$$

and therefore those of $f_{2n}(z)$ lie in

$$|z| \leqq 2\sqrt{n(2n-1)} \sim 2\sqrt{2n}.$$

4. (*a*) The triangle joining (0, 0), (0, 1), (1, 0).
(*b*) The whole complex plane.

5. We must show that if a point set S is contained in a circle, then so is its convex hull. But the circle is a convex set containing S. The convex hull of S is contained on every convex set containing S, and hence in the circle.

6. Suppose Budan's theorem gives exact information on any interval (a, b). Making $a \to -\infty$, $b \to \infty$, $V(a) - V(b) \to n$; hence $f(z)$ has n real zeros.

Conversely, let $f(z)$ have n real zeros, and let (a, b) be given. Let x_0 be a point where $f^{(p)}(x_0) = 0$. Since, by Gauss-Lucas' theorem $f^{(p-1)}(x)$ has only real zeros $f^{(p-1)}(x)$ must be positive at its maxima and negative at its minima. In either case clearly, $f^{(p-1)}(x_0)f^{(p+1)}(x_0) < 0$, for $f^{(p+1)}(x)$ is the second derivative of $f^{(p-1)}(x)$. Hence the sequence

$$f(x), \ldots, f^{(n)}(x)$$

is in this case a Sturm sequence and therefore gives exact information on every interval. This argument ignores certain possible degeneracies, which are easily dealt with by taking limits through nondegenerate cases.

7. This is just a rephrasing of exercise 5 of Chapter 2.

8. As in the proof of Theorem 8, $V(x)$ is unaffected by the zeros of members of the Sturm sequence other than $P(x)$ itself. At any zero of $P(x)$ we make the table

Left of x_0		Right of x_0	
$P(x)$	$Q(x)$	$P(x)$	$Q(x)$
+	+	−	+
+	−	−	−
−	+	+	+
−	−	+	−

and observe that $V(x)$ increases by one if $P(x)/Q(x)$ changes from $+\infty$ to $-\infty$, and decreases by one otherwise. Hence $V(\infty) - V(-\infty)$ is the excess of "plus-to-minus" points over "minus-to-plus" points, which was to be shown.

The phrase "combining these results" refers to the following operation: On the interval $(\xi_k + \varepsilon, \xi_{k+1} - \varepsilon)$,

$$\Delta \arg f(x) = \frac{\pi}{2}[\operatorname{sgn} \psi(\xi_{k+1} - \varepsilon) - \operatorname{sgn} \psi(\xi_k + \varepsilon)].$$

Furthermore, on $(-\infty, \xi_1 - \varepsilon)$,

$$\Delta \arg f(x) = \frac{\pi}{2} \operatorname{sgn} \psi(\xi_1 - \varepsilon)$$

since ξ_1 is the smallest zero of $f(x)$, and $f(x)$ cannot change sign in $(-\infty, \xi_1 - \varepsilon)$. Similarly, in $(\xi_m + \varepsilon, \infty)$,

$$\Delta \arg f(x) = -\frac{\pi}{2} \operatorname{sgn} \psi(\xi_m + \varepsilon).$$

Adding all these,

$$\begin{aligned}\Delta_R \arg f(x) &= \frac{\pi}{2}\left\{\sum_{k=1}^{m-1} [\operatorname{sgn} \psi(\xi_{k+1} - \varepsilon) - \operatorname{sgn} \psi(\xi_k + \varepsilon)]\right. \\ &\quad \left. + \operatorname{sgn} \psi(\xi_1 - \varepsilon) - \operatorname{sgn} \psi(\xi_m + \varepsilon)\right\} \\ &= \pi \sum_{k=1}^{m} \left\{\frac{\operatorname{sgn} \psi(\xi_k - \varepsilon) - \operatorname{sgn} \psi(\xi_k + \varepsilon)}{2}\right\}\end{aligned}$$

as claimed.

9. We have

$$f(\rho z) = a_0 + a_1 \rho z + \cdots + a_n \rho^n z^n.$$

Replacing a_j by $\rho^j a_j$ in (16), the zeros of $f(\rho z)$, which are the zeros of $f(z)$ divided by ρ, lie in the circle

$$|z| \leqq 1 + \max \left\{\left|\frac{a_{n-i}}{a_n}\right| \rho^{-i}\right\}$$

and the result follows.

10. If $f(z) = P(z) + iQ(z)$, $P(z)$, $Q(z)$ are real, then $f(z) = 0$, z real, imply $P(z) = Q(z) = 0$. Hence the real zeros of $f(z)$ are the zeros of the real polynomial $q(z)$, which is the highest common factor of $P(z)$ and $Q(z)$. This highest common factor can be found in the usual way from the negative-remainder division algorithm, and the number of zeros of $q(z)$ in (a, b) in the usual way from Sturm's theorem for real polynomials.

11. The number of zeros of $f(z)$ in $|z| < R$ is the number of zeros of $f(Rz)$ in $|z| < 1$. Now the mapping

$$w = \frac{1 + z}{1 - z}$$

maps the unit circle onto the right half-plane. Hence the number of zeros of $f(Rz)$ in $|z| < 1$ is the number of zeros of

$$f\left(R\frac{w - 1}{w + 1}\right)$$

in the right half-plane. This is not a polynomial, but

$$(w + 1)^n f\left(R\frac{w - 1}{w + 1}\right)$$

is, and has the same zeros. Rotating by 90° and replacing w by z, we deduce that $f(z)$ has precisely as many zeros in $|z| < R$ as

$$g(z) = (1 - iz)^n f\left(R\frac{1 + iz}{iz - 1}\right)$$

has in the upper half-plane.

For the particular $f(z)$ given, we find

$$g(z) = (1 - iz)^3 f\left(\frac{1 + iz}{iz - 1}\right) = -2i(z^3 + iz^2 + z - 3i).$$

The Sturm sequence is

$$z^3 + z,\ z^2 - 3,\ -4z,\ 3$$

and the required number of zeros is

$$\begin{aligned} p &= \tfrac{3}{2} + \tfrac{1}{2}V(+, +, -, +) - \tfrac{1}{2}V(-, +, +, +) \\ &= \tfrac{3}{2} + 1 - \tfrac{1}{2} \\ &= 2. \end{aligned}$$

CHAPTER 4

1. (*a*) False $\qquad \sqrt{2x^2 + 1} \sim \sqrt{2}x \qquad (x \to \infty)$
(*b*) True
(*c*) True
(*d*) True, since

$$\left(1 + \frac{1}{n}\right)^{\sqrt{n}} \leqq \left(1 + \frac{1}{n}\right)^n < e.$$

2. Though

$$x^2 + x \sim x^2 \qquad (x \to \infty)$$

it is not true that

$$e^{x^2+x} \sim e^{x^2} \qquad (x \to \infty).$$

However, if

$$f(x) = g(x) + o(1) \qquad (x \to \infty)$$

then

$$\begin{aligned} e^{f(x)} &= e^{g(x)+o(1)} \\ &= e^{g(x)}e^{o(1)} \\ &\sim e^{g(x)} \qquad (x \to \infty). \end{aligned}$$

3. We have

$$\begin{aligned} \log f(x) &- \log g(x) \\ &= \log \frac{f(x)}{g(x)} \\ &\to \log 1 \qquad (x \to \infty) \\ &= 0. \end{aligned}$$

4. Regarding x as a parameter, put

$$\varphi(y) = \frac{x}{y^2} e^{-x/y^2}$$

in the restricted form of the Euler-MacLaurin sum formula. Then

$$\sum_{\nu=1}^{n} \frac{x}{\nu^2} e^{-x/\nu^2} = \int_1^n \frac{x}{y^2} e^{-x/y^2}\, dy + \tfrac{1}{2} x e^{-x} + \frac{1}{2}\frac{x}{n^2} e^{-x/n^2} + \int_1^n (y - [y] - \tfrac{1}{2})\, \varphi'(y)\, dy.$$

Making $n \to \infty$,

$$\sum_{\nu=1}^{\infty} \frac{x}{\nu^2} e^{-x/\nu^2} = \int_1^\infty \frac{x}{y^2} e^{-x/y^2}\, dy + \tfrac{1}{2} x e^{-x} + \int_1^\infty (y - [y] - \tfrac{1}{2})\, \varphi'(y)\, dy.$$

Now

$$\begin{aligned}
\left| \int_1^\infty (y - [y] - \tfrac{1}{2})\, \varphi'(y)\, dy \right| &= \left| \int_1^\infty (y - [y] - \tfrac{1}{2}) e^{-x/y^2} \left(\frac{x}{y^2} - 1\right) \frac{2x}{y^3}\, dy \right| \\
&= \left| \int_0^{\sqrt{x}} \left(\frac{\sqrt{x}}{t} - \left[\frac{\sqrt{x}}{t}\right] - \frac{1}{2} \right) e^{-t^2} (t^2 - 1) 2t\, dt \right| \\
&\leqq \text{const.} \int_0^{\sqrt{x}} 2t^3 e^{-t^2}\, dt \\
&= O(1) \qquad (x \to \infty)
\end{aligned}$$

Hence

$$\begin{aligned}
\sum_{n=1}^{\infty} \frac{x}{n^2} e^{-x/n^2} &= \int_1^\infty \frac{x}{y^2} e^{-x/y^2}\, dy + O(1) \qquad (x \to \infty) \\
&= \sqrt{x} \int_0^{\sqrt{x}} e^{-t^2}\, dt + O(1) \\
&= \sqrt{x} \int_0^\infty e^{-z^2}\, dt - \sqrt{x} \int_{\sqrt{x}}^\infty e^{-t^2}\, dt + O(1).
\end{aligned}$$

Finally,

$$\sqrt{x} \int_{\sqrt{x}}^\infty e^{-t^2}\, dt \leqq \sqrt{x} \int_{\sqrt{x}}^\infty e^{-t}\, dt = \sqrt{x}\, e^{-\sqrt{x}} = o(1) \qquad (x \to \infty)$$

completing the proof.

5. Since

$$J_n(x) = \frac{1}{n!} \left(\frac{x}{2}\right)^n \sum_{k=0}^{\infty} \frac{(-1)^k (x/2)^{2k}}{k!\, (n+1) \cdots (n+k)}$$

we have

$$\begin{aligned}
J_n(\sqrt{n}) &= \frac{1}{n!} \left(\frac{\sqrt{n}}{2}\right)^n \sum_{k=0}^{\infty} \left(-\frac{1}{4}\right)^k \frac{1}{k!} \frac{n^k}{(n+1) \cdots (n+k)} \\
&\sim \frac{1}{n!} \left(\frac{\sqrt{n}}{2}\right)^n \sum_{k=0}^{\infty} \left(-\frac{1}{4}\right)^k \frac{1}{k!} \\
&= \frac{1}{n!} \left(\frac{\sqrt{n}}{2}\right)^n e^{-1/4}
\end{aligned}$$

and the result follows from Stirling's formula (72).

6. First, from (45), it is clear that

$$\zeta(2n) \sim 1 \qquad (n \to \infty)$$

and from (50),

$$B_{2n} \sim (-1)^{n+1}(2n)!\,\frac{2}{(2\pi)^{2n}} \qquad (n \to \infty)$$

whence the result follows from Stirling's formula (72).

7. A direct application of the Euler-MacLaurin formula to $f(x) = x^{-1}$ gives

$$1 + \frac{1}{2} + \cdots + \frac{1}{n} \approx \log n + \gamma + \frac{1}{2n} - \frac{B_2}{2n^2} - \frac{B_4}{4n^4} - \cdots \qquad (n \to \infty).$$

8. First,

$$\begin{aligned}\frac{1 \cdot 3 \cdot 5 \cdots (2n-1)}{2 \cdot 4 \cdots (2n)} &= \frac{(2n-1)!}{2^n n!\,(2 \cdot 4 \cdots (2n-2))} \\ &= \frac{(2n-1)!}{2^{2n-1} n!\,(n-1)!} \\ &= \frac{(2n)!}{4^n (n!)^2} \sim 4^{-n}\left\{\sqrt{4\pi n}\left(\frac{2n}{e}\right)^{2n}\right\}\left\{\sqrt{2\pi n}\left(\frac{n}{e}\right)^{n}\right\}^{-2} \\ &= \frac{1}{\sqrt{n\pi}} \qquad (n \to \infty).\end{aligned}$$

Therefore

$$\left(\frac{1 \cdot 3 \cdot 5 \cdots (2n-1)}{2 \cdot 4 \cdots 2n}\right)^2 \sim \frac{1}{n\pi} \qquad (n \to \infty)$$

and

$$G_n \sim \sum_{\nu=1}^{n} \frac{1}{\nu\pi} \sim \frac{1}{\pi}\log n \qquad (n \to \infty)$$

according to exercise 14.

9. Suppose first that $A = 1$. If $\varepsilon > 0$ is given, choose δ such that

$$|f(t) - f(1)| \leqq \varepsilon$$

for $1 - \delta \leqq t \leqq 1$. Then

$$\int_0^1 t^n f(t)\,dt = \int_0^{1-\delta} t^n f(t)\,dt + \int_{1-\delta}^1 t^n f(t)\,dt.$$

But

$$\begin{aligned}\left|\int_0^{1-\delta} t^n f(t)\,dt\right| &\leqq M \int_0^{1-\delta} t^n\,dt \\ &= M\,\frac{(1-\delta)^{n+1}}{n+1} \\ &= o(n^{-1}) \qquad (n \to \infty)\end{aligned}$$

and

$$(f(1) - \varepsilon)\int_{1-\delta}^1 t^n\,dt \leqq \int_{1-\delta}^1 t^n f(t)\,dt \leqq (f(1) + \varepsilon)\int_{1-\delta}^1 t^n\,dt$$

or

$$\frac{f(1) - 2\varepsilon}{n + 1} \leq \int_{1-\delta}^{1} t^n f(t)\, dt \leq \frac{f(1) + 2\varepsilon}{n + 1}$$

for all sufficiently large n, and the result follows. The general problem can be reduced to the case $A = 1$ by letting $t = At'$ in the integral.

10. First,

$$J_0(x) = \frac{1}{2\pi}\int_0^{2\pi} e^{ix\sin\theta}\, d\theta$$

which is of the form (118) with $\phi(\theta) = \sin\theta$. This function has two stationary points, at $\pi/2$, $3\pi/2$, and its second derivative vanishes at $0, \pi$, violating one of the hypothesis of Theorem 5. Hence we must consider

$$\frac{1}{2\pi}\left\{\int_{\delta}^{\pi-\delta} + \int_{\pi-\delta}^{\pi+\delta} + \int_{\pi+\delta}^{2\pi-\delta} + \int_{2\pi-\delta}^{2\pi+\delta}\right\} e^{ix\sin\theta}\, d\theta.$$

The first and third of these integrals each contain one stationary point, and $\phi''(\theta)$ is of constant sign in each of them. Therefore, by (118),

$$\frac{1}{2\pi}\int_{\delta}^{\pi-\delta} e^{ix\sin\theta}\, d\theta = \frac{e^{-i(\pi/4)+ix}}{\sqrt{2\pi x}} + O(x^{-3/5}) \qquad (x \to \infty)$$

and

$$\frac{1}{2\pi}\int_{\pi+\delta}^{2\pi-\delta} e^{ix\sin\theta}\, d\theta = \frac{e^{i(\pi/4)-ix}}{\sqrt{2\pi x}} + O(x^{-3/5}) \qquad (x \to \infty).$$

For the second of the four integrals, take $h(y) = x\sin y$ in Lemma 3. Then, on $(\pi - \delta, \pi + \delta)$, $|h'(y)| = |x\cos y| \geq x|\cos(\pi - \delta)| = m$. Thus, by Lemma 3, keeping δ fixed, the second integral (and similarly the fourth) is $O(x^{-1})(x \to \infty)$. Hence

$$\frac{1}{2\pi}\int_0^{2\pi} e^{ix\sin\theta}\, d\theta = \frac{1}{\sqrt{2\pi x}}\{e^{i(x-(\pi/4))} + e^{-i(x-(\pi/4))}\} + O(x^{-3/5}) \qquad (x \to \infty)$$

and

$$J_0(x) = \sqrt{\frac{2}{\pi x}}\cos\left(x - \frac{\pi}{4}\right) + O(x^{-3/5}) \qquad (x \to \infty).$$

11. (*a*) If

$$f(z) = \sum_{\nu=0}^{\infty} \frac{\alpha_\nu}{\nu!} z^\nu$$

then

$$e^z f(z) = \sum_{\nu=0}^{\infty} \frac{\sum_{\mu=1}^{\nu} \binom{\nu}{\mu} \alpha_\mu}{\nu!} z^\nu$$

$$= 1 + \sum_{\nu=1}^{\infty} \frac{\alpha_\nu + \alpha_\nu}{\nu!} z^\nu$$

$$= 1 + 2\{f(z) - 1\}$$

and

$$f(z) = (2 - e^z)^{-1}.$$

(*b*) The function $f(z)$ clearly has a singularity at $z = \log 2$, and this is, therefore, the radius of convergence of its power series. In other words,

$$\frac{1}{\log 2} = \limsup_{n\to\infty} \left(\frac{\alpha_n}{n!}\right)^{1/n}$$

Hence, if $\varepsilon > 0$ is given, for all sufficiently large n we have

$$\left(\frac{\alpha_n}{n!}\right)^{1/n} \leqq \left(\frac{1}{\log 2} + \varepsilon\right)$$

or

$$\alpha_n \leqq \left(\frac{1}{\log 2} + \varepsilon\right)^n n! \qquad (n \geqq n_0)$$

as required. If $\epsilon < 0$, then for infinitely many n,

$$\left(\frac{\alpha_n}{n!}\right)^{1/n} \geqq \frac{1}{\log 2} - \varepsilon$$

proving the second part.

12. If

$$f(z) = \sum_{\nu=0}^{\infty} \frac{a_\nu}{\nu!} z^\nu$$

then

$$f^{(k)}(z) = \sum_{\nu=0}^{\infty} \frac{a_\nu}{\nu!} \nu(\nu - 1)\cdots(\nu - k + 1)z^{\nu-k}$$

$$= \sum_{\nu=k}^{\infty} \frac{a_\nu}{(\nu - k)!} z^{\nu-k} = \sum_{\nu=0}^{\infty} \frac{a_{\nu+k}}{\nu!} z^\nu.$$

Finally,

$$e^{\alpha z} f(z) = \sum_{\nu=0}^{\infty} \frac{\alpha^\nu z^\nu}{\nu!} \sum_{\nu=0}^{\infty} \frac{a_\nu z^\nu}{\nu!}$$

$$= \sum_{\nu=0}^{\infty} z^\nu \sum_{\mu=0}^{\nu} \frac{\alpha^\mu}{\mu!} \frac{a_{\nu-\mu}}{(\nu - \mu)!}$$

$$= \sum_{\nu=0}^{\infty} \frac{z^\nu}{\nu!} \sum_{\mu=0}^{\nu} a_{\nu-\mu} \alpha^\mu \binom{\nu}{\mu}$$

13. The generating function is

$$F(t) = \sum_{n=0}^{\infty} \frac{\lambda_n}{n!} t^n = \int_0^\pi \sum_{n=0}^{\infty} \frac{(xt)^n}{n!} \sin x \, dx$$

$$= \int_0^\pi e^{xt} \sin x \, dx = \frac{1 + e^{\pi t}}{1 + t^2}$$

$$= 2 + \pi t - \left(2 - \frac{\pi^2}{2!}\right)t^2 - \left(\pi - \frac{\pi^3}{3!}\right)t^3 + \left(2 - \frac{\pi^2}{2!} + \frac{\pi^4}{4!}\right)t^4 + \cdots$$

The coefficient of t^{2n+1} is

$$\frac{\lambda_{2n+1}}{(2n+1)!} = (-1)^n\left\{\pi - \frac{\pi^3}{3!} + \cdots + (-1)^n \frac{\pi^{2n+1}}{(2n+1)!}\right\}$$

$$= (-1)^n\left\{\sin \pi + (-1)^n \frac{\pi^{2n+3}}{(2n+3)!} + (-1)^{n+1} \frac{\pi^{2n+5}}{(2n+5)!} + \cdots\right\}$$

$$= \frac{\pi^{2n+3}}{(2n+3)!}\left\{1 - \frac{\pi^2}{(2n+4)(2n+5)} + \cdots\right\}.$$

Multiplying by $(2n + 1)!$ and replacing $2n + 1$ by n throughout, the result follows for odd n, and similarly for even n.

14. Since

$$\lim_{\nu\to\infty} \frac{a_\nu}{b_\nu} = 1$$

if $\varepsilon > 0$ is given, there is an $n_0 = n_0(\varepsilon)$ such that for $\nu \geqq n_0$

$$\left|\frac{a_\nu}{b_\nu} - 1\right| \leqq \frac{\varepsilon}{2}$$

i.e.,

$$\left(1 - \frac{\varepsilon}{2}\right)b_\nu \leqq a_\nu \leqq \left(1 + \frac{\varepsilon}{2}\right)b_\nu \qquad (\nu \geqq n_0).$$

Now, for $n > n_0$,

$$\sum_{\nu=1}^{n} a_\nu = \sum_{\nu=1}^{n_0-1} a_\nu + \sum_{\nu=n_0}^{n} a_\nu = K(\varepsilon) + \sum_{\nu=n_0}^{n} a_\nu, \qquad \text{say.}$$

Hence

$$K(\varepsilon) + \left(1 - \frac{\varepsilon}{2}\right)\sum_{\nu=n_0}^{n} b_\nu \leqq \sum_{\nu=1}^{n} a_\nu \leqq K(\varepsilon) + \left(1 + \frac{\varepsilon}{2}\right)\sum_{\nu=n_0}^{n} b_\nu$$

for $n > n_0$, and if we put $L(\varepsilon) = \sum_{\nu=1}^{n_0-1} b_\nu$

$$K(\varepsilon) - \left(1 - \frac{\varepsilon}{2}\right)L(\varepsilon) + \left(1 - \frac{\varepsilon}{2}\right)\sum_{\nu=1}^{n} b_\nu$$

$$\leqq \sum_{\nu=1}^{n} a_\nu \leqq K(\varepsilon) - \left(1 + \frac{\varepsilon}{2}\right)L(\varepsilon) + \left(1 + \frac{\varepsilon}{2}\right)\sum_{\nu=1}^{n} b_\nu.$$

Therefore,

$$1 - \frac{\varepsilon}{2} + \frac{M_-(\varepsilon)}{\sum_{\nu=1}^{n} b_\nu} \leqq \frac{\sum_{\nu=1}^{n} a_\nu}{\sum_{\nu=1}^{n} b_\nu} \leqq \left(1 + \frac{\varepsilon}{2}\right) + \frac{M_+(\varepsilon)}{\sum_{\nu=1}^{n} b_\nu}$$

for $n > n_0$, where we have put $M_\pm(\varepsilon) = K(\varepsilon) - (1 \pm \epsilon/2)L(\varepsilon)$. Keeping ε fixed, by hypothesis III, for all large enough n the ratio in the middle will lie between $1 - \varepsilon$ and $1 + \varepsilon$, which was to be shown.

CHAPTER 5

1. In all cases, no Lipschitz condition is satisfied in any open set containing the initial data point. This is obvious in (5), (6). In (7) what we have to show is that $\sqrt{y}$ satisfies no Lipschitz condition in an interval containing the origin. Otherwise, however, there would be a constant L and an interval $[0, \delta]$, $\delta > 0$, such that

$$\sqrt{y_2} - \sqrt{y_1} \leqq L(y_2 - y_1)$$

for y_1, y_2 in $[0, \delta]$. But then, with $y_1 = 0$, we have $\sqrt{y_2} \leqq Ly$ or $L \geqq y^{-\frac{1}{2}}$ for all y_2 in $[0, \delta]$, which is absurd. The situation in (8) is identical.

2. Clearly $y'(x) > 0$ always; hence $y(x)$ increases and is therefore always positive. Thus $e^{-xy(x)} \leqq 1$ always, and the given equation is dominated by $y'(x) = 1$, $y(1) = 1$, whose solution is $y(x) = x$.

3. Theorem 1 gives first

$$M(a, c) = 2a(1 + c)^2$$

then

$$b = \max_{a,c>0} \min \left(a, \frac{c}{2a(1 + c)^2}\right) = \frac{1}{2\sqrt{2}}.$$

From Theorem 3 we find

$$\begin{aligned} \varphi(y) &= \max |2xs^2| \qquad \text{on} \qquad |x| \leqq a; \quad 1 - |y| \leqq s \leqq 1 + |y| \\ &= 2a(1 + |y|)^2 \end{aligned}$$

and from equation (27) we may take

$$\begin{aligned} b &= \max_{(a,c>0)} \min \left(a, \int_0^c \frac{dy}{\varphi(y)}\right) \\ &= \max_{(a,c>0)} \min \left(a, \frac{c}{2a(c + 1)}\right) \\ &= \max_{c>0} \sqrt{\frac{c}{2(c + 1)}} = \frac{1}{\sqrt{2}}. \end{aligned}$$

The true solution is $y(x) = (1 + x^2)^{-1}$; hence, on the real axis, the solution is everywhere nonsingular. That is, the best possible value of b is $+\infty$. The reason for the large discrepancy lies with the poles of $y(x)$ on the *imaginary axis*. Indeed, if $f(x, y)$ is analytic, Picards's theorem goes through virtually unchanged, to provide an existence theorem in the complex domain. Hence the number b should be interpreted, in such cases, as an estimate of the radius of convergence of the power series development of the solution.

4. In the general formula

$$u_{n+1} = a_0 u_n + \cdots + a_{-p} u_{n-p} + h[b_1 f(x_{n+1}, u_{n+1}) + \cdots + b_{-p} f(x_{n-p}, u_{n-p})]$$

put $x_{n+1} = h$, $x_n = 0$, $x_{n-1} = -h, \ldots, x_{n-p} = -ph$, $u(x) = x^r$, $f(x, u) = rx^{r-1}$. Then

$$h^r = a_0 \cdot 0^r + \cdots + a_{-p}(-ph)^r + hr[b_1 h^{r-1} + \cdots + b_{-p}(-ph)^{r-1}].$$

Cancelling h^r, the condition is

$$\sum_{j=0}^{p} a_{-j}(-j)^r + r \sum_{j=-1}^{p} b_{-j}(-j)^{r-1} = 1 \qquad (r = 0, 1, \ldots, k;\ 0^0 = 1).$$

5. Read the Euler-MacLaurin sum formula ((54), Chapter 4) from right to left. It says

$$\int_1^n f(x)\,dx = \tfrac{1}{2}(f(1) + f(2)) + \tfrac{1}{2}(f(2) + f(3)) + \cdots + \tfrac{1}{2}(f(n-1) + f(n))$$
$$- \int_1^n (x - [x] - \tfrac{1}{2}) f'(x)\,dx.$$

6. (*a*) The general formula is

$$u_{n+1} = a_0 u_n + a_{-1} u_{n-1} + h\{b_1 f(x_{n+1}, u_{n+1}) + b_0 f(x_n, u_n) + b_{-1} f(x_{n-1}, u_{n-1})\}$$

Imposing the first two exactness conditions of exercise 4,

$$a_0 + a_{-1} = 1, \qquad -a_{-1} + b_1 + b_0 + b_{-1} = 1$$

we take $a_0 = t$, $a_{-1} = 1 - t$, $b_1 = 2 - t - (b_0 + b_{-1})$, and the most general formula exact for $1, x$ is

$$u_{n+1} = tu_n + (1 - t)u_{n-1} + h\{[2 - t - (b_0 + b_{-1})]$$
$$+ b_0 f(x_n, u_n) + b_{-1} f(x_{n-1}, u_{n-1}).$$

The stability condition is that the roots of

$$\frac{r^2 - tr - (1 - t)}{r - 1} = r + (1 - t) = 0$$

should lie in $|z| \geqq \frac{1}{2}$, which requires $\frac{1}{2} \leqq t \leqq \frac{3}{2}$.

(*b*) The formula of part (*a*) contains three parameters, b_0, b_{-1}, t, the first two being unrestricted, the last satisfying $\frac{1}{2} \leqq t \leqq \frac{3}{2}$. The conditions of exactness for x^2, x^3, x^4 are, after some simplifications,

$$b_0 + 2b_{-1} = 2 - \frac{t}{2}$$

$$b_0 = \frac{4 - 2t}{3}$$

$$b_0 + 2b_{-1} = 2 - \frac{5t}{4}.$$

The last of these contradicts the first unless $t = 0$, which is outside $[\frac{1}{2}, \frac{3}{2}]$. Hence the largest k for which there is a stable formula exact for $1, x, \ldots, x^k$ is $k = 3$. Taking t as parameter and eliminating b_0, b_{-1} from the first two equations above, all such formulas are

$$u_{n+1} = tu_n + (1 - t)u_{n-1} + h\left[\left(\frac{1}{3} + \frac{t}{12}\right) f(x_{n+1}, u_{n+1}) + \frac{1 + t}{3} f(x_n, u_n) + \left(\frac{1}{3} - \frac{5t}{12}\right) f(x_{n-1}, u_{n-1})\right]$$

where $\frac{1}{2} \leqq t \leqq \frac{3}{2}$.

7. Put $z = \frac{1}{2} + it$ in (120), getting

$$\Gamma\left(\frac{1}{2} + it\right)\Gamma\left(\frac{1}{2} - it\right) = \frac{\pi}{\sin \pi(\frac{1}{2} + it)} = \frac{\pi}{\cosh \pi t},$$

hence

$$\left|\Gamma\left(\frac{1}{2} + it\right)\right| = \sqrt{\frac{\pi}{\cosh \pi t}} \qquad (t \text{ real}).$$

8. Let $\phi(z) = [z\Gamma(z)]^{-1}$. Then

$$\phi(z) = e^{\gamma z} \prod_{n=1}^{\infty} \left[\left(1 + \frac{z}{n}\right) e^{-z/n}\right]$$

and clearly $\phi(0) = 1$. Logarithmic differentiation gives

$$\frac{\phi'(z)}{\varphi(z)} = \gamma + \sum_{n=1}^{\infty} \left[\frac{1}{n + z} - \frac{1}{n}\right]$$

whence $\phi'(0) = \gamma$. Differentiating once more,

$$\left(\frac{\phi'(z)}{\phi(z)}\right)' = \frac{\phi(z)\phi''(z) - [\phi'(z)]^2}{[\phi(z)]^2}$$

$$= -\sum_{n=1}^{\infty} \frac{1}{(n + z)^2}.$$

Putting $z = 0$, this gives

$$\phi''(0) = \gamma^2 - \sum_{n=1}^{\infty} \frac{1}{n^2}$$

$$= \gamma^2 - \zeta(2) = \gamma^2 - \frac{\pi^2}{6}.$$

Taylor series expansion of $\phi(z)$ is

$$\phi(z) = 1 + \gamma z + \frac{1}{2}\left(\gamma^2 - \frac{\pi^2}{6}\right) z^2 + \cdots$$

in which the coefficient of z^2 is negative.

9. Put $z = \frac{1}{2}$ in (118), using (129).

10. We have, using (130),

$$J_{1/2}(x) = \sum_{\nu=0}^{\infty} \frac{(-1)^\nu \left(\frac{x}{2}\right)^{2\nu + 1/2}}{\nu!\, \Gamma(\nu + 1 + \frac{1}{2})}$$

$$= \sum_{\nu=0}^{\infty} \frac{(-1)^\nu \left(\frac{x}{2}\right)^{2\nu + 1/2} 2^{\nu+1}}{\nu!\, 1 \cdot 3 \cdot 5 \cdots (2\nu + 1)\sqrt{\pi}}$$

$$= \sum_{\nu=0}^{\infty} \frac{(-1)^\nu \left(\frac{x}{2}\right)^{2\nu + 1/2} 2^{\nu+1} 2^\nu \nu!}{\nu!\, \sqrt{\pi}\, (2\nu + 1)!}$$

$$= \sqrt{\frac{2x}{\pi}} \sum_{\nu=0}^{\infty} \frac{(-1)^\nu x^{2\nu}}{(2\nu + 1)!}$$

$$= \sqrt{\frac{2}{\pi x}} \sin x.$$

11. (*a*) We find

$$\exp\left[\frac{x}{2}\left(t - \frac{1}{t}\right)\right] \exp\left[\frac{y}{2}\left(t - \frac{1}{t}\right)\right]$$

$$= \exp\left[\frac{x+y}{2}\left(t - \frac{1}{t}\right)\right]$$

$$= \sum_{n=-\infty}^{\infty} J_n(x+y)t^n$$

$$= \sum_{n=-\infty}^{\infty} J_n(x)t^n \sum_{n=-\infty}^{\infty} J_n(y)t^n$$

$$= t^0[J_0(x)J_0(y) - 2J_1(x)J_1(y) + 2J_2(x)J_2(y) - \cdots]$$

+ terms involving t.

(*b*) Putting $y = -x$,

$$J_0(0) = 1 = J_0(x)J_0(-x) - 2J_1(x)J_1(-x) + \cdots$$

$$= J_0^2(x) + 2J_1^2(x) + 2J_2^2(x) + \ldots$$

Naturally, each term is $\leqq 1$, since all are positive.

CHAPTER 6

1. (*a*) The question is: For what w does the equation

$$f(z) = z^2 - \left(2 + \frac{1}{w}\right)z + 1 = 0$$

have a root in the unit circle? Using the solution of exercise (11), Chapter 3, the number of zeros of $f(z)$ in the unit circle is the same as the number of zeros of

$$g(z) = (1 - iz)^2 f\left(\frac{1+iz}{iz-1}\right)$$

$$= z^2\frac{1}{w} + 4 + \frac{1}{w}$$

in the upper half-plane. But the zeros of $g(z)$ are the numbers

$$\pm\sqrt{-4w-1}$$

one of which must lie in the upper half-plane unless $-4w - 1$ is a positive real number, that is, unless w is real and less than $-\frac{1}{4}$. Hence the function in question maps the unit circle 1-1 onto the full w-plane cut along the ray $-\infty < w < -\frac{1}{4}$ on the negative real axis.

(*b*) The full right half-plane. To verify this, show that $\operatorname{Re} w > 0$ in $|z| < 1$, then that w omits no such values.

2. (*a*) We showed in the previous solution that $f(z)$ has as many zeros in $|z| < 1$ as $g(z)$ has in $\mathbf{Im}\, z > 0$, but that number is one.
(*b*) If

$$\frac{1+z_1}{1-z_1} = \frac{1+z_2}{1-z_2}$$

then $$1 - z_2 + z_1 - z_1z_2 = 1 - z_1 + z_2 - z_1z_2, \text{ or } z_1 = z_2.$$

3. If $f(g(z_1)) = f(g(z_2))$ then $g(z_1) = g(z_2)$ and so $z_1 = z_2$.

4. If $f(z)$ is univalent in $|z| < R$, then $f(Rz)$ is univalent in $|z| < 1$; hence so is

$$\frac{f(Rz)}{nR} = \frac{\left(1 + nRz + \dfrac{n(n-1)}{2}R^2z^2 + \cdots\right) - 1}{nR}$$

$$= z + \frac{n-1}{2}Rz^2 + \cdots$$

By theorem 12,

$$\left|\frac{n-1}{2}R\right| \leqq 2$$

or

$$R \leqq \frac{4}{n-1}$$

which is less than unity when $n \geqq 6$

5. We have to show that

$$w = \log\frac{z-1}{z+1}$$

maps the upper half-plane onto the strip $0 < \mathbf{Im}\, w < \pi$. This is evident from the equation preceding (25).

6. If $f_1(z)$ maps the unit circle onto the upper half-plane and $f_2(z)$ maps the upper half-plane onto the strip in question, then $f_2(f_1(z))$ is the function we seek. But from exercise 1,

$$f_1(z) = i\frac{1+z}{1-z}$$

while we have already shown that

$$f_2(z) = \sin^{-1} z.$$

Thus

$$f(z) = \sin^{-1}\frac{i+iz}{1-z}$$

is the required function.

7. According to exercise 1(*b*), the function

$$w = \frac{\dfrac{1+z}{1-z} - 1}{2} = \frac{z}{1-z} = z + z^2 + \cdots$$

maps the unit circle onto the convex set $\operatorname{Re} w > -\frac{1}{2}$. All of its coefficients $a_\nu = 1 (\nu = 1, 2, \ldots)$.

The function

$$w = \frac{1}{2}\frac{1+z}{1-z} = \frac{1}{2} + z + z^2 + \cdots$$

has positive real part, and the sign of equality holds in (31) for each $n = 1, 2, 3, \ldots$

8. According to exercise (1), the function

$$\frac{z}{(1-z)^2} = z + 2z^2 + 3z^3 + \cdots$$

is univalent in $|z| < 1$, and the sign of equality holds in (33), (34) for each $n = 1, 2, 3, \ldots$.

9. No, for

$$\frac{f(z)}{6} = \tfrac{1}{2} + \tfrac{1}{6}z + \tfrac{7}{6}z^2 + \cdots$$

would violate Theorem 10.

10. For a convex map we have, for $|z| \leqq r$,

$$|f(z)| = |z + a_2 z^2 + \cdots| \leqq r + r^2 + r^3 + \cdots = \frac{r}{1-r}.$$

Hence the maximum modulus is $O((1-r)^{-1})$ as $r \to 1^-$. The function in question does not satisfy this condition.

For (*b*), (*c*), we have for such functions

$$|f(z)| \leqq r + 2r^2 + \cdots = \frac{r}{(1-r)^2}$$

and this is not violated; so the answer to (*b*), (*c*) is "possibly."

11. From equation (20),

$$\nabla^2\Phi(x, y) = (u_x^2 + u_y^2)\,\nabla^2\phi(x, y)$$

which will be positive if $\phi(x, y)$ is subharmonic.

12. If $f_0(z)$ is any function with the desired mapping property, then

$$f(z) = e^{i\varphi}\frac{f_0(z) - w_0}{1 - \overline{w}_0 f_0(z)}$$

is the most general such function, where $|w_0| < 1$. Now, for the extremal function, we have $f(2i) = 0$. Furthermore, we may choose

$$f_0(z) = \frac{z-i}{z+i}$$

by rotating and inverting the result of exercise 1(*b*). To make $f(z)$ vanish at $2i$, then, we take

$$f(z) = \frac{f_0(z) - f_0(2i)}{1 - \overline{f_0(2i)}\, f_0(z)}$$

$$= \frac{\dfrac{z-i}{z+i} - \dfrac{1}{3}}{1 - \dfrac{1}{3}\dfrac{z-i}{z+i}}$$

$$= \frac{z-2i}{z+2i}.$$

The maximum value of the derivative in the whole class of functions is therefore

$$|f'(2i)| = \tfrac{1}{4}.$$

CHAPTER 7

1. By differentiation, the function has no interior extrema. On the boundary joining (0, 0) to (1, 0) the function is $f(x, 0) = x$ whose maximum is at (1, 0). On the vertical boundary $f(0, y) = y^2$ whose maximum is at (0, 1), whereas on the other boundary $f(x, 1 - x) = x^2 - x + 1$ which has a minimum at $x = \frac{1}{2}$, maxima at the end points. Hence, in this case (not in general for nonlinear functions) it is enough to examine the vertices, and the required maximum is 1.

2. According to equation (38) of Chapter 1, the required polynomial is the sum of the first $n + 1$ terms of the formal expansion of $f(x)$ in a series of Legendre polynomials.

3. If

$$f(z) = z + a_2 z^2 + \cdots + a_n z^n$$

is univalent in $|z| < 1$, then

$$f'(z) = 1 + 2a_2 z + \cdots + n a_n z^{n-1}$$

is not zero there. Hence the product of the zeros of $f'(z)$, namely $1/na_n$, exceeds unity in modulus. Hence $|a_n| \leqq 1/n$. The polynomial

$$f_0(z) = z + \frac{z^n}{n}$$

satisfies the conditions of the problem.

4. By Schwarz's inequality (Theorem 1 of Chapter 1),

$$(\mathbf{x}, \mathbf{x})(\mathbf{a}, \mathbf{a}) \geqq (\mathbf{a}, \mathbf{x})^2 = 1$$

from which

$$(\mathbf{x}, \mathbf{x}) \geqq (\mathbf{a}, \mathbf{a})^{-1}$$

with equality if and only if $\mathbf{x} = (\mathbf{a}, \mathbf{a})^{-2}\mathbf{a}$. The problem can also be done with Lagrange multipliers, but this would amount to another proof of Schwarz's inequality.

5. (*a*) The functional is

$$J[y] = \frac{1}{x - x_0}\int_{x_0}^{x} y'(x)\,dx$$
$$= \frac{y(x) - y_0}{x - x_0}$$

where $(x, y(x))$ lies on C. The value of $J[y]$ depends only on the endpoints, and thus we may restrict attentions to straight lines. To find the straight line of least slope is easy: start with a vertical line through P (slope $-\infty$). Rotate the line clockwise about P until the curve C is touched. The position of the line segment $\overline{PC}$ then solves the problem.

6. By minimizing

$$\int F(x, y, y')\,dx$$

subject to the side condition

$$\int G(x, y, y') = 1.$$

7. As in equation (20), we minimize

$$\int_0^b \sqrt{\frac{1 + y'^2}{y}}\,dx$$

with the conditions $y(0) = y_0$, $y(b) = 0$, and

$$\int_0^b \sqrt{1 + y'^2}\,dx = L.$$

The Euler equations are then

$$\frac{d}{dx}\frac{\partial}{\partial y'}\left\{\sqrt{\frac{1 + y'^2}{y}} + \lambda\sqrt{1 + y'^2}\right\} - \frac{\partial}{\partial y}\left\{\sqrt{\frac{1 + y'^2}{y}} + \lambda\sqrt{1 + y'^2}\right\} = 0.$$

Since the integrand was independent of x, a first integral is

$$y'\frac{\partial}{\partial y'}\left\{\sqrt{\frac{1 + y'^2}{y}} + \lambda\sqrt{1 + y'^2}\right\} - \left\{\sqrt{\frac{1 + y'^2}{y}} + \lambda\sqrt{1 + y'^2}\right\} = c_1$$

which simplifies to

$$\frac{dy}{\sqrt{\left(\lambda + \dfrac{1}{\sqrt{y}}\right)^2 - c_1^2}} = \frac{dx}{c_1}$$

and the problem is reduced to a quadrature.

8. The problem is incompletely formulated as it stands, for the boundary values were not given. We have to discover the kind of boundary conditions which are natural to the problem. Let us use F_1, F_2, F_3 to denote the derivatives of F with respect to its first, second, and third arguments, respectively.

Choosing a variation $\eta(x)$,

$$I[y + \varepsilon\eta] = \int_a^b F(x, y(x) + \varepsilon\eta(x), y(x + h) + \varepsilon\eta(x + h))\, dx$$

and

$$I'[y + \varepsilon\eta]_{\varepsilon=0} = \int_a^b \{F_2(x, y(x), y(x + h))\eta(x) + F_3(x, y(x), y(x + h))\eta(x + h)\}\, dx$$

$$= \int_a^b F_2(x, y(x), y(x + h))\eta(x)\, dx + \int_{a+h}^{b+h} F_3(x - h, y(x - h), y(x))\eta(x)\, dx.$$

In order to collect these two integrals into one we clearly need $\eta(x)$ to vanish on $[a, a + h)$ and on $(b, b + h]$. This means that $y(x)$ is given on those two intervals. If this is the case, the Euler equation is

$$F_2(x, y(x), y(x + h)) + F_3(x - h, y(x - h), y(x)) = 0$$

where the values of $y(x)$ are prescribed on $a \leqq x \leqq a + h$ and $b \leqq x \leqq b + h$.

9. We first let $x_1 = x$, $x_2 = y$, and introduce slack variables x_3, x_4, x_5, x_6 by

$$-x_1 + x_2 + x_3 = 1 \qquad -\tfrac{1}{2}x_1 - x_2 + x_4 = -2$$

$$-\tfrac{1}{2}x_1 + x_2 + x_5 = 3 \qquad x_1 + x_6 = 5$$

with the constraints $x_i \geqq 0$ $(i = 1, \ldots, 6)$. The vertex $(\frac{2}{3}, \frac{5}{3})$ corresponds to $x_1 > 0$, $x_2 > 0$, $x_5 > 0$, $x_6 > 0$; $x_3 = x_4 = 0$. The matrix of (60) is

$$\begin{pmatrix} -1 & 1 & 1 & 0 & 0 & 0 & 1 \\ -\frac{1}{2} & -1 & 0 & 1 & 0 & 0 & -2 \\ -\frac{1}{2} & 1 & 0 & 0 & 1 & 0 & 3 \\ 1 & 0 & 0 & 0 & 0 & 1 & 5 \end{pmatrix}$$

and the columns to be used for the basis are the first, second, fifth, and sixth.

If we now express each column as a linear combination of these four, as in (63), we find the matrix of γ_{ij} to be

$$\begin{matrix} \text{row } 1 \\ \text{row } 2 \\ \text{row } 5 \\ \text{row } 6 \end{matrix} \begin{pmatrix} 1 & 0 & -\frac{2}{3} & -\frac{2}{3} & 0 & 0 \\ 0 & 1 & \frac{1}{3} & -\frac{2}{3} & 0 & 0 \\ 0 & 0 & -\frac{2}{3} & \frac{1}{3} & 1 & 0 \\ 0 & 0 & \frac{2}{3} & \frac{2}{3} & 0 & 1 \end{pmatrix}$$

From (64), $\varphi_j = \gamma_{1j} + 2\gamma_{2j}$, and so $\varphi_1 = 1, \varphi_2 = 1, \varphi_3 = 0, \varphi_4 = -2, \varphi_5 = \varphi_6 = 0$.

The variables for which $c_j > \phi_j$ are clearly the second and fourth. Using the current values of the variables $x_1 = \frac{2}{3}$, $x_2 = \frac{5}{3}$, $x_3 = 0$, $x_4 = 0$, $x_5 = \frac{5}{3}$, $x_6 = \frac{13}{3}$ in (69), we find

$$(c_2 - \phi_2) \min_i \left(\frac{x_i}{\gamma_{i2}}\right) = \frac{5}{3}$$

$$(c_4 - \phi_4) \min_i \left(\frac{x_i}{\gamma_{i4}}\right) = 10.$$

The fourth variable is then to be entered in the basis with the value $\theta = 5$; the other variables now have the values

$$x_1 = \tfrac{2}{3} - 5(-\tfrac{2}{3}) = 4$$

$$x_2 = \tfrac{5}{3} - 5(-\tfrac{2}{3}) = 5$$

$$x_5 = \tfrac{5}{3} - 5 \cdot \tfrac{1}{3} = 0$$

$$x_6 = \tfrac{13}{3} - 5 \cdot \tfrac{2}{3} = 1.$$

Thus, at the beginning of the second iteration, the values of the six variables are 4, 5, 0, 5, 0, 1, and the value of the function we seek to maximize has increased from 4 to 14. Inspection of a graph will show that we have moved to a vertex next to the one which actually solves the problem. Hence the next iteration, which we omit here, will be the last.

10. If the line in question is $y = \alpha x + \beta$, then we have the conditions

$$M = e^0 - \beta = -\{e^\xi - \alpha\xi - \beta\} = e - (\alpha + \beta)$$

and

$$e^\xi - \alpha = 0.$$

We find readily that

$$\alpha = e - 1$$

$$\beta = \tfrac{1}{2}\{1 + (e - 1)[1 - \log(e - 1)].$$

Books referred to in the text

1. Bernstein, S., *Leçons sur les Propriétés Extrémales et la Meilleure Approximation des Fonctions Analytiques d'une Variable Réele*, Gauthier-Villars, Paris, 1926.
2. Bliss, G. A., *Lectures on the Calculus of Variations*, University of Chicago Press, 1946.
3. de Bruijn, N. G., *Asymptotic Methods in Analysis, Interscience Publishers, New York*, 1958.
4. Caratheodory, C., *Conformal Representation*, Cambridge Tracts, 28, Cambridge University Press, New York, 1932.
5. Churchill, R. V., *Introduction to Complex Variables and Applications*, McGraw-Hill Book Co., New York, 1948.
6. Coddington, E. A. and N. Levinson, *Theory of Ordinary Differential Equations*, McGraw-Hill Book Co., New York, 1955.
7. Collatz, L., *The Numerical Treatment of Differential Equations*, Springer-Verlag, Berlin, 1960.
8. Courant, R., and D. Hilbert, *Methods of Mathematical Physics*, Interscience Publishers, New York, 1953.
9. Dickson, L. E., *A First Course in the Theory of Equations*, John Wiley and Sons, New York, 1922.
10. Dieudonné, J., *La Theorie Analytique des Polynomes d'une Variable*, Memorial des Sciences Mathematiques, vol. 93, Paris, 1938.
11. Erdélyi, A., *Asymptotic Expansions*, Dover Publications, New York, 1956.
12. Erdélyi, A., et al., *Bateman Manuscript Project*, McGraw-Hill Book Co., New York, 1954.
13. Fox, L., *The Numerical Solution of Two-Point Boundary Problems*, Clarendon Press, Oxford, 1957.
14. Gantmacher, F. R., *Applications of the Theory of Matrices*, Chelsea Press, New York 1960.

15. Goursat, E. J. B., *A Course in Mathematical Analysis*, Ginn and Co., New York, 1945.
16. Halmos, P. R., *Finite Dimensional Vector Spaces*, Princeton University Press, 1948.
17. Halmos, P. R., *Introduction to Hilbert Space*, Chelsea Publishing Co., New York, 1957.
18. Hayman, W. K., *Multivalent Functions*, Cambridge Tracts, 48, Cambridge University Press, New York, 1958.
19. Hildebrand, F. B., *An Introduction to Numerical Analysis*, McGraw-Hill Book Co., New York, 1956.
20. Hildebrand, F. B., *Methods of Applied Mathematics*, Prentice-Hall, New York, 1952.
21. Householder, A., *Principles of Numerical Analysis*, McGraw-Hill Book Co., New York, 1953.
22. Ince, E. L., *Ordinary Differential Equations*, Dover Publications, New York, 1944.
23. Jeffreys, H., and B. S. Jeffreys, *Methods of Mathematical Physics*, Cambridge University Press, New York, 1950.
24. Knopp, K., *Theory and Application of Infinite Series*, Blackie and Son, London, 1928.
25. Kober, H., *A Dictionary of Conformal Representations*, Dover Publications, New York, 1952.
26. Kopal, Z., *Numerical Analysis*, John Wiley and Sons, New York, 1955.
27. MacDuffee, C. C., *Theory of Matrices*, Chelsea Publishing Co., New York, 1946.
28. Marden, M., *The Geometry of the Zeros of a Polynomial in a Complex Variable*, American Mathematical Society Survey III, New York, 1949.
29. Milne, W. E., *The Numerical Solution of Differential Equations*, John Wiley and Sons, New York, 1953.
30. Murray, F. J., and K. S. Miller, *Existence Theorems*, New York University Press, 1954.
31. Nehari, Z., *Conformal Mapping*, McGraw-Hill Book Co., New York, 1952.
32. Pólya, G. and G. Szegö, *Aufgaben und Lehrsätze aus der Analysis*, Springer-Verlag, Berlin, 1954.
33. Rainville, E., *Special Functions*, The Macmillan Co., New York, 1960.
34. Ralston, A., and H. S. Wilf, *Mathematical Methods for Digital Computers*, John Wiley and Sons, New York, 1960.
35. Riesz, F. and B. v. sz. Nagy, *Leçons d'Analyse Fonctionelle*, Akademiai Kiado, Budapest, 1952.
36. Shohat, J. et al., *A Bibliography of Orthogonal Polynomials*, National Research Council, 1940.
37. Szegö, G., *Orthogonal Polynomials*, American Mathematical Society Colloquium Publications, vol. 23, 1939.
38. Titchmarsh, E. C., *The Theory of the Riemann Zeta Function*, Oxford University Press, New York, 1951.
39. Tricomi, F. G., *Vorlesungen über Orthogonalreihen*, Springer-Verlag, Berlin, 1955.
40. Watson, G. N., *A Treatise on the Theory of Bessel Functions*, The Macmillan Co., New York, 1944.
41. Whittaker, E. J. and G. N. Watson, *Modern Analysis*, Cambridge University Press, New York, 1927.
42. Zygmund, A., *Trigonometric Series*, Cambridge University Press, New York, 1959.
43. *Contributions to the Calculus of Variations*, University of Chicago Press, 1931.
44. *Tables of Functions and Zeros of Functions*, National Bureau of Standards Applied Mathematics Series, no. 37, 1954.

Original works cited in the text

S. Bernstein

1. Sur les formules de quadrature de Cotes et de Tschebycheff, *C. R. de L'Academie des Sciences de L'URSS*, **14** (1937), 323–326.

N. Bourbaki

1. *Éléments de Mathématique*, XII, livre IV, chap. IV, *Equations Differentielles*, Hermann and Cie, Paris, 1951.

A. Brauer

1. Limits for the Characteristic Roots of a Matrix, *Duke Math. J.*, **13** (1946), 387–395.

A. Cauchy

1. *J. École Poly.*, **25** (1837), 176.

E. B. Christoffel

1. Über die Gaussische Quadratur und eine Verallgemeinerung derselben, *J. für Math.*, **55** (1858), 61–62.

G. Dantzig

1. The Maximization of a Linear Function of Variables Subject to Linear Inequalities, *Activity Analysis of Production and Allocation*, Cowles Commission Monograph 13, John Wiley and Sons, New York, 1951.

G. Darboux

1. Mémoire sur l'approximation des fonctions de très grands nombres, *J. de Mathématiques*, **4** (1878), 5–56.

P. Erdös and P. Turán

1. On the Distribution of Roots of Polynomials, *Annals of Math.*, **51** (1950), 105–119.

L. Euler

1. *Commentarii Acad. Sci. Imp. Petropolitanae*, vol. 11 (1739).

L. Fejér

1. Untersuchungen über die Fourierschen Reihen, *Math. Ann.*, **58** (1903), 51.
2. Nombre des changements de signe d'une fonction dans un intervalle et ses moments, *C. R. Acad. Sci.*, **158** (1914), 1328–1331.

M. Fekete

1. Sur une limite inférieure des changements de signe d'une fonction dans un intervalle, *C. R. Acad. Sci.*, **158** (1914), 1256–1258.

G. Frobenius

1. Über Matrizen aus positiven Elementen, *Sitz. Akad. Wiss. Berlin*, Phys. Math. Klasse, 1908, pp. 471–476; 1909, pp. 514–158.
2. Über Matrizen aus nicht negativen Elementen, *Sitz. Akad. Wiss. Berlin*, Phys. Math. Klasse, 1912, pp. 456–477.

M. Fujiwara

1. Über die obere Schranke des absoluten Betrages der Wurzeln einer algebraischen Gleichung, *Tôhoku Math. J.*, **10** (1916), 167–171.

C. F. Gauss

1. *Collected Works*, vol. 10, part II, pp. 189–191.
2. Methodus nova integralium valores per approximationem inveniendi, *Werke*, vol. 3, pp. 163–196.

S. Gerschgorin

1. *Bull. Acad. Sci. de l'URSS*, Leningrad, Classe Math., 7ᵉ série, 1931, pp. 749–754.

T. H. Gronwall

1. *Proc. National Academy of Sciences*, U.S.A., **6** (1920), 300–302.

J. Hadamard

1. *Leçons sur la Propagation des Ondes*, Collège de France, Paris (1903) pp. 13–14.

W. K. Hayman

1. A Generalisation of Stirling's Formula, *Jour. für Math.*, **196** (1956), 67–95.

M. R. Hestenes and W. Karush

1. A Method of Gradients for the Calculation of the Characteristic Roots and Vectors of a Real Symmetric Matrix, *J. Res. Nat'l. Bur. of Stds.*, vol. 47, pp. 45–61.

A. Hurwitz

1. Über die Bedingungen unter welchen eine Gleichung nur Wurzeln mit negativen reelen Theilen besitzt, *Math. Ann.*, **46** (1895), 273–284.

C. J. G. Jacobi

1. Über Gauss' neue Methode, die Werthe der Integrale näherungsweise zu finden, *J. für Math.*, **1** (1826), pp. 301–308.

P. Koebe

1. *J. für die Reine und Angewandte Mathematik*, **145** (1915), 177–225.

T. Kojima

1. On the limits of the roots of an algebraic equation, *Tôhoku Math. J.*, **11** (1917), 119–127.

E. Landau

1. Abschätzung der Koeffizientensumme einer Potenzreihe, *Arch. für Math.*, **21** 42–50.

P. S. Laplace

1. *Oeuvres*, vol. 7, Gauthier-Villars, Paris, 1886, p. 89.

L. Lévy

1. *C. R. Acad. Sci.*, **93** (1881), 707–708.

K. Löwner

1. Untersuchungen über die Verzerrung bei konformen Abbildungen des Einheitskreises $|z| \leqq 1$, die durch Funktionen mit nicht verschwindender Ableitung geliefert werden, Leipzig Ber., **69** (1917), 89–106.

F. Lucas

1. Géométrie des Polynômes, *J. École Poly.*, **46** (1879), 1–33.

C. MacLaurin

1. *Treatise of Fluxions*, Edinburgh, 1742.

P. Montel

1. *Leçons sur les fonctions univalentes ou multivalentes*, Paris, 1933.

M. A. Pellet

1. Sur un mode de separation des racines des équations et la formule de Lagrange, *Bull. des Sciences Math.*, **5** (1881), 393–395.

O. Perron

1. Grundlagen für eine Theorie des Jacobischen Kettenbruchalgorithmus, *Math. Ann.*, **64,** (1907), 1–76.
2. Zur Theorie der Matrizen, *Math. Ann.* **64** (1907), 248–263.

W. W. Rogosinski

1. Über positive harmonische Sinusentwicklungen, *Jahrsber. deutsch. Math. Ver.*, **40** (1931), 33–35.

E. J. Routh

1. *Dynamics of a System of Rigid Bodies, London*, 1905.

J. Sherman and W. J. Morrison

1. Adjustment of an Inverse Matrix Corresponding to Changes in the Elements of a Given Column or a Given Row of the Original Matrix, *Ann. Math. Stat.*, **20** (1949), 621.

L. Throumolopoulos

1. On the modulus of the roots of polynomials, *Bull. Greek Math. Soc.*, **23** (1947), 18–20.

J. L. Ullman

1. The Tschebycheff Method of Approximate Integration, Abstract, *Notices of the Amer. Math. Soc.*, **8** (1961), 49.

H. Wielandt

1. Unzerlegbare, nicht negative Matrizen, *Math. Z.*, **52** (1950), 642–648.

H. S. Wilf

1. Perron-Frobenius Theory and the Zeros of Polynomials, *Proc. Amer. Math. Soc.*, **12** (1961), 247–250.
2. The Possibility of Tschebycheff Quadrature on Infinite Intervals, *Proc. Nat'l. Acad. Sci. U.S.A.*, **47** (1961), 209–213.

J. E. Wilkins Jr.

1. The Average of the Reciprocal of a Function, *Proc. Amer. Math. Soc.*, **6** (1955), 806–815.

A. Wintner

1. On the Process of Successive Approximation in Initial Value Problems, *Annali di Matematica Pura ed Applicata*, **41** (1956), 343–357.

M. Woodbury

1. *Inverting Modified Matrices*, Memo Rpt. 42, Stat. Res. Group, Princeton, 1950.

Index

A CATALOGUE OF SELECTED DOVER BOOKS IN ALL FIELDS OF INTEREST

A CATALOGUE OF SELECTED DOVER BOOKS IN ALL FIELDS OF INTEREST

AUSTRIAN COOKING AND BAKING, Gretel Beer. Authentic thick soups, wiener schnitzel, veal goulash, more, plus dumplings, puff pastries, nut cakes, sacher tortes, other great Austrian desserts. 224pp. USO 23220-4 Pa. $2.50

CHEESES OF THE WORLD, U.S.D.A. Dictionary of cheeses containing descriptions of over 400 varieties of cheese from common Cheddar to exotic Surati. Up to two pages are given to important cheeses like Camembert, Cottage, Edam, etc. 151pp. 22831-2 Pa. $1.50

TRITTON'S GUIDE TO BETTER WINE AND BEER MAKING FOR BEGINNERS, S.M. Tritton. All you need to know to make family-sized quantities of over 100 types of grape, fruit, herb, vegetable wines; plus beers, mead, cider, more. 11 illustrations. 157pp. USO 22528-3 Pa. $2.25

DECORATIVE LABELS FOR HOME CANNING, PRESERVING, AND OTHER HOUSEHOLD AND GIFT USES, Theodore Menten. 128 gummed, perforated labels, beautifully printed in 2 colors. 12 versions in traditional, Art Nouveau, Art Deco styles. Adhere to metal, glass, wood, most plastics. 24pp. 8¼ x 11. 23219-0 Pa. $2.00

FIVE ACRES AND INDEPENDENCE, Maurice G. Kains. Great back-to-the-land classic explains basics of self-sufficient farming: economics, plants, crops, animals, orchards, soils, land selection, host of other necessary things. Do not confuse with skimpy faddist literature; Kains was one of America's greatest agriculturalists. 95 illustrations. 397pp. 20974-1 Pa. $3.00

GROWING VEGETABLES IN THE HOME GARDEN, U.S. Dept. of Agriculture. Basic information on site, soil conditions, selection of vegetables, planting, cultivation, gathering. Up-to-date, concise, authoritative. Covers 60 vegetables. 30 illustrations. 123pp. 23167-4 Pa. $1.35

FRUITS FOR THE HOME GARDEN, Dr. U.P. Hedrick. A chapter covering each type of garden fruit, advice on plant care, soils, grafting, pruning, sprays, transplanting, and much more! Very full. 53 illustrations. 175pp. 22944-0 Pa. $2.50

GARDENING ON SANDY SOIL IN NORTH TEMPERATE AREAS, Christine Kelway. Is your soil too light, too sandy? Improve your soil, select plants that survive under such conditions. Both vegetables and flowers. 42 photos. 148pp. USO 23199-2 Pa. $2.50

THE FRAGRANT GARDEN: A BOOK ABOUT SWEET SCENTED FLOWERS AND LEAVES, Louise Beebe Wilder. Fullest, best book on growing plants for their fragrances. Descriptions of hundreds of plants, both well-known and overlooked. 407pp. 23071-6 Pa. $4.00

EASY GARDENING WITH DROUGHT-RESISTANT PLANTS, Arno and Irene Nehrling. Authoritative guide to gardening with plants that require a minimum of water: seashore, desert, and rock gardens; house plants; annuals and perennials; much more. 190 illustrations. 320pp. 23230-1 Pa. $3.50

THE MAGIC MOVING PICTURE BOOK, Bliss, Sands & Co. The pictures in this book move! Volcanoes erupt, a house burns, a serpentine dancer wiggles her way through a number. By using a specially ruled acetate screen provided, you can obtain these and 15 other startling effects. Originally "The Motograph Moving Picture Book." 32pp. 8¼ x 11. 23224-7 Pa. $1.75

STRING FIGURES AND HOW TO MAKE THEM, Caroline F. Jayne. Fullest, clearest instructions on string figures from around world: Eskimo, Navajo, Lapp, Europe, more. Cats cradle, moving spear, lightning, stars. Introduction by A.C. Haddon. 950 illustrations. 407pp. 20152-X Pa. $3.50

PAPER FOLDING FOR BEGINNERS, William D. Murray and Francis J. Rigney. Clearest book on market for making origami sail boats, roosters, frogs that move legs, cups, bonbon boxes. 40 projects. More than 275 illustrations. Photographs. 94pp. 20713-7 Pa. $1.25

INDIAN SIGN LANGUAGE, William Tomkins. Over 525 signs developed by Sioux, Blackfoot, Cheyenne, Arapahoe and other tribes. Written instructions and diagrams: how to make words, construct sentences. Also 290 pictographs of Sioux and Ojibway tribes. 111pp. 6⅛ x 9¼. 22029-X Pa. $1.50

BOOMERANGS: HOW TO MAKE AND THROW THEM, Bernard S. Mason. Easy to make and throw, dozens of designs: cross-stick, pinwheel, boomabird, tumblestick, Australian curved stick boomerang. Complete throwing instructions. All safe. 99pp. 23028-7 Pa. $1.75

25 KITES THAT FLY, Leslie Hunt. Full, easy to follow instructions for kites made from inexpensive materials. Many novelties. Reeling, raising, designing your own. 70 illustrations. 110pp. 22550-X Pa. $1.25

TRICKS AND GAMES ON THE POOL TABLE, Fred Herrmann. 79 tricks and games, some solitaires, some for 2 or more players, some competitive; mystifying shots and throws, unusual carom, tricks involving cork, coins, a hat, more. 77 figures. 95pp. 21814-7 Pa. $1.25

WOODCRAFT AND CAMPING, Bernard S. Mason. How to make a quick emergency shelter, select woods that will burn immediately, make do with limited supplies, etc. Also making many things out of wood, rawhide, bark, at camp. Formerly titled Woodcraft. 295 illustrations. 580pp. 21951-8 Pa. $4.00

AN INTRODUCTION TO CHESS MOVES AND TACTICS SIMPLY EXPLAINED, Leonard Barden. Informal intermediate introduction: reasons for moves, tactics, openings, traps, positional play, endgame. Isolates patterns. 102pp. USO 21210-6 Pa. $1.35

LASKER'S MANUAL OF CHESS, Dr. Emanuel Lasker. Great world champion offers very thorough coverage of all aspects of chess. Combinations, position play, openings, endgame, aesthetics of chess, philosophy of struggle, much more. Filled with analyzed games. 390pp. 20640-8 Pa. $4.00

EAST O' THE SUN AND WEST O' THE MOON, George W. Dasent. Considered the best of all translations of these Norwegian folk tales, this collection has been enjoyed by generations of children (and folklorists too). Includes True and Untrue, Why the Sea is Salt, East O' the Sun and West O' the Moon, Why the Bear is Stumpy-Tailed, Boots and the Troll, The Cock and the Hen, Rich Peter the Pedlar, and 52 more. The only edition with all 59 tales. 77 illustrations by Erik Werenskiold and Theodor Kittelsen. xv + 418pp. 22521-6 Paperbound **$4.00**

GOOPS AND HOW TO BE THEM, Gelett Burgess. Classic of tongue-in-cheek humor, masquerading as etiquette book. 87 verses, twice as many cartoons, show mischievous Goops as they demonstrate to children virtues of table manners, neatness, courtesy, etc. Favorite for generations. viii + 88pp. 6½ x 9¼.
22233-0 Paperbound **$2.00**

ALICE'S ADVENTURES UNDER GROUND, Lewis Carroll. The first version, quite different from the final *Alice in Wonderland,* printed out by Carroll himself with his own illustrations. Complete facsimile of the "million dollar" manuscript Carroll gave to Alice Liddell in 1864. Introduction by Martin Gardner. viii + 96pp. Title and dedication pages in color. 21482-6 Paperbound **$1.50**

THE BROWNIES, THEIR BOOK, Palmer Cox. Small as mice, cunning as foxes, exuberant and full of mischief, the Brownies go to the zoo, toy shop, seashore, circus, etc., in 24 verse adventures and 266 illustrations. Long a favorite, since their first appearance in St. Nicholas Magazine. xi + 144pp. 6⅝ x 9¼.
21265-3 Paperbound **$2.50**

SONGS OF CHILDHOOD, Walter De La Mare. Published (under the pseudonym Walter Ramal) when De La Mare was only 29, this charming collection has long been a favorite children's book. A facsimile of the first edition in paper, the 47 poems capture the simplicity of the nursery rhyme and the ballad, including such lyrics as I Met Eve, Tartary, The Silver Penny. vii + 106pp. (USO) 21972-0 Paperbound **$2.00**

THE COMPLETE NONSENSE OF EDWARD LEAR, Edward Lear. The finest 19th-century humorist-cartoonist in full: all nonsense limericks, zany alphabets, Owl and Pussycat, songs, nonsense botany, and more than 500 illustrations by Lear himself. Edited by Holbrook Jackson. xxix + 287pp. (USO) 20167-8 Paperbound **$3.00**

BILLY WHISKERS: THE AUTOBIOGRAPHY OF A GOAT, Frances Trego Montgomery. A favorite of children since the early 20th century, here are the escapades of that rambunctious, irresistible and mischievous goat—Billy Whiskers. Much in the spirit of *Peck's Bad Boy,* this is a book that children never tire of reading or hearing. All the original familiar illustrations by W. H. Fry are included: 6 color plates, 18 black and white drawings. 159pp. 22345-0 Paperbound **$2.75**

MOTHER GOOSE MELODIES. Faithful republication of the fabulously rare Munroe and Francis "copyright 1833" Boston edition—the most important Mother Goose collection, usually referred to as the "original." Familiar rhymes plus many rare ones, with wonderful old woodcut illustrations. Edited by E. F. Bleiler. 128pp. 4½ x 6⅜. 22577-1 Paperbound **$1.50**

HOW TO SOLVE CHESS PROBLEMS, Kenneth S. Howard. Practical suggestions on problem solving for very beginners. 58 two-move problems, 46 3-movers, 8 4-movers for practice, plus hints. 171pp. 20748-X Pa. $2.00

A GUIDE TO FAIRY CHESS, Anthony Dickins. 3-D chess, 4-D chess, chess on a cylindrical board, reflecting pieces that bounce off edges, cooperative chess, retrograde chess, maximummers, much more. Most based on work of great Dawson. Full handbook, 100 problems. 66pp. 7⁷/₈ x 10¾. 22687-5 Pa. $2.00

WIN AT BACKGAMMON, Millard Hopper. Best opening moves, running game, blocking game, back game, tables of odds, etc. Hopper makes the game clear enough for anyone to play, and win. 43 diagrams. 111pp. 22894-0 Pa. $1.50

BIDDING A BRIDGE HAND, Terence Reese. Master player "thinks out loud" the binding of 75 hands that defy point count systems. Organized by bidding problem—no-fit situations, overbidding, underbidding, cueing your defense, etc. 254pp. EBE 22830-4 Pa. $3.00

THE PRECISION BIDDING SYSTEM IN BRIDGE, C.C. Wei, edited by Alan Truscott. Inventor of precision bidding presents average hands and hands from actual play, including games from 1969 Bermuda Bowl where system emerged. 114 exercises. 116pp. 21171-1 Pa. $1.75

LEARN MAGIC, Henry Hay. 20 simple, easy-to-follow lessons on magic for the new magician: illusions, card tricks, silks, sleights of hand, coin manipulations, escapes, and more —all with a minimum amount of equipment. Final chapter explains the great stage illusions. 92 illustrations. 285pp. 21238-6 Pa. $2.95

THE NEW MAGICIAN'S MANUAL, Walter B. Gibson. Step-by-step instructions and clear illustrations guide the novice in mastering 36 tricks; much equipment supplied on 16 pages of cut-out materials. 36 additional tricks. 64 illustrations. 159pp. 6⁵/₈ x 10. 23113-5 Pa. $3.00

PROFESSIONAL MAGIC FOR AMATEURS, Walter B. Gibson. 50 easy, effective tricks used by professionals —cards, string, tumblers, handkerchiefs, mental magic, etc. 63 illustrations. 223pp. 23012-0 Pa. $2.50

CARD MANIPULATIONS, Jean Hugard. Very rich collection of manipulations; has taught thousands of fine magicians tricks that are really workable, eye-catching. Easily followed, serious work. Over 200 illustrations. 163pp. 20539-8 Pa. $2.00

ABBOTT'S ENCYCLOPEDIA OF ROPE TRICKS FOR MAGICIANS, Stewart James. Complete reference book for amateur and professional magicians containing more than 150 tricks involving knots, penetrations, cut and restored rope, etc. 510 illustrations. Reprint of 3rd edition. 400pp. 23206-9 Pa. $3.50

THE SECRETS OF HOUDINI, J.C. Cannell. Classic study of Houdini's incredible magic, exposing closely-kept professional secrets and revealing, in general terms, the whole art of stage magic. 67 illustrations. 279pp. 22913-0 Pa. $2.50

AGAINST THE GRAIN (A REBOURS), Joris K. Huysmans. Filled with weird images, evidences of a bizarre imagination, exotic experiments with hallucinatory drugs, rich tastes and smells and the diversions of its sybarite hero Duc Jean des Esseintes, this classic novel pushed 19th-century literary decadence to its limits. Full unabridged edition. Do not confuse this with abridged editions generally sold. Introduction by Havelock Ellis. xlix + 206pp. 22190-3 Paperbound **$2.50**

VARIORUM SHAKESPEARE: HAMLET. Edited by Horace H. Furness; a landmark of American scholarship. Exhaustive footnotes and appendices treat all doubtful words and phrases, as well as suggested critical emendations throughout the play's history. First volume contains editor's own text, collated with all Quartos and Folios. Second volume contains full first Quarto, translations of Shakespeare's sources (Belleforest, and Saxo Grammaticus), Der Bestrafte Brudermord, and many essays on critical and historical points of interest by major authorities of past and present. Includes details of staging and costuming over the years. By far the best edition available for serious students of Shakespeare. Total of xx + 905pp.
21004-9, 21005-7, 2 volumes, Paperbound **$11.00**

A LIFE OF WILLIAM SHAKESPEARE, Sir Sidney Lee. This is the standard life of Shakespeare, summarizing everything known about Shakespeare and his plays. Incredibly rich in material, broad in coverage, clear and judicious, it has served thousands as the best introduction to Shakespeare. 1931 edition. 9 plates. xxix + 792pp. 21967-4 Paperbound $4.50

MASTERS OF THE DRAMA, John Gassner. Most comprehensive history of the drama in print, covering every tradition from Greeks to modern Europe and America, including India, Far East, etc. Covers more than 800 dramatists, 2000 plays, with biographical material, plot summaries, theatre history, criticism, etc. "Best of its kind in English," *New Republic*. 77 illustrations. xxii + 890pp.
20100-7 Clothbound **$10.00**

THE EVOLUTION OF THE ENGLISH LANGUAGE, George McKnight. The growth of English, from the 14th century to the present. Unusual, non-technical account presents basic information in very interesting form: sound shifts, change in grammar and syntax, vocabulary growth, similar topics. Abundantly illustrated with quotations. Formerly *Modern English in the Making*. xii + 590pp.
21932-1 Paperbound **$4.00**

AN ETYMOLOGICAL DICTIONARY OF MODERN ENGLISH, Ernest Weekley. Fullest, richest work of its sort, by foremost British lexicographer. Detailed word histories, including many colloquial and archaic words; extensive quotations. Do not confuse this with the Concise Etymological Dictionary, which is much abridged. Total of xxvii + 830pp. 6½ x 9¼.
21873-2, 21874-0 Two volumes, Paperbound **$10.00**

FLATLAND: A ROMANCE OF MANY DIMENSIONS, E. A. Abbott. Classic of science-fiction explores ramifications of life in a two-dimensional world, and what happens when a three-dimensional being intrudes. Amusing reading, but also useful as introduction to thought about hyperspace. Introduction by Banesh Hoffmann. 16 illustrations. xx + 103pp. 20001-9 Paperbound **$1.50**

THE RED FAIRY BOOK, Andrew Lang. Lang's color fairy books have long been children's favorites. This volume includes Rapunzel, Jack and the Bean-stalk and 35 other stories, familiar and unfamiliar. 4 plates, 93 illustrations x + 367pp.
21673-X Paperbound **$3.00**

THE BLUE FAIRY BOOK, Andrew Lang. Lang's tales come from all countries and all times. Here are 37 tales from Grimm, the Arabian Nights, Greek Mythology, and other fascinating sources. 8 plates, 130 illustrations. xi + 390pp.
21437-0 Paperbound **$3.50**

HOUSEHOLD STORIES BY THE BROTHERS GRIMM. Classic English-language edition of the well-known tales — Rumpelstiltskin, Snow White, Hansel and Gretel, The Twelve Brothers, Faithful John, Rapunzel, Tom Thumb (52 stories in all). Translated into simple, straightforward English by Lucy Crane. Ornamented with headpieces, vignettes, elaborate decorative initials and a dozen full-page illustrations by Walter Crane. x + 269pp. 21080-4 Paperbound **$3.00**

THE MERRY ADVENTURES OF ROBIN HOOD, Howard Pyle. The finest modern versions of the traditional ballads and tales about the great English outlaw. Howard Pyle's complete prose version, with every word, every illustration of the first edition. Do not confuse this facsimile of the original (1883) with modern editions that change text or illustrations. 23 plates plus many page decorations. xxii + 296pp.
22043-5 Paperbound **$4.00**

THE STORY OF KING ARTHUR AND HIS KNIGHTS, Howard Pyle. The finest children's version of the life of King Arthur; brilliantly retold by Pyle, with 48 of his most imaginative illustrations. xviii + 313pp. $6\frac{1}{8}$ x $9\frac{1}{4}$.
21445-1 Paperbound **$3.50**

THE WONDERFUL WIZARD OF OZ, L. Frank Baum. America's finest children's book in facsimile of first edition with all Denslow illustrations in full color. The edition a child should have. Introduction by Martin Gardner. 23 color plates, scores of drawings. iv + 267pp. 20691-2 Paperbound **$3.00**

THE MARVELOUS LAND OF OZ, L. Frank Baum. The second Oz book, every bit as imaginative as the Wizard. The hero is a boy named Tip, but the Scarecrow and the Tin Woodman are back, as is the Oz magic. 16 color plates, 120 drawings by John R. Neill. 287pp. 20692-0 Paperbound **$3.00**

THE MAGICAL MONARCH OF MO, L. Frank Baum. Remarkable adventures in a land even stranger than Oz. The best of Baum's books not in the Oz series. 15 color plates and dozens of drawings by Frank Verbeck. xviii + 237pp.
21892-9 Paperbound **$2.95**

THE BAD CHILD'S BOOK OF BEASTS, MORE BEASTS FOR WORSE CHILDREN, A MORAL ALPHABET, Hilaire Belloc. Three complete humor classics in one volume. Be kind to the frog, and do not call him names . . . and 28 other whimsical animals. Familiar favorites and some not so well known. Illustrated by Basil Blackwell. 156pp. (USO) 20749-8 Paperbound **$2.00**

VISUAL ILLUSIONS: THEIR CAUSES, CHARACTERISTICS, AND APPLICATIONS, Matthew Luckiesh. Thorough description and discussion of optical illusion, geometric and perspective, particularly; size and shape distortions, illusions of color, of motion; natural illusions; use of illusion in art and magic, industry, etc. Most useful today with op art, also for classical art. Scores of effects illustrated. Introduction by William H. Ittleson. 100 illustrations. xxi + 252pp.
21530-X Paperbound **$2.50**

A HANDBOOK OF ANATOMY FOR ART STUDENTS, Arthur Thomson. Thorough, virtually exhaustive coverage of skeletal structure, musculature, etc. Full text, supplemented by anatomical diagrams and drawings and by photographs of undraped figures. Unique in its comparison of male and female forms, pointing out differences of contour, texture, form. 211 figures, 40 drawings, 86 photographs. xx + 459pp. 5⅜ x 8⅜. 21163-0 Paperbound **$5.00**

150 MASTERPIECES OF DRAWING, Selected by Anthony Toney. Full page reproductions of drawings from the early 16th to the end of the 18th century, all beautifully reproduced: Rembrandt, Michelangelo, Dürer, Fragonard, Urs, Graf, Wouwerman, many others. First-rate browsing book, model book for artists. xviii + 150pp. 8⅜ x 11¼. 21032-4 Paperbound **$4.00**

THE LATER WORK OF AUBREY BEARDSLEY, Aubrey Beardsley. Exotic, erotic, ironic masterpieces in full maturity: Comedy Ballet, Venus and Tannhauser, Pierrot, Lysistrata, Rape of the Lock, Savoy material, Ali Baba, Volpone, etc. This material revolutionized the art world, and is still powerful, fresh, brilliant. With *The Early Work,* all Beardsley's finest work. 174 plates, 2 in color. xiv + 176pp. 8⅛ x 11.
21817-1 Paperbound **$4.00**

DRAWINGS OF REMBRANDT, Rembrandt van Rijn. Complete reproduction of fabulously rare edition by Lippmann and Hofstede de Groot, completely reedited, updated, improved by Prof. Seymour Slive, Fogg Museum. Portraits, Biblical sketches, landscapes, Oriental types, nudes, episodes from classical mythology—All Rembrandt's fertile genius. Also selection of drawings by his pupils and followers. "Stunning volumes," *Saturday Review.* 550 illustrations. lxxviii + 552pp. 9⅛ x 12¼. 21485-0, 21486-9 Two volumes, Paperbound **$12.00**

THE DISASTERS OF WAR, Francisco Goya. One of the masterpieces of Western civilization—83 etchings that record Goya's shattering, bitter reaction to the Napoleonic war that swept through Spain after the insurrection of 1808 and to war in general. Reprint of the first edition, with three additional plates from Boston's Museum of Fine Arts. All plates facsimile size. Introduction by Philip Hofer, Fogg Museum. v + 97pp. 9⅜ x 8¼. 21872-4 Paperbound **$3.00**

GRAPHIC WORKS OF ODILON REDON. Largest collection of Redon's graphic works ever assembled: 172 lithographs, 28 etchings and engravings, 9 drawings. These include some of his most famous works. All the plates from *Odilon Redon: oeuvre graphique complet,* plus additional plates. New introduction and caption translations by Alfred Werner. 209 illustrations. xxvii + 209pp. 9⅛ x 12¼.
21966-8 Paperbound **$6.00**

The Art Deco Style, ed. by Theodore Menten. Furniture, jewelry, metalwork, ceramics, fabrics, lighting fixtures, interior decors, exteriors, graphics from pure French sources. Best sampling around. Over 400 photographs. 183pp. 8$\frac{3}{8}$ x 11$\frac{1}{4}$.
22824-X Pa. $4.00

The Gentleman and Cabinet Maker's Director, Thomas Chippendale. Full reprint, 1762 style book, most influential of all time; chairs, tables, sofas, mirrors, cabinets, etc. 200 plates, plus 24 photographs of surviving pieces. 249pp. 9$\frac{7}{8}$ x 12$\frac{3}{4}$.
21601-2 Pa. $6.00

Pine Furniture of Early New England, Russell H. Kettell. Basic book. Thorough historical text, plus 200 illustrations of boxes, highboys, candlesticks, desks, etc. 477pp. 7$\frac{7}{8}$ x 10$\frac{3}{4}$.
20145-7 Clothbd. $12.50

Oriental Rugs, Antique and Modern, Walter A. Hawley. Persia, Turkey, Caucasus, Central Asia, China, other traditions. Best general survey of all aspects: styles and periods, manufacture, uses, symbols and their interpretation, and identification. 96 illustrations, 11 in color. 320pp. 6$\frac{1}{8}$ x 9$\frac{1}{4}$.
22366-3 Pa. $5.00

Decorative Antique Ironwork, Henry R. d'Allemagne. Photographs of 4500 iron artifacts from world's finest collection, Rouen. Hinges, locks, candelabra, weapons, lighting devices, clocks, tools, from Roman times to mid-19th century. Nothing else comparable to it. 420pp. 9 x 12.
22082-6 Pa. $8.50

The Complete Book of Doll Making and Collecting, Catherine Christopher. Instructions, patterns for dozens of dolls, from rag doll on up to elaborate, historically accurate figures. Mould faces, sew clothing, make doll houses, etc. Also collecting information. Many illustrations. 288pp. 6 x 9. 22066-4 Pa. $3.00

Antique Paper Dolls: 1915-1920, edited by Arnold Arnold. 7 antique cut-out dolls and 24 costumes from 1915-1920, selected by Arnold Arnold from his collection of rare children's books and entertainments, all in full color. 32pp. 9$\frac{1}{4}$ x 12$\frac{1}{4}$.
23176-3 Pa. $2.00

Antique Paper Dolls: The Edwardian Era, Epinal. Full-color reproductions of two historic series of paper dolls that show clothing styles in 1908 and at the beginning of the First World War. 8 two-sided, stand-up dolls and 32 complete, two-sided costumes. Full instructions for assembling included. 32pp. 9$\frac{1}{4}$ x 12$\frac{1}{4}$.
23175-5 Pa. $2.00

A History of Costume, Carl Köhler, Emma von Sichardt. Egypt, Babylon, Greece up through 19th century Europe; based on surviving pieces, art works, etc. Full text and 595 illustrations, including many clear, measured patterns for reproducing historic costume. Practical. 464pp.
21030-8 Pa. $4.00

Early American Locomotives, John H. White, Jr. Finest locomotive engravings from late 19th century: historical (1804-1874), main-line (after 1870), special, foreign, etc. 147 plates. 200pp. 11$\frac{3}{8}$ x 8$\frac{1}{4}$.
22772-3 Pa. $3.50

DECORATIVE ALPHABETS AND INITIALS, edited by Alexander Nesbitt. 91 complete alphabets (medieval to modern), 3924 decorative initials, including Victorian novelty and Art Nouveau. 192pp. 7¾ x 10¾. 20544-4 Pa. $4.00

CALLIGRAPHY, Arthur Baker. Over 100 original alphabets from the hand of our greatest living calligrapher: simple, bold, fine-line, richly ornamented, etc. —all strikingly original and different, a fusion of many influences and styles. 155pp. 11⅜ x 8¼. 22895-9 Pa. $4.50

MONOGRAMS AND ALPHABETIC DEVICES, edited by Hayward and Blanche Cirker. Over 2500 combinations, names, crests in very varied styles: script engraving, ornate Victorian, simple Roman, and many others. 226pp. 8⅛ x 11. 22330-2 Pa. $5.00

THE BOOK OF SIGNS, Rudolf Koch. Famed German type designer renders 493 symbols: religious, alchemical, imperial, runes, property marks, etc. Timeless. 104pp. 6⅛ x 9¼. 20162-7 Pa. $1.75

200 DECORATIVE TITLE PAGES, edited by Alexander Nesbitt. 1478 to late 1920's. Baskerville, Dürer, Beardsley, W. Morris, Pyle, many others in most varied techniques. For posters, programs, other uses. 222pp. 8⅜ x 11¼. 21264-5 Pa. $5.00

DICTIONARY OF AMERICAN PORTRAITS, edited by Hayward and Blanche Cirker. 4000 important Americans, earliest times to 1905, mostly in clear line. Politicians, writers, soldiers, scientists, inventors, industrialists, Indians, Blacks, women, outlaws, etc. Identificatory information. 756pp. 9¼ x 12¾. 21823-6 Clothbd. $30.00

ART FORMS IN NATURE, Ernst Haeckel. Multitude of strangely beautiful natural forms: Radiolaria, Foraminifera, jellyfishes, fungi, turtles, bats, etc. All 100 plates of the 19th century evolutionist's Kunstformen der Natur (1904). 100pp. 9⅜ x 12¼. 22987-4 Pa. $4.00

DECOUPAGE: THE BIG PICTURE SOURCEBOOK, Eleanor Rawlings. Make hundreds of beautiful objects, over 550 florals, animals, letters, shells, period costumes, frames, etc. selected by foremost practitioner. Printed on one side of page. 8 color plates. Instructions. 176pp. 9 3/16 x 12¼. 23182-8 Pa. $5.00

AMERICAN FOLK DECORATION, Jean Lipman, Eve Meulendyke. Thorough coverage of all aspects of wood, tin, leather, paper, cloth decoration — scapes, humans, trees, flowers, geometrics — and how to make them. Full instructions. 233 illustrations, 5 in color. 163pp. 8⅜ x 11¼. 22217-9 Pa. $3.95

WHITTLING AND WOODCARVING, E.J. Tangerman. Best book on market; clear, full. If you can cut a potato, you can carve toys, puzzles, chains, caricatures, masks, patterns, frames, decorate surfaces, etc. Also covers serious wood sculpture. Over 200 photos. 293pp. 20965-2 Pa. $3.00

MODERN CHESS STRATEGY, Ludek Pachman. The use of the queen, the active king, exchanges, pawn play, the center, weak squares, etc. Section on rook alone worth price of the book. Stress on the moderns. Often considered the most important book on strategy. 314pp. 20290-9 Pa. $3.50

CHESS STRATEGY, Edward Lasker. One of half-dozen great theoretical works in chess, shows principles of action above and beyond moves. Acclaimed by Capablanca, Keres, etc. 282pp. USO 20528-2 Pa. $3.00

CHESS PRAXIS, THE PRAXIS OF MY SYSTEM, Aron Nimzovich. Founder of hypermodern chess explains his profound, influential theories that have dominated much of 20th century chess. 109 illustrative games. 369pp. 20296-8 Pa. $3.50

HOW TO PLAY THE CHESS OPENINGS, Eugene Znosko-Borovsky. Clear, profound examinations of just what each opening is intended to do and how opponent can counter. Many sample games, questions and answers. 147pp. 22795-2 Pa. $2.00

THE ART OF CHESS COMBINATION, Eugene Znosko-Borovsky. Modern explanation of principles, varieties, techniques and ideas behind them, illustrated with many examples from great players. 212pp. 20583-5 Pa. $2.50

COMBINATIONS: THE HEART OF CHESS, Irving Chernev. Step-by-step explanation of intricacies of combinative play. 356 combinations by Tarrasch, Botvinnik, Keres, Steinitz, Anderssen, Morphy, Marshall, Capablanca, others, all annotated. 245 pp. 21744-2 Pa. $3.00

HOW TO PLAY CHESS ENDINGS, Eugene Znosko-Borovsky. Thorough instruction manual by fine teacher analyzes each piece individually; many common endgame situations. Examines games by Steinitz, Alekhine, Lasker, others. Emphasis on understanding. 288pp. 21170-3 Pa. $2.75

MORPHY'S GAMES OF CHESS, Philip W. Sergeant. Romantic history, 54 games of greatest player of all time against Anderssen, Bird, Paulsen, Harrwitz; 52 games at odds; 52 blindfold; 100 consultation, informal, other games. Analyses by Anderssen, Steinitz, Morphy himself. 352pp. 20386-7 Pa. $4.00

500 MASTER GAMES OF CHESS, S. Tartakower, J. du Mont. Vast collection of great chess games from 1798-1938, with much material nowhere else readily available. Fully annotated, arranged by opening for easier study. 665pp. 23208-5 Pa. $6.00

THE SOVIET SCHOOL OF CHESS, Alexander Kotov and M. Yudovich. Authoritative work on modern Russian chess. History, conceptual background. 128 fully annotated games (most unavailable elsewhere) by Botvinnik, Keres, Smyslov, Tal, Petrosian, Spassky, more. 390pp. 20026-4 Pa. $3.95

WONDERS AND CURIOSITIES OF CHESS, Irving Chernev. A lifetime's accumulation of such wonders and curiosities as the longest won game, shortest game, chess problem with mate in 1220 moves, and much more unusual material —356 items in all, over 160 complete games. 146 diagrams. 203pp. 23007-4 Pa. $3.50

SLEEPING BEAUTY, illustrated by Arthur Rackham. Perhaps the fullest, most delightful version ever, told by C.S. Evans. Rackham's best work. 49 illustrations. 110pp. 7⅞ x 10¾. 22756-1 Pa. $2.00

THE WONDERFUL WIZARD OF OZ, L. Frank Baum. Facsimile in full color of America's finest children's classic. Introduction by Martin Gardner. 143 illustrations by W.W. Denslow. 267pp. 20691-2 Pa. $3.00

GOOPS AND HOW TO BE THEM, Gelett Burgess. Classic tongue-in-cheek masquerading as etiquette book. 87 verses, 170 cartoons as Goops demonstrate virtues of table manners, neatness, courtesy, more. 88pp. 6½ x 9¼. 22233-0 Pa. $2.00

THE BROWNIES, THEIR BOOK, Palmer Cox. Small as mice, cunning as foxes, exuberant, mischievous, Brownies go to zoo, toy shop, seashore, circus, more. 24 verse adventures. 266 illustrations. 144pp. 6⅝ x 9¼. 21265-3 Pa. $2.50

BILLY WHISKERS: THE AUTOBIOGRAPHY OF A GOAT, Frances Trego Montgomery. Escapades of that rambunctious goat. Favorite from turn of the century America. 24 illustrations. 259pp. 22345-0 Pa. $2.75

THE ROCKET BOOK, Peter Newell. Fritz, janitor's kid, sets off rocket in basement of apartment house; an ingenious hole punched through every page traces course of rocket. 22 duotone drawings, verses. 48pp. 6⅞ x 8⅜. 22044-3 Pa. $1.50

PECK'S BAD BOY AND HIS PA, George W. Peck. Complete double-volume of great American childhood classic. Hennery's ingenious pranks against outraged pomposity of pa and the grocery man. 97 illustrations. Introduction by E.F. Bleiler. 347pp. 20497-9 Pa. $2.50

THE TALE OF PETER RABBIT, Beatrix Potter. The inimitable Peter's terrifying adventure in Mr. McGregor's garden, with all 27 wonderful, full-color Potter illustrations. 55pp. 4¼ x 5½. USO 22827-4 Pa. $1.00

THE TALE OF MRS. TIGGY-WINKLE, Beatrix Potter. Your child will love this story about a very special hedgehog and all 27 wonderful, full-color Potter illustrations. 57pp. 4¼ x 5½. USO 20546-0 Pa. $1.00

THE TALE OF BENJAMIN BUNNY, Beatrix Potter. Peter Rabbit's cousin coaxes him back into Mr. McGregor's garden for a whole new set of adventures. A favorite with children. All 27 full-color illustrations. 59pp. 4¼ x 5½. USO 21102-9 Pa. $1.00

THE MERRY ADVENTURES OF ROBIN HOOD, Howard Pyle. Facsimile of original (1883) edition, finest modern version of English outlaw's adventures. 23 illustrations by Pyle. 296pp. 6½ x 9¼. 22043-5 Pa. $4.00

TWO LITTLE SAVAGES, Ernest Thompson Seton. Adventures of two boys who lived as Indians; explaining Indian ways, woodlore, pioneer methods. 293 illustrations. 286pp. 20985-7 Pa. $3.00

JEWISH GREETING CARDS, Ed Sibbett, Jr. 16 cards to cut and color. Three say "Happy Chanukah," one "Happy New Year," others have no message, show stars of David, Torahs, wine cups, other traditional themes. 16 envelopes. 8¼ x 11.
23225-5 Pa. $2.00

AUBREY BEARDSLEY GREETING CARD BOOK, Aubrey Beardsley. Edited by Theodore Menten. 16 elegant yet inexpensive greeting cards let you combine your own sentiments with subtle Art Nouveau lines. 16 different Aubrey Beardsley designs that you can color or not, as you wish. 16 envelopes. 64pp. 8¼ x 11.
23173-9 Pa. $2.00

RECREATIONS IN THE THEORY OF NUMBERS, Albert Beiler. Number theory, an inexhaustible source of puzzles, recreations, for beginners and advanced. Divisors, perfect numbers. scales of notation, etc. 349pp. 21096-0 Pa. $4.00

AMUSEMENTS IN MATHEMATICS, Henry E. Dudeney. One of largest puzzle collections, based on algebra, arithmetic, permutations, probability, plane figure dissection, properties of numbers, by one of world's foremost puzzlists. Solutions. 450 illustrations. 258pp. 20473-1 Pa. $3.00

MATHEMATICS, MAGIC AND MYSTERY, Martin Gardner. Puzzle editor for Scientific American explains math behind: card tricks, stage mind reading, coin and match tricks, counting out games, geometric dissections. Probability, sets, theory of numbers, clearly explained. Plus more than 400 tricks, guaranteed to work. 135 illustrations. 176pp. 20335-2 Pa. $2.00

BEST MATHEMATICAL PUZZLES OF SAM LOYD, edited by Martin Gardner. Bizarre, original, whimsical puzzles by America's greatest puzzler. From fabulously rare Cyclopedia, including famous 14-15 puzzles, the Horse of a Different Color, 115 more. Elementary math. 150 illustrations. 167pp. 20498-7 Pa. $2.50

MATHEMATICAL PUZZLES FOR BEGINNERS AND ENTHUSIASTS, Geoffrey Mott-Smith. 189 puzzles from easy to difficult involving arithmetic, logic, algebra, properties of digits, probability. Explanation of math behind puzzles. 135 illustrations. 248pp. 20198-8 Pa. $2.75

BIG BOOK OF MAZES AND LABYRINTHS, Walter Shepherd. Classical, solid, and ripple mazes; short path and avoidance labyrinths; more —50 mazes and labyrinths in all. 12 other figures. Full solutions. 112pp. 8⅛ x 11. 22951-3 Pa. $2.00

COIN GAMES AND PUZZLES, Maxey Brooke. 60 puzzles, games and stunts — from Japan, Korea, Africa and the ancient world, by Dudeney and the other great puzzlers, as well as Maxey Brooke's own creations. Full solutions. 67 illustrations. 94pp. 22893-2 Pa. $1.50

HAND SHADOWS TO BE THROWN UPON THE WALL, Henry Bursill. Wonderful Victorian novelty tells how to make flying birds, dog, goose, deer, and 14 others. 32pp. 6½ x 9¼. 21779-5 Pa. $1.25

THE JOURNAL OF HENRY D. THOREAU, edited by Bradford Torrey, F.H. Allen. Complete reprinting of 14 volumes, 1837-1861, over two million words; the sourcebooks for Walden, etc. Definitive. All original sketches, plus 75 photographs. Introduction by Walter Harding. Total of 1804pp. 8½ x 12¼.
20312-3, 20313-1 Clothbd., Two vol. set $50.00

MASTERS OF THE DRAMA, John Gassner. Most comprehensive history of the drama, every tradition from Greeks to modern Europe and America, including Orient. Covers 800 dramatists, 2000 plays; biography, plot summaries, criticism, theatre history, etc. 77 illustrations. 890pp. 20100-7 Clothbd. $10.00

GHOST AND HORROR STORIES OF AMBROSE BIERCE, Ambrose Bierce. 23 modern horror stories: The Eyes of the Panther, The Damned Thing, etc., plus the dream-essay Visions of the Night. Edited by E.F. Bleiler. 199pp. 20767-6 Pa. $2.00

BEST GHOST STORIES, Algernon Blackwood. 13 great stories by foremost British 20th century supernaturalist. The Willows, The Wendigo, Ancient Sorceries, others. Edited by E.F. Bleiler. 366pp. USO 22977-7 Pa. $3.00

THE BEST TALES OF HOFFMANN, E.T.A. Hoffmann. 10 of Hoffmann's most important stories, in modern re-editings of standard translations: Nutcracker and the King of Mice, The Golden Flowerpot, etc. 7 illustrations by Hoffmann. Edited by E.F. Bleiler. 458pp. 21793-0 Pa. $3.95

BEST GHOST STORIES OF J.S. LEFANU, J. Sheridan LeFanu. 16 stories by greatest Victorian master: Green Tea, Carmilla, Haunted Baronet, The Familiar, etc. Mostly unavailable elsewhere. Edited by E.F. Bleiler. 8 illustrations. 467pp.
20415-4 Pa. $4.00

SUPERNATURAL HORROR IN LITERATURE, H.P. Lovecraft. Great modern American supernaturalist brilliantly surveys history of genre to 1930's, summarizing, evaluating scores of books. Necessary for every student, lover of form. Introduction by E.F. Bleiler. 111pp. 20105-8 Pa. $1.50

THREE GOTHIC NOVELS, ed. by E.F. Bleiler. Full texts Castle of Otranto, Walpole; Vathek, Beckford; The Vampyre, Polidori; Fragment of a Novel, Lord Byron. 331pp. 21232-7 Pa. $3.00

SEVEN SCIENCE FICTION NOVELS, H.G. Wells. Full novels. First Men in the Moon, Island of Dr. Moreau, War of the Worlds, Food of the Gods, Invisible Man, Time Machine, In the Days of the Comet. A basic science-fiction library. 1015pp.
USO 20264-X Clothbd. $6.00

LADY AUDLEY'S SECRET, Mary E. Braddon. Great Victorian mystery classic, beautifully plotted, suspenseful; praised by Thackeray, Boucher, Starrett, others. What happened to beautiful, vicious Lady Audley's husband? Introduction by Norman Donaldson. 286pp. 23011-2 Pa. $3.00

150 MASTERPIECES OF DRAWING, edited by Anthony Toney. 150 plates, early 15th century to end of 18th century; Rembrandt, Michelangelo, Dürer, Fragonard, Watteau, Wouwerman, many others. 150pp. 8³/8 x 11¼. 21032-4 Pa. $4.00

THE GOLDEN AGE OF THE POSTER, Hayward and Blanche Cirker. 70 extraordinary posters in full colors, from Maîtres de l'Affiche, Mucha, Lautrec, Bradley, Cheret, Beardsley, many others. 9³/8 x 12¼. 22753-7 Pa. $4.95
21718-3 Clothbd. $7.95

SIMPLICISSIMUS, selection, translations and text by Stanley Appelbaum. 180 satirical drawings, 16 in full color, from the famous German weekly magazine in the years 1896 to 1926. 24 artists included: Grosz, Kley, Pascin, Kubin, Kollwitz, plus Heine, Thöny, Bruno Paul, others. 172pp. 8½ x 12¼. 23098-8 Pa. $5.00
23099-6 Clothbd. $10.00

THE EARLY WORK OF AUBREY BEARDSLEY, Aubrey Beardsley. 157 plates, 2 in color: Manon Lescaut, Madame Bovary, Morte d'Arthur, Salome, other. Introduction by H. Marillier. 175pp. 8½ x 11. 21816-3 Pa. $4.00

THE LATER WORK OF AUBREY BEARDSLEY, Aubrey Beardsley. Exotic masterpieces of full maturity: Venus and Tannhäuser, Lysistrata, Rape of the Lock, Volpone, Savoy material, etc. 174 plates, 2 in color. 176pp. 8½ x 11. 21817-1 Pa. $4.00

DRAWINGS OF WILLIAM BLAKE, William Blake. 92 plates from Book of Job, Divine Comedy, Paradise Lost, visionary heads, mythological figures, Laocoön, etc. Selection, introduction, commentary by Sir Geoffrey Keynes. 178pp. 8½ x 11.
22303-5 Pa. $3.50

LONDON: A PILGRIMAGE, Gustave Doré, Blanchard Jerrold. Squalor, riches, misery, beauty of mid-Victorian metropolis; 55 wonderful plates, 125 other illustrations, full social, cultural text by Jerrold. 191pp. of text. 8¹/8 x 11.
22306-X Pa. $5.00

THE COMPLETE WOODCUTS OF ALBRECHT DÜRER, edited by Dr. W. Kurth. 346 in all: Old Testament, St. Jerome, Passion, Life of Virgin, Apocalypse, many others. Introduction by Campbell Dodgson. 285pp. 8½ x 12¼. 21097-9 Pa. $6.00

THE DISASTERS OF WAR, Francisco Goya. 83 etchings record horrors of Napoleonic wars in Spain and war in general. Reprint of 1st edition, plus 3 additional plates. Introduction by Philip Hofer. 97pp. 9³/8 x 8¼. 21872-4 Pa. $3.00

ENGRAVINGS OF HOGARTH, William Hogarth. 101 of Hogarth's greatest works: Rake's Progress, Harlot's Progress, Illustrations for Hudibras, Midnight Modern Conversation, Before and After, Beer Street and Gin Lane, many more. Full commentary. 256pp. 11 x 14. 22479-1 Pa. $7.00
23023-6 Clothbd. $13.50

PRIMITIVE ART, Franz Boas. Great anthropologist on ceramics, textiles, wood, stone, metal, etc.; patterns, technology, symbols, styles. All areas, but fullest on Northwest Coast Indians. 350 illustrations. 378pp. 20025-6 Pa. $3.75